Heterojunctions and Semiconductor Superlattices

Heterojunctions and Semiconductor Superlattices

Proceedings of the Winter School
Les Houches, France, March 12–21 1985

Editors: G. Allan, G. Bastard, N. Boccara,
M. Lannoo, and M. Voos

With 168 Figures

Springer-Verlag Berlin Heidelberg New York
London Paris Tokyo

Dr. Guy Allan
Dr. Michel Lannoo

Laboratoire de Physique des Solides, ISEN, 41 Boulevard Vauban,
F-59046 Lille Cedex, France

Dr. Gérald Bastard
Dr. Michel Voos

Laboratoire de Physique, Ecole Normale Supérieure, 24 rue Lhomond,
F-75231 Paris Cedex 05, France

Professor Nino Boccara

Université Scientifique et Médicale de Grenoble, Centre de Physique, Côte des Chavants,
F-74310 Les Houches, France

Organizing Committee:
M. Bensoussan, J. P. Gaillard, H. Launois, H. Martinot, J. B. Theeten, and C. Weisbuch

---

ISBN 3-540-16259-3 Springer-Verlag Berlin Heidelberg New York
ISBN 0-387-16259-3 Springer-Verlag New York Berlin Heidelberg

Printed in Germany

Offset printing: Weihert-Druck GmbH, D-6100 Darmstadt
2153/3150-543210

Preface

The Winter School held in Les Houches on March 12-21, 1985 was devoted to Semiconductor Heterojunctions and Superlattices, a topic which is recognized as being now one of the most interesting and active fields in semiconductor physics. In fact, following the pioneering work of Esaki and Tsu in 1970, the study of these two-dimensional semiconductor heterostructures has developed rapidly, both from the point of view of basic physics and of applications. For instance, modulation-doped heterojunctions are nowadays currently used to investigate the quantum Hall effect and to make very fast transistors.

This book contains the lectures presented at this Winter School, showing in particular that many aspects of semiconductor heterojunctions and superlattices were treated, extending from the fabrication of these two-dimensional systems to their basic properties and applications in micro- and opto-electronics. Among the subjects which were covered, one can quote as examples: molecular beam epitaxy and metallorganic chemical vapor deposition of semiconductor compounds; band structure of superlattices; properties of electrons in heterojunctions, including the fractional quantum Hall effect; optical properties of two-dimensional heterostructures; quantum well lasers; and two-dimensional electron gas field effect transistors.

It is clear that two-dimensional semiconductor systems are raising a great deal of interest in many industrial and university laboratories. From the number of applications which were received and from the reactions of the participants, it can certainly be asserted that this School corresponded to a need and came at the right time.

Finally, we wish to thank warmly all the lecturers and participants, as well as the Organizing Committee, who made this Winter School a success. We are also very grateful to the Board of the Physics Center of Les Houches, and to the following organizations and companies for their efficient support: Bull, CIT-Alcatel, CNRS, DRET, Electricité de France, IBM-France, L.E.P., LETI-Ceng, Ministère de l'Education Nationale, Ministère de la Recherche et de la Technologie, RTC-La Radiotechnique COMPELEC, Thomson-CSF and the Université Scientifique et Médicale de Grenoble.

Gif-sur-Yvette
Lille, Paris, March 1986

N. Boccara G. Allan M. Lannoo
G. Bastard M. Voos

Contents

Part I

Introduction

Advances in Semiconductor Superlattices and Quantum Wells

L. Esaki

IBM Thomas J. Watson Research Center, Yorktown Heights, NY 10598, USA

In 1969, research on synthesized semiconductor superlattices was initiated with a proposal for a one-dimensional periodic structure consisting of alternating ultra-thin layers by ESAKI and TSU [1,2]. Two types of superlattices , doping and compositional, were envisioned. The idea of the superlattice occurred to us while investigating the possible observation of resonant tunneling through double and multiple potential-barriers [3]. In general, if characteristic dimensions such as superlattice periods and widths of potential wells in semiconductor nanostructures are reduced to less than the electron mean free path, the entire electron system will enter a quantum regime of reduced dimensionality with the presence of nearly-ideal hetero-interfaces. Our effort for the semiconductor superlattice [4] is viewed as a search for novel phenomena in such a regime with engineered structures.

It was recognized at the beginning that, while the structure was undoubtedly of considerable interest, the engineering of such a crystal consisting of ultra-thin layers would be a formidable task. Nevertheless, the proposal of a semiconductor superlattice inspired a number of material scientists [5-8]. Indeed, steady improvements in thin-film growth techniques such as MBE or MOCVD during the last decade have made possible high-quality heterostructures having designed potential profiles and impurity distributions with a dimensional control close to interatomic spacing and with virtually defect-free interfaces in a lattice-matched case such as GaAs-$Ga_{1-x}Al_xAs$. Such great precision, indeed, has allowed access to a new quantum regime. In this review, I will present the evolution of semiconductor superlattices and quantum wells, surveying significant milestones chosen from the research occurring over the past fifteen years.

1. Superlattice Band Model and Experiment (1969-1972)

A synthesized superlattice possesses unusual electronic properties of quasi-two-dimensional character. The introduction of the superlattice potential perturbs the host bandstructure in such a manner as to give rise to narrow subbands separated by forbidden regions, analogous to the Kronig-Penney band model [9]. The electron dynamics in the superlattice direction, with a simplified path integration method [10], was analyzed for conduction electrons in narrow subbands. This calculation predicted an unusual current-voltage characteristic including a differential negative resistance. In 1972, ESAKI et al.[11] found that a MBE-Grown GaAs-GaAlAs superlattice [12] exhibited a negative resistance in its transport properties, which was, for the first time, interpreted in terms of the superlattice effect.

It is worthwhile mentioning here that, in 1974, GNUTZMANN and CLAUSEKER [13] pointed out an interesting possibility; namely, the occurrence of a direct-gap superlattice made of indirect-gap host materials, because of Brillouin-zone folding which was later reexamined by MADHUKAR [14]. The idea suggests the synthesis of new optical materials.

2. Multibarrier Tunneling and Quantum Wells : Theory and Experiment (1973-1974)

In 1973, TSU and ESAKI [15] computed the resonant transmission coefficient T*T for double, triple, and quintuple barrier structures from the tunneling point of view, leading to the derivation of the current-voltage characteristics. The superlattice band model previously presented assumed an infinite periodic structure, whereas, in reality, not only a finite number of periods is prepared with alternating epitaxy, but also the electron mean free path is limited. Thus, this multibarrier tunneling model provided useful insight into the transport mechanism. In early 1974, CHANG, ESAKI and TSU [16] observed resonant tunneling in double-barriers, and subsequently, ESAKI and CHANG [17] measured quantum transport properties for a superlattice having a tight-binding potential. The resonance is achieved at such applied voltages as to align the Fermi level of the electrode with the bound states in the well. Their energies can be obtained from resonant voltages. A number of double barriers were MBE-grown with four different well widths, 35, 40, 50 and 65Å, for which measured values for the bound states were found to agree with the calculation.

These experiments probably constitute the first clear observation of man-made bound states in both single and multiple potential wells. Such achievements can be viewed as a laboratory practice of elementary one-dimensional quantum physics described in textbooks (Do-It-Yourself Quantum Mechanics!). SOLLNER et al. [18] recently demonstrated dramatically improved I-V characteristics in resonant double-barrier tunneling. Such an improvement clearly endorses the evolution of MBE in the last decade.

3. Optical Absorption for Quantum Wells and Superlattices (1974-1975)

DINGLE et al. [19,20] observed pronounced structure in the optical absorption spectrum, representing bound states in isolated [19] and double quantum wells [20]. For the former, GaAs well widths in the range between 70Å and 500Å were prepared. The GaAs wells were separated by $Ga_{1-x}Al_xAs$ barriers which were normally thicker than 250Å. In low-temperature measurements for such structures, several exciton peaks, associated with different bound-electron and bound-hole states, were resolved. For the latter study, a series of structures, with GaAs well widths in the range between 50Å and 200Å and $Ga_{1-x}Al_xAs$ ($0.19 < x < 0.27$) barrier widths between 12Å and 18Å, were grown by MBE on the GaAs substrates. The spectra at low temperatures clearly indicated the evolution of resonantly split discrete states into the lowest subband of a superlattice.

4. Raman scattering (1976-1980)

MANUEL et al. [21] reported the observation of enhancement in the Raman cross-section for photon energies near electronic resonance in GaAs-$Ga_{1-x}Al_xAs$ superlattices of a variety of configurations. The results elucidated the two-dimensionality in the electronic structure. Later, however, the significance of resonant inelastic light scattering as a spectroscopic tool was pointed out by BURSTEIN et al.[22], claiming that the method yields separate spectra of single particle and collective excitations, which will lead to the determination of electronic energy levels in quantum wells as well as Coulomb interactions. Subsequently, ABSTREITER et al. [23] and PINCZUK et al. [24] observed light scattering by intersubband single particle excitations between discrete energy levels, of two-dimensional electrons in GaAs-$Ga_{1-x}Al_xAs$ heterojunctions and quantum wells.

Meanwhile, COLVARD et al.[25] reported the observation of Raman scattering from folded acoustic longitudinal phonons in a GaAs (13.6Å)-AlAs(11.4Å) superlattice. Before this observation, Narayanamurti et al.[26] showed selective transmission of high-frequency phonons due to narrow band reflection determined by the superlattice period.

5. Modulation Doping (1978) and Subsequent Developments

In the original article [1], the spatial separation of carriers and their parent impurities was proposed in order to reduce impurity scattering by means of concentrated doping in the regions of the potential hills. In 1978, DINGLE et al [27] successfully implemented such a concept in modulation-doped GaAs-GaAlAs superlattices, achieving electron mobilities which far exceed the Brooks-Herring predictions. Soon after, STÖRMER et al.[28] reported a two-dimensional electron gas at modulation-doped GaAs-GaAlAs heterostructures. Such heterostructures were used to fabricate a new high-speed field-effect-transistor [29,30] called MODFET. Hall mobilities in the dark at 4.2K for such confined electrons recently exceeded 1000000 cm^2/V.sec. [31,32]. Subsequently, a similar technique provided a two-dimensional hole gas at the hetero-interface [33]. Such a hole gas not only revealed characteristic quantum effects [34] but also was found useful for p-channel MODFETs [35]. More recently, WANG et al.[36] reported a hole mobility as high as 97000cm^2/V.sec at 4.2K and deduced a valence-band offset of 210$\pm$30meV for $Ga_{0.5}Al_{0.5}As$-GaAs heterojunctions, corresponding to $\Delta E_c/\Delta E_g$ = 0.62$\pm$0.05.

6. Quantized Hall Effect (1980-1981) and Discovery of Fractional Filling (1982)

In 1980, V. KLITZING et al.[37] demonstrated the interesting proposition that quantized Hall resistance in Si MOSFETs can be used for precision determination of the fine structure constant α. Subsequently, TSUI and GOSSARD [38] found the modulation-doped GaAs-GaAlAs heterostructures quite desirable for the same purpose, primarily because of their high electron mobilities, which led to the determination of α with great accuracy [39]. The quantized Hall effect in the two-dimensional electron or hole [34,40] system is observable at sufficiently high magnetic fields and low temperatures; in such a range of magnetic fields as to locate the Fermi level in the localized states between the extended states, the magneto-resistance ρ_{xx} vanishes and the Hall resistance ρ_{xy} goes through plateau. This surprising result can be understood by the argument that the localized states do not take part in quantum transport [41].

Recently, TSUI, STÖRMER and GOSSARD [42] discovered a striking phonomenon : the existence of an anomalous quantized Hall effect, a Hall plateau in ρ_{xy} and a dip in ρ_{xx}, at a fractional filling factor of 1/3 in the extreme quantum limit at temperatures lower than 4.2K. This discovery has spurred a large number of experimental and theoretical studies. LAUGHLIN [43], as an explanation of such a fractional filling, presented variational wave functions which describe the condensation of a two-dimensional electron gas into a new state of matter, an incompressible quantum fluid; the elementary excitations of such quantum fluid are fractionally charged.

MENDEZ et al.[44] performed magnetotransport measurements at 0.51K (later 68mK) [45] and up to 28 T for a dilute two-dimensional electron gas with a concentration of $6x10^{10}cm^{-2}$ in a GaAs-GaAlAs heterojunction. The magneto-resistance indicated a substantial deviation from linearity above 18 T and exhibited no additional features for filling factors beyond 1/5, which suggested a transition to a crystalline state. Meantime, Lam and Girvin [46] calculated the critical Landau-level filling factor for transition from Laughlin's liquid state to a Wigner crystal in comparing the energies of these states. The result appears to be consistent with the experimental observation.

7. Variety of Heterojunctions and Superlattices and Relevant Topics

A major portion of the studies reported up to now have been carried out with the GaAs-$Ga_{1-x}Al_xAs$ system. However, a variety of other systems, notably InAs-GaSb(-AlSb), InAlAs-InGaAs, [47], InP-lattice matched alloys [48,49,50,51], Ge-GaAs [52,53], CdTe-HgTe [54,55,56], PbTe-PbSnTe [57,58], and lattice-mismatched pairs of III-V compounds [59,60], (strained-layer superlattices) have also been seriously explored from scientific as well as technical aspects.

Semiconductor hetero-interfaces exhibit the abrupt discontinuity in the local band structure, usually associated with a gradual band bending in its neighborhood which reflects space-charge. Hetero-interfaces may be classified into four kinds : type I, type II-staggered, type II-misaligned, and type III. The conduction band discontinuity ΔE_c is equal to the difference in the electron affinities of the two semiconductors. Type I is applied to the GaAs-AlAs, GaSb-AlSb, GaAs-GaP systems, etc., where their energy-gap difference $\Delta E_g = \Delta E_c + \Delta E_v$. On the other hand, type IIs are applied to pairs of InAs-GaSb, $(InAs)_{1-x}(GaAs)_x$-$(GaSb)_{1-y}(GaAs)_y$ [61], InP-$Al_{0.48}In_{0.52}As$ [51], etc., where electrons and holes are confined in the different semiconductors at their heterojunctions and superlattices. Particularly in type II-misaligned, the top of the valence band in GaSb is located above the bottom of the conduction band in InAs by the amount of E_s, differing from type II-staggered. In this classification, we add a unique member of the family, HgTe-CdTe [56], as type III, where HgTe is a zero-gap semiconductor due to the inversion of the relative positions of Γ_6 and Γ_8 edges. The Γ_8 light-hole band in CdTe becomes the conduction band in HgTe.

The bandedge discontinuities at the hetero-interfaces obviously command all properties of quantum wells and superlattices, and thus constitute the most relevant parameters for device design [62] as well. Recently, considerable efforts have been made to understand the electronic structure at interfaces [63] or heterojunctions [64,65,66]. Even in an ideal situation, the discontinuity provides formidable tasks in theoretical handling : Propagation and evanescent Bloch waves should be matched across the interface, satisfying continuity conditions on the envelope wave functions [67,68]. The fundamental understanding, as well as the experimental determination for such parameters as ΔE_c and ΔE_v, however, is still not satisfactory, even in the most-studied GaAs-$Ga_{1-x}Al_xAs$ system. Values of $\Delta E_c/\Delta E_g$, 85%, and $\Delta E_v/\Delta E_g$, 15% [20], determined in 1975, apparently, are to be revised to about 60% and 40%, respectively [36,69,70,71,72].

Another breed called "n-i-p-i", the outgrowth of a doping superlattice, was pursued by DÖHLER [73] and PLOOG et al. [74]. DÖHLER et al. [75] observed in GaAs doping superlattices that the photon energies in luminescence were varied by the laser excitation intensity. Superlattice structures were also made of amorphous semiconductors. ABELES and TIEDJE [76] pioneered the development of such structures consisting of alternating layers of hydrogenated amorphous silicon, germanium, silicon nitride, and silicon carbide. ESAKI et al [77] proposed the introduction of a third constituent such as AlSb in the InAs-GaSb system. Such a triple-constituent system leads to a new concept of man-made polytype superlattices, which offers an additional degree of freedom.

In heterostructures, it is certainly desirable to select a pair of materials closely lattice-matched in order to minimize defect formation or stress. However, heterostructures lattice-mismatched to a limited extent can be grown with essentially no misfit dislocations, if the layers are sufficiently thin, because the mismatch is accommodated by uniform lattice strain [78]. On the basis of such premise, OSBOURN [59] and his co-workers [60] prepared a number of strained-layer superlattices from lattice-mismatch pairs, claiming their relatively high-quality superlattices to be suitable for opto-devices. It is certainly true that, without the requirement of lattice matching, the number of available pairs for superlattice formation is greatly inflated.

VOISIN et al. [79] made optical absorption measurements on GaSb-AlSb superlattices, of which the spectrum exhibits the two-dimensional density of states and pronounced free exciton peaks. In comparison with the reported optical absorption spectrum [20] for the GaAs-GaAlAs case, it was noticed that the first two exciton peaks appear to be exchanged. Using an effective mass theory, such observation was interpreted as the occurrence of the reversal of the heavy- and light-hole bands, due to the strain effect induced by a lattice mismatch equal to 0.65%.

8. InAs-GaSb Superlattices (1977-1982) and Quantum Wells (1982-)

In 1977, while searching for a new type of superlattice, InAs-GaSb was selected, because of its extraordinary bandedge relationship at the interface, called type II-misaligned. It was observed that, in the study of $(InAs)_{1-x}(GaAs)_x$-$(GaSb)_{1-y}(GaAs)_y$ p-n heterojunctions [61], the rectifying characteristic changes to nonrectification as both x and y approach zero, implying the change-over from the "staggered" heterojunction to the "misaligned" one. At the misaligned-bandgap interface, electrons which "flood" from the GaSb valence band to the InAs conduction band, leaving holes behind, produce a dipole layer consisting of two-dimensional electron and hole gases.

First, SAI-HALASZ et al. [80] made a one-dimensional calculation and, subsequently, the LCAO band calculation [80] was performed for InAs-GaSb superlattices. The calculated subband structure is strongly dependent upon the period : The energy gap decreases with increase in the period, becoming zero at 170Å, corresponding to a semiconductor-to-semimetal transition. In those calculations, the misaligned magnitude, E_s, was set at 0.15eV : a value which had been derived from analysis of optical absorption [82]. Recently, ALTARELLI [83] performed self-consistent electronic structure calculations in the envelope-function approximation with a three-band k.p formalism for this superlattice.

The electron concentration in superlattices was measured as a function of InAs layer thickness [84]; it exhibited a sudden increase of an order-of-magnitude in the neighborhood of 100Å. Such increase indicates the onset of electron-transfer from GaSb to InAs which is in good agreement with theoretical prediction. Far-infrared magneto-absorption experiments [85,86] were performed at 1.6K for semimetallic superlattices,which confirmed their negative energy gap.

MBE-grown GaSb-InAs-GaSb quantum wells have been investigated, where the unique bandedge relationship allows the coexistence of electrons and holes across the two interfaces. Before an experimental approach, BASTARD et al. [87] had performed self-consistent calculations for the electronic properties for such quantum wells, predicting the existence of a semiconductor-to-semimetal transition as a result of electron-transfer from GaSb when the InAs quantum well thickness reaches a threshold, somewhat similar to the mechanism in the InAs-GaSb superlattices. Such a transition was confirmed experimentally by MENDEZ et al. [88]. The threshold thickness was found to be about 60Å. The electron mobility in InAs quantum wells of 150-200Å ranges between 1.5 and $2.0\times10^5 cm^2/V.sec$ at 4.2K. Such values of the mobility were compared with theoretical calculations [89].

Measurements of magnetoresistance ρ_{xx} and the Hall resistance ρ_{xy} caused a few surprises : No holes are evident in ρ_{xy} whereas the increase of ρ_{xx} is proportional to B^2 at low fields, suggesting a two-carrier conduction mechanism; there is a nearly 20% decrease of electrons at high fields; the most of all is the advent of anomalous peaks adjacent to ordinary peaks in Shubnikov-de Haas oscillations. WASHBURN et al.[90] observed that such anomalous peaks weaken with decreasing temperature and finally vanish at 19mK.

Such phenomena appear to be too unusual to be interpreted in terms of the known effects at this time. Nevertheless, we can present the following considerations : First of all, the electron number, n, in the InAs layer, may be balanced not by the hole number, p, alone, but rather by p + N_D, where N_D is the concentration of ionized donor states in the neighborhood of the interfaces. The InAs well is also susceptible to unwanted modulation-doping from GaSb. It seems possible that p is a minor fraction of n, which means a rather large deviation from the theoretical calculations [87]. Secondly, although these holes are contributing to conduction at low magnetic fields, they will form electron-hole bound states, possibly induced by a certain magnetic field strength, 5T or so. The reduction of n may be indicative of the formation of such two-particle states which manifest themselves as anomalous peaks, but cease to contribute to conduction at very low temperatures. Along with this line, the binding energy for such an excitonic state is derived to be 6.4meV from the difference of the positions between the normal and anomalous peaks. The obtained value is about three times the effective Rydberg of InAs (1.9meV), comparable to the hydrogenic binding energy [91] in the two-dimensional system, but larger than the previously calculated exciton energy [92] at no magnetic field condition. Limited experiments have also been carried out for AlSb-InAs-AlSb quantum wells [93].

We have witnessed the remarkable development of an interdisciplinary nature on this subject. Such development can be considered to be a *renaissance* in semiconductor research. Indeed, a variety of *engineered* structures exhibited extraordinary transport and optical properties; some of them, such as ultrahigh carrier mobilities, semimetallic coexistence of electrons and holes, etc., may not even exist in any *natural* crystal. Thus, this new degree of freedom offered in semiconductor research *through advanced material engineering* has inspired many ingeneous experiments, resulting in observations of not only predicted effects but also totally unknown phenomena such as fractional quantization,which require novel interpretations. Activities in this new frontier of semiconductor physics, in turn, give immeasurable stimulus to device physics, leading to novel devices such as MODFETs [29,30], MQW-lasers [94], advanced APDs [95,96] and real-space electron-transfer devices [97], or provoking new ideas [98,99] for applications.

This article concludes with some comments on the prospective research;

1. The route to one dimension is still wide open. A number of attempts, such as GaAs quantum well wires [100] or pinched accumulation layers in Si MOSFET [101], have already been made. A true one-dimensional electron gas, however, can be created with a combination of a superlattice potential and the surface inversion.

2. Our early efforts focussed on transport properties in the direction of the one-dimensional periodic potential including resonant tunneling. Such efforts apparently demand the stringent requirements for lateral uniformity as well as the defect density in grown wafers. With improved techniques there will be more promising studies in this direction.

3. The electronic properties of two-dimensional holes associated with the complexity of the valence-band structure at the hetero-interfaces deserve further scrutiny.

4. The discovery of the fractionally quantized Hall effect is probably one of the most significant events in this field. More efforts will be needed for in-depth understanding of the quantum fluid, including a transition to a Wigner crystal [44,45,46].

5. In two dimensions, there is no true metallic conduction [102] : all of the two-dimensional electron gas systems exhibit a logarithmic increase of the resistance as the temperature decreases [103], which is characteristic of the weak localization. Thus, localization studies are pertinent to the physics of reduced dimensionality.

6. Still required are innovations in theory and experiment for accurate determination of the bandedge offsets at the hetero-interfaces.

7. It is worth pursuing multilayer structures which exhibit electric field-induced effects in either optical [104] or transport properties. Such structures clearly possess device potentials.

Hopefully, this presentation, which cannot possibly cover every landmark, provides some flavor of the excitement in this field.

References

1. L. Esaki, R. Tsu : IBM Research Note RC-2418 (1969)
2. L. Esaki, R. Tsu : IBM J. Res. Develop. 14, 61 (1970)
3. D. Bohm : *Quantum Theory* (Prentice Hall, Englewood Cliffs, N.J. 1951) p. 283
4. L. Esaki : *Les Prix Nobel en 1973*, Imprimerie Royale P.A. Norstedt & Söner, Stockholm 1974, p. 66; Science 183, 1149 (1974)
5. A.E. Blakeslee, C.F. Aliotta : IBM J. Res. Develop. 14, 686 (1970)
6. L. Esaki, L.L. Chang, R. Tsu : In Proc. 12th Int. Conf. Low Temp. Phys., Kyoto, Japan, September 1970. (Keigaku Tokyo, Japan) p. 551

7. A.Y. Cho : Appl. Phys. Lett. 19, 467 (1971)
8. J.M. Woodall : J. Cryst. Growth 12, 32 (1972)
9. R. de L. Kronig, W.J. Penney : Proc. Roy. Soc. A130, 499 (1930)
10. R.G. Chambers : Proc. Phys. Soc. (London) A65, 458 (1952)
11. L. Esaki, L.L. Chang, W.E. Howard, V.L. Rideout : Proc. 11th Int. Conf. Phys. Semicond. Warsaw, Poland, 1972, edited by the Polish Academy of Sciences (PWN-Polish Scientific Publishers, Warsaw, Poland 1972) p. 431
12. L.L. Chang, L. Esaki, W.E. Howard, R. Ludeke : J. Vac. Sci. Technol. 10, 11 (1973); L.L. Chang, L. Esaki, W.E. Howard, R. Ludeke, G. Schul : J. Vac. Sci. Technol. 10, 655 (1973)
13. U. Gnutzmann and K. Clauseker : Appl. Phys. 3, 9 (1974)
14. A. Madhukar : J. Vac. Sci. Technol. 20, 149 (1982)
15. R. Tsu, L. Esaki : Appl. Phys. Lett. 22, 562 (1973)
16. L.L. Chang, L. Esaki, R. Tsu : Appl. Phys. Lett. 24, 593 (1974)
17. L. Esaki, L.L. Chang : Phys. Rev. Lett. 33, 495 (1974)
18. T.C.L.G. Sollner, W.E. Goodhue, P.E. Tannenwald, C.D. Parker, D.D. Peck : Appl. Phys. Lett. 43, 588 (1983)
19. R. Dingle, W. Wiegmann, C.H. Henry : Phys. Rev. Lett. 33, 827 (1974)
20. R. Dingle, A.C. Gossard, W. Wiegmann : Phys. Rev. Lett. 34, 1327 (1975)
21. P. Manuel, G.A. Sai-Halasz, L.L. Chang, C.-A. Chang, L. Esaki : Phys. Rev. Lett. 25, 1701 (1976)
22. E. Burstein, A. Pinczuk, S. Buchner : Physics of Semiconductors 1978, Institute of Physics Conference Series 43, (London 1979), p. 1231
23. G. Abstreiter, K. Ploog : Phys. Rev. Lett. 42, 1308 (1979)
24. A. Pinczuk, H.L. Störmer, R. Dingle, J.M. Worlock, W. Wiegmann, A.C. Gossard : Solid State Commun. 32, 1001 (1979)
25. C. Colvard, R. Merlin, M.V. Klein, A.C. Gossard : Phys. Rev. Lett. 45, 298 (1980)
26. V. Narayanamurti, H.L. Störmer, M.A. Chin, A.C. Gossard, W. Wiegmann : Phys. Rev. Lett. 43, 2012 (1979)
27. R. Dingle, H.L. Störmer, A.C. Gossard, W. Wiegmann : Appl. Phys. Lett. 33, 665 (1978)
28. H.L. Störmer, R. Dingle, A.C. Gossard, W. Wiegmann, M.D. Sturge : Solid State Comm. 29, 705 (1979)
29. T. Mimura, S. Hiyamizu, T. Fujii, K. Nanbu : Jpn. J. Appl. Phys. 19, L225 (1980)
30. D. Delagebeaudeuf, P. Delescluse, P. Etienne, M. Laviron, J. Chaplart, N.T. Linh : Electron. Lett. 16, 667 (1980)
31. M. Heiblum, E.E. Mendez, F. Stern : Appl. Phys. Lett. 44, 1064 (1984)
32. E.E. Mendez, P.J. Price, M. Heiblum : to appear in Appl. Phys. Lett.
33. H.L. Störmer, W.T. Tsand : Appl. Phys. Lett. 36, 685 (1980)
34. H.L. Störmer, Z. Schlesinger, A. Chang, D.C. Tsui, A.C. Gossard, W. Wiegman : Phys. Rev. Lett. 51, 126 (1983)
35. H.L. Störmer, K. Baldwin, A.C. Gossard, W. Wiegmann : Appl. Phys. Lett. 44, 1062 (1984)
36. W.I. Wang, E.E. Mendez, F. Stern : Appl. Phys. Lett. 45, 639 (1984)
37. K.v. Klitzing, G. Dorda, M. Pepper : Phys. Rev. Lett. 45, 494 (1980)
38. D.C. Tsui, A.C. Gossard : Appl. Phys. Lett. 38, 550 (1981)
39. D.C. Tsui, A.C. Gossard, B.F. Field, M.E. Cage, R.F. Dziuba : Phys. Rev. Lett. 48, 3 (1982)
40. E.E. Mendez, W.I. Wang, L.L. Chang, L. Esaki : Phys. Rev. B 28, 4886 (1983)
41. T. Ando, Y. Uemura : J. Phys. Soc. Jpn. 36, 959 (1974)
42. D.C. Tsui, H.L. Störmer, A.C. Gossard : Phys. Rev. Lett. 48, 1559 (1982)
43. R.B. Laughlin : Phys. Rev. Lett. 50, 1395 (1983)
44. E.E. Mendez, M. Heiblum, L.L. Chang, L. Esaki : Phys. Rev. B, 28, 4886 (1983)
45. E.E. Mendez, L.L. Chang, M. Heiblum, L. Esaki, M. Naughton, K. Martin, J. Brooks : to appear in Phys. Rev. B
46. Pui K. Lam, S.M. Girvin : Phys. Rev. B, 30, 473 (1984)
47. R. People, K.W. Wecht, K. Alavi, A.Y. Cho : Appl. Phys. Lett. 43, 118 (1983)
48. K.Y. Cheng, A.Y. Cho, W.R. Wagner : Appl. Phys. Lett. 39, 607 (1981)
49. M. Razeghi, J.P. Duchemin : J. Vac. Sci. Technol. B, 1, 262 (1983)
50. M. Voos : J. Vac. Sci. Technol. B, 1, 404 (1983)

51. E.J. Caine, S. Subbanna, H. Kröemer, J.L. Merz, A.Y. Cho : to appear in Appl. Phys. Lett.
52. P.M. Petroff, A.C. Gossard, A. Savage, W. Wiegmann : J. Cryst. Growth 46, 172 (1979)
53. C-A. Chang, A. Segmüller, L.L. Chang, L. Esaki : Appl. Phys. Lett. 38, 912 (1981)
54. J.N. Schulman, T.C. McGill : Appl. Phys. Lett. 34, 883 (1979)
55. G. Bastard : Phys. Rev. B, 25, 7584 (1982)
56. Y. Guldner, G. Bastard, J.P. Vieren, M. Voos, J.P. Faurie, A. Million, Phys. Rev. Lett. 51, 907 (1983)
57. H. Kinoshita, H. Fujiyasu : J. Appl. Phys. 51, 5845 (1980)
58. E.F. Fantner, G. Bauer : *In Two-Dimensional Systems, Heterostructures, and Superlattices*, ed. by G. Bauer, F. Kuchar, H. Heinrich, Springer Ser. Solid-State Sci.Vol. 53, (Springer, Berlin, Heidelberg (1984), p. 207
59. G.C. Osbourn : J. Appl. Phys. 53, 1586 (1982)
60. G.C. Osbourn, R.M. Biefeld, P.L. Gourley : Appl. Phys. Lett. 41, 172 (1982)
61. H. Sakaki, L.L. Chang, R. Ludeke, C.-A. Chang, G.A. Sai-Halasz, L. Esaki : Appl. Phys. Lett. 31, 211 (1977)
62. H. Kroemer : Surf. Sci. 132, 543 (1983)
63. M.L. Cohen : Advances in Electronics and Electron Physics, Vol. 51, 1 (Academic, New York 1980)
64. W. Harrison : J. Vac. Sci. Technol. 14, 1016 (1977)
65. W.R. Frensley, H. Kröemer : Phys. Rev. B16, 2642 (1977)
66. J. Tersoff : to appear in Phys. Rev. B
67. G. Bastard : Phys. Rev. B24, 5693 (1981)
68. S.R. White, L.J. Sham : Phys. Rev. Lett. 47, 879 (1981)
69. H. Kröemer, Wu-Yi Chen, J.S. Harris, Jr., D.D. Edwall : Appl. Phys. Lett. 36, 295 (1980)
70. R.C. Miller, A.C. Gossard, D.A. Kleinman, O. Munteanu : Phys. Rev. B29, 3740 (1984)
71. R.C. Miller, D.A. Kleinman, A.C. Gossard : Phys. Rev. B29, 7085 (1984)
72. T.W. Hickmott, P.M. Solomon, R. Fisher, H. Morkoc : to appear in J. Appl. Phys.
73. G.H. Döhler, Phys. Status Solidi (b) 52, 79, 533 (1972)
74. K. Ploog, A. Fischer, G.H. Döhler, H. Künzel : In *Gallium Arsenide and Related Compounds 1980*, Institute of Physics Conference Series N° 56, ed. by H.W. Thim (Institute of Physics, London 1981) p. 721
75. G.H. Döhler, H. Kunzel, D. Olego, K. Ploog, P. Ruden, H.J. Stolz, G. Abstreiter, Phys. Rev. Lett. 47, 864 (1981)
76. A. Abeles, T. Tiedje : Phys. Rev. Lett. 51, 2003 (1983)
77. L. Esaki, L.L. Chang, E.E. Mendez : Jpn. J. Appl. Phys. 20, L529 (1981)
78. J.H. van der Merwe : J. Appl. Phys. 34, 117 (1963)
79. P. Voisin, C. Delalande, M. Voos, L.L. Chang, A. Segmüller, C.A. Chang, L. Esaki, to appear in Phys. Rev. B
80. G.A. Sai-Halasz, R. Tsu, L. Esaki : Appl. Phys. Lett. 30, 651 (1977)
81. G.A. Sai-Halasz, L. Esaki, W.A. Harrison : Phys. Rev. B18, 2812 (1978)
82. G.A. Sai-Halasz, L.L. Chang, J-M. Welter, C.-A. Chang, L. Esaki : Solid State Commun. 25, 935 (1978)
83. M. Altarelli : Phys. Rev. B28, 842 (1983)
84. L.L. Chang, N.J. Kawai, G.A. Sai-Halasz, R. Ludeke, L. Esaki : Appl. Phys. Lett. 35, 939 (1979)
85. Y. Guldner, J.P. Vieren, P. Voisin, M. Voos, L.L. Chang, L. Esaki : Phys. Rev. Lett. 45, 1719 (1980)
86. J.C. Maan, Y. Gulder, J.P. Vieren, P. Voisin, M. Voos, L.L. Chang, Esaki : Solid State Commun. 39, 683 (1981)
87. G. Bastard, E.E. Mendez, L.L. Chang, L. Esaki : J. Vac. Sci. Technol. 21, 531 (1982)
88. E.E. Mendez, L.L. Chang, C-A. Chang, L.F. Alexander, L. Esaki : Surf. Sci. 142, 215 (1984)
89. E.E. Mendez, G. Bastard, L.L. Chang, C-A. Chang, L. Esaki : Bull. Am. Phys. Soc. 29, 471 (1984)
90. W. Washburn, R.A. Webb, E.E. Mendez, L.L. Chang, L. Esaki : to appear in Phys. Rev. B

91. G. Bastard : Phys. Rev. B24, 4714 (1981)
92. G. Bastard, E.E. Mendez, L.L. Chang, L. Esaki : Phys. Rev. B 26, 1974 (1982)
93. C-A. Chang, E.E. Mendez, L.L. Chang, L. Esaki : Surf. Sci. 142, 598 (1984).
94. W.T. Tsang, Appl. Phys. Lett. 39, 786 (1981)
95. F. Capasso : J. Vac. Sci. Technol. B1, 457 (1983)
96. T. Tanoue, H. Sakaki : Appl. Phys. Lett. 41, 67 (1982).
97. K. Hess, M. Morkoc, H. Shichijo, B.G. Streetman : Appl. Phys. Lett. 35, 469 (1979)
98. J.J. Quinn, U. Strom, L.L. Chang : Solid State Commun. 45, 111 (1983)
99. T. Nakagawa, N.J. Kawai, K. Ohta, M. Kawashima : Electronics Lett. 19, 822 (1983). Phys. Rev. B29, 3752 (1984)
100. P.M. Petroff, A.C. Gossard, R.A. Logan, W. Wiegmann : Appl. Phys. Lett. 41, 635 (1982)
101. A.B. Fowler, A. Hartstein, R.A. Webb : Phys. Rev. Lett. 48, 196 (1982)
102. E. Abrahams, P.W. Anderson, D.C. Licciardello, T.V. Ramakrishnan, Phys. Rev. Lett. 42, 673 (1979)
103. S. Washburn, R.A. Webb, E.E. Mendez, L.L. Chang, L. Esaki : Phys. Rev. B29, 3752 (1984)
104. E.E. Mendez, G. Bastard, L.L. Chang, L. Esaki : Phys. Rev. B26, 7101 (1982)

Part II

Theory

Band Structure, Impurities and Excitons in Superlattices

M. Altarelli

Max-Planck-Institut für Festkörperforschung,
D-7000 Stuttgart 80, Fed. Rep. of Germany* and
Hochfeld-Magnetlabor, BP 166 X, F-38042 Grenoble, France

In these lectures I shall review (at an introductory level) the present knowledge of electronic states in semiconductor heterostructures. Although comparison with experiment will be illustrated as often as possible, the emphasis is on the theoretical description through the envelope-function approximation. This approach to the calculation of electronic levels has proven to be simple, accurate and versatile and, therefore, superior to other methods. Fully microscopic calculations become indeed intractable, for quantum wells and superlattices with thicknesses of experimental interest. Empirical tight binding calculations are on the other hand simple, but can hardly be extended to include external fields, or to account for charge rearrangements in a self-consistent way.

Electronic states in three types of heterostructures will be considered : superlattices, isolated quantum wells and heterojunctions (Fig.1). However, the latter two can be regarded as limiting cases of the first, when the thickness of the layers becomes much larger than some characteristic wavefunction decay length or screening length. When the thickness of the "barrier" material becomes large, one recovers the isolated quantum well limit; when also the "well" material is very thick one obtains isolated heterojunctions. Therefore, at least in principle, a complete treatment for superlattices includes the other cases of interest.

We shall proceed by discussing first the motion of electron and holes in a slowly varying field, via the effective-mass equation, and then considering the appropriate boundary conditions on the envelope-functions (i.e. effective-mass wavefunctions) at a sharp interface between the semiconductors. The actual band structure of III-V semiconductors, in particular the valence band degeneracy and anisotropy, complicates the structure of the equations. We shall, however, see that it is possible to obtain numerical solutions for the band structure of superlattices that compare favorably with experimental information. The inclusion of perpendicular magnetic fields will then be discussed. Finally, a short review of calculations of impurity and exciton states in quantum wells will be given.

1. The Envelope-Function Approximation

1.1 The Effective-Mass Equation

The envelope-function approximation is an effective-mass theory, familiar from the treatment of shallow impurities [1]. Consider the motion of an electron in a semiconductor in presence of some additional potential $U(\vec{r})$. The Schrödinger equation can be written :

$$\left[\frac{p^2}{2m_o} + V_{per}(\vec{r}) + U(\vec{r}) - E \right] \psi(\vec{r}) = 0 \qquad (1)$$

Here m_o is the free-electron mass, V_{per} is the periodic potenial of the perfect bulk semiconductor and $U(\vec{r})$ is the additional potential, which we assume to be slowly varying and weak, in the sense precisely stated below. For example, U(z) could be the band-bending potential on the right side of Fig. 1(c) (z being the

---

* Present address

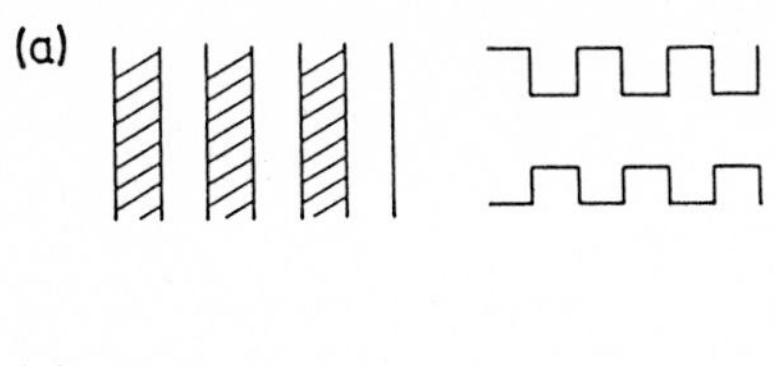

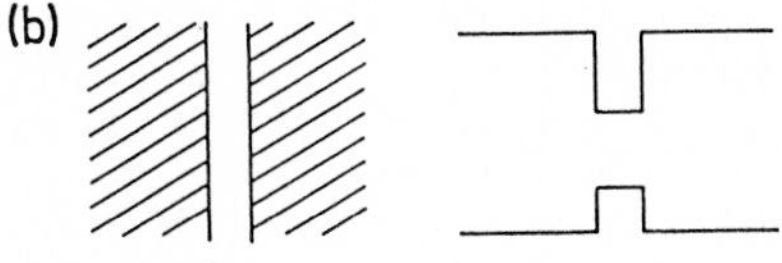

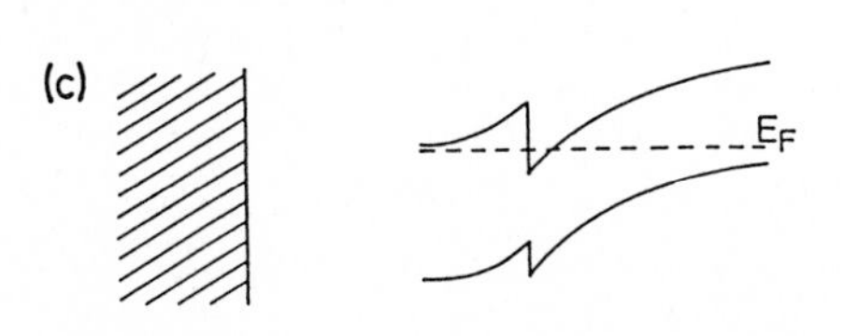

Fig.1 Schematic representation of three types of heterostructures and, on the right, the corresponding band-edge profile. (a) superlattice; (b) isolated quantum well; (c) single heterojunction between an n-type material (on the left side of the junction) and a p-type one (on the right). E_F denotes the Fermi level.

coordinate normal to the interface), arising from the depleted acceptors. Eq. (1) would then describe the motion of an electron on the right side of the junction. A similar equation, but with different V_{per} and U would hold on the left side, and we shall worry later about the matching of the solutions across the interface. If the potential $U(\vec{r})$ is vanishing, we know the solutions of Eq. (1) to be the Bloch functions $\psi_{n\vec{k}}(\vec{r})$, with :

$$\left[\frac{p^2}{2m_o} + V_{per}(\vec{r}) \right] \psi_{n\vec{k}}(\vec{r}) = E_n(\vec{k})\, \psi_{n\vec{k}}(\vec{r}) \tag{2}$$

and

$$\psi_{n\vec{k}}(\vec{r}) = e^{i\vec{k}\vec{r}} u_{n\vec{k}}(\vec{r}), \quad (u_n \text{ periodic}) \tag{2'}$$

In Eq. (2,2') $\vec{k}$ is restricted to the first Brillouin zone. For $U(\vec{r}) \neq 0$ we expand the solutions of (1) in the form

$$\psi(\vec{r}) = \sum_{n\vec{k}}\sum \phi_n(\vec{k})\, \psi_{n\vec{k}}(\vec{r}) \tag{3}$$

Replacing this in (1) gives, for the coefficient functions $\phi_n(\vec{k})$:

$$[E_n(\vec{k}) - E]\, \phi_n(\vec{k}) + \sum_{n'\vec{k}'} \langle \psi_{n\vec{k}} | U(\vec{r}) | \psi_{n'\vec{k}'} \rangle\, \phi_{n'}(\vec{k}') = 0 \tag{4}$$

where the potential matrix element is :

$$\int \psi^*_{n\vec{k}}(\vec{r})\, U(\vec{r}) \psi_{n'\vec{k}'}(\vec{r})\, d^3r = \int e^{-i(\vec{k}-\vec{k}')\vec{r}} u^*_{n\vec{k}}(\vec{r}) u_{n'\vec{k}'}(\vec{r}) U(\vec{r})\, d^3r$$

$$= \sum_{\vec{G}} \tilde{U}(\vec{k}'-\vec{G})\, c(n\vec{k}, n'\vec{k};\vec{G}) \tag{5}$$

The last step is derived by expanding the function $u^*_{n\vec{k}} u_{n'\vec{k}'}$, which is periodic, in a Fourier series of reciprocal lattice vectors :

$$u^*_{n\vec{k}}(\vec{r})u_{n'\vec{k}'}(\vec{r}) = \frac{1}{(2\pi)^3}\sum_{\vec{G}} c(n\vec{k}, n'\vec{k}'; \vec{G})e^{i\vec{G}.\vec{r}} \tag{6}$$

so that the d^3r integration in Eq. (5) yields the Fourier transform $\tilde{U}$ of U. To proceed further we make use of the assumption that $U(\vec{r})$ is slowly varying on the length scale of the unit cell. This means that the Fourier transform $\tilde{U}$ is very small unless its argument is much smaller than the Brillouin zone. Let us consider, as we shall almost invariably do in the following, states with energy close to the bottom of the conduction band or, equivalently, to the top of the valence band, both of which we assume located at $\vec{k} = 0$, as in GaAs and in most III-V semiconductors. Then, only $\vec{k}$ vectors in a small region of the zone are mixed by U and contribute to the expansion (3) for a given state. We can then ignore all terms with $\vec{G} \neq 0$ and write

$$< \psi_{n\vec{k}}|U(\vec{r})|\psi_{n'\vec{k}'} > \simeq \tilde{U}(\vec{k}-\vec{k}')\ c(n\vec{k}, n'\vec{k}'\ ;\ 0) \tag{7}$$

In the evaluation of $c(n\vec{k}, n'\vec{k}'; 0)$ we may use the fact that $\vec{k}$ and $\vec{k}'$ are near $\vec{k} = 0$ and express $u_{n\vec{k}}$ and $u_{n'\vec{k}'}$ in terms of the $\vec{k} = 0$ periodic functions u_{m0}, by means of the standard $\vec{k}.\vec{p}$ perturbation theory :

$$u_{n\vec{k}}(\vec{r}) = u_{n0}(\vec{r}) + \sum_{m\neq n} \frac{\vec{k}.\vec{p}_{mn}}{m_o(E_n(0) - E_m(0))} u_{m0}(r) + \ldots.. \tag{8}$$

The terms omitted are of order $(\vec{k}.\vec{p}_{mn}/m_o(E_m(0)-E_n(0))^2$ and smaller. If the band n is well separated by all others, then the denominators are always much larger than the numerators, for small $\vec{k}$, and the approximation is very good. If on the other hand there are small energy gaps or even degeneracy at $\vec{k} = 0$ of several bands, the theory must be modified, as discussed in Section 1.3. Until then, we deal with the case of a non-degenerate band separated by large gaps from all others. To the same order of accuracy, we find from Eq. (8) that

$$c(n\vec{k};\ n'\vec{k}\ ;\ 0) = \delta nn' + \frac{(\vec{k}-\vec{k}').\vec{p}_{nn'}}{m_0(E_n(o)-E_{n'}(o))}(1-\delta_{nn'}) \tag{9}$$

Therefore from Eq. (7)

$$< \psi_{n\vec{k}}|U(r)|\psi_{n'\vec{k}'} > \simeq \tilde{U}(\vec{k}-\vec{k}')(\delta_{nn'} + A_{nn'}(1-\delta_{nn'})) \tag{10}$$

where $A_{nn'}$ is the off-diagonal contribution from the second piece of Eq. (9). If the potential is weak, however, this off-diagonal term can be ignored; it gives, indeed, when treated as a perturbation, a second-order correction to the energy, and a correction of order

$$\frac{(\vec{k}-\vec{k}').\vec{p}_{nn'}}{m_0(E_n(o)-E_{n'}(0))} \cdot \frac{\tilde{U}(\vec{k}-\vec{k}')}{(E_n(o)-E_{n'}(o))} u_{n'} \tag{11}$$

to the wavefunction, i.e. smaller by a factor $\sim U/\Delta E$ than the first order $\vec{k}.\vec{p}$ correction of Eq. (8). Here ΔE is a typical energy gap of the system ($\sim$1eV) and U a matrix element of the potential; by "weak" potential we mean that the ratio $U/\Delta E$ is small.

Eq. (4) reduces finally to

$$[E_n(\vec{k}) - E]\ \phi_n(\vec{k}) + \sum_{k'} \tilde{U}(\vec{k}-\vec{k}')\phi_n(\vec{k}') = 0 \tag{12}$$

In the small region around $\vec{k} = 0$ of interest here, $E(\vec{k})$ is also well approximated by the 2nd order k.p expansion, which in general reads

$$E_n(\vec{k}) = E_n(o) + \sum_{\alpha,\beta=1}^{3} \frac{\hbar^2}{2m_0} (\delta\alpha\beta + \frac{1}{m_0} \sum_{m\neq n} \frac{p^{\alpha}_{nm}p^{\beta}_{mn} + p^{\beta}_{nm}p^{\alpha}_{mn}}{E_n(o) - E_m(o)}) k_\alpha k_\beta \qquad (13)$$

where α, β run over x,y and z. In the case of a simple, isotropic non-degenerate band extremum (e.g. the conduction band minimum of GaAs-related direct semiconductors) Eq. (13) has nonvanishing diagonal contributions only, i.e. it reduces to :

$$E_n(\vec{k}) = E_n(o) + \frac{\hbar^2}{2m^*} k^2 \quad \text{where}$$

$$\frac{1}{m^*} = \frac{1}{m_0} + \frac{2}{m_0^2} \sum_{m\neq n} \frac{p^{\alpha}_{nm}p^{\alpha}_{mn}}{E_n(o)-E_m(o)} \qquad (13')$$

(α = x or y or z)
is the inverse effective mass of band n.

Taking a Fourier transform of (2) and using Eq. (13) and (13') as though they were valid in the whole k-space, one finds :

$$[-\frac{\hbar^2}{2m^*} \nabla^2 + U(\vec{r})]\, F(\vec{r}) = (E - E_n(o))\, F(\vec{r}) \qquad (14)$$

$F(\vec{r})$ being the Fourier transform of $\phi_n(\vec{k})$. Eq. (14) is the effective-mass equation familiar from the theory of shallow impurities. It is important for our purposes to stress the meaning of the "effective-mass" or "envelope" function $F(\vec{r})$. The total wavefunction, in this approximation, is, from Eq. (3) and (8) :

$$\psi(\vec{r}) = \sum_{\vec{k}} \phi_n(\vec{k})\psi_{n\vec{k}}(\vec{r}) =$$

$$= \sum_k \psi_n(\vec{k})\, e^{i\vec{k}\vec{r}} [u_{no}(r) + \sum_{m\neq n} \frac{\vec{k}.\vec{p}_{mn}}{m_o(E_n(o) - E_m(o)} u_{mo}(r)]$$

$$= F(\vec{r})u_{no}(\vec{r}) + \sum_{m\neq n} \frac{-i(\vec{\nabla}F(\vec{r})) \cdot \vec{p}_{mn}}{m_o(E_n(o) - E_m(o))} u_{mo}(\vec{r}) \qquad (15)$$

Eq. (15) shows that to lowest order $F(\vec{r})$ is a slowly varying "envelope" modulating the rapidly varying Bloch part $u_{no}(\vec{r})$. The following correction term shows that there is a contribution from other $\vec{k}$ = 0 Bloch functions, proportional to the gradient of the envelope $F(\vec{r})$. Eq. (15) is most often quoted without the gradient term. However, it is important not to forget this term if one is interested in investigating the *derivative* of the wave function $\psi(\vec{r})$, as we shall do next to investigate the boundary conditions for the effective-mass equation at the sharp boundary between two semiconductors. This was actually the reason for this pedantic rederivation of the effective-mass equation.

1.2 Boundary conditions for the envelope function

The effective-mass equation (14) has the remarkable feature that all reference to the microscopic structure of the host semiconductor is condensed in the effective-mass m* and the band edge energy E_n (o). This is possible when the potential $U(\vec{r})$ is weak and slowly varying. The two parameters m* and $E_n(o)$ assume different values in the two semiconductors, say A and B, making up an interface system. Given the high quality of state-of-the-art heterostructures, the transition region includes only a few atomic layers. One could think of writing a more general form of Eq. (14), in which the effective-mass and the band edge vary as a function of z. The z-dependent $E_n(o)$ can be attached to $U(\vec{r})$ to form a new effective potential $U(\vec{r}) + E_n(o,z)$. The modified equation (14) would then read :

$$[- \frac{\hbar^2}{2} \vec{\nabla} \cdot (\frac{1}{m^*(z)} \vec{\nabla}) + U(\vec{r}) + E_n(o,z)] F(\vec{r}) = E F(\vec{r}) \tag{16}$$

The kinetic energy term has been rewritten, for a z-dependent mass, in a way which restores the hermitian character of the Hamiltonian, following Harrison [3] and Ben Daniel and Duke [4]. For z well to the left of the interface, $m^* = m^*_A$, $E_n(o) = E_n^A(o)$, and for z well on the right side $m^* = m^*_B$ and $E_n(o) = E_n^B(o)$.

We cannot take Eq. (16) seriously, however, because the variation in $E_n(o,z)$ between $E_n^A(o)$ and $E_n^B(o)$, which is typically as large as 0.1-1 eV, takes place over a few lattice distances. The potential term in Eq. (16), therefore, varies much too rapidly for the effective-mass formalism to be valid. Nevertheless , we can learn something about the boundary conditions from this differential equation. Taking the limit in which the effective-mass and bandedge energy variations occur over an infinitesimal thickness 2ε and integrating Eq. (16) between $z = -\varepsilon$ and $z = +\varepsilon$ ~~and z = +ε~~ we obtain

$$F^A(-\varepsilon) = F^B(+\varepsilon)$$
$$\frac{1}{m^*_A} \frac{\partial}{\partial z} F^A)_{-\varepsilon} = \frac{1}{m^*_B} \frac{\partial}{\partial z} F^B)_{+\varepsilon} \tag{17}$$

In order for the boundary conditions (17) on the envelope functions to make physical sense, we must see their implications on the total wavefunction as described by Eq.(15). In order for ψ to be continuous when F is, one must assume

$$u^A_{no} \simeq u^B_{no} \tag{18}$$

and that the second term of the wavefunction (15) is small, i.e. that the $\vec{k}$-dependence of the Bloch function $u_{n\vec{k}}$ about $\vec{k} = 0$ be weak, as emphasized by Ben Daniel and Duke [4]. Eq. (18) is plausible, in the III-V semiconductor family, as long as we are considering the same band edge on both sides (e.g. the conduction band direct minimum). Then, by looking at pseudopotential wavefunctions, one sees that Eq. (18) is reasonably verified.

Consider now the probability current operator. The existence of stationary states (probability density constant in time) implies that the z-component of the current be the same on all planes parallel to the interface, therefore, also that its average over a microscopic volume Ω, including one or few unit cells, be the same on both sides of the interface. Let us calculate this average on the A side. To do this, we use the wavefunction as given by Eq. (15) and recall that

$$\int_{cell} u^*_{no} \frac{\partial}{\partial z} u_{mo} = \frac{i}{\hbar} p_{nm} \tag{19}$$

and that this matrix element vanishes for m = n, for symmetry reasons [5]. We find, for the average current J

$$\bar{J}_A = \frac{\hbar}{m_o} \operatorname{Im} \int_\Omega d^3r \, \psi^*_A \frac{\partial}{\partial z} \psi_A = \frac{\hbar}{m^*_A} \operatorname{Im} (F^*_A(o) \frac{\partial}{\partial z} F_A(o)) \tag{20}$$

where use was made of Eq. (13') and (15). Therefore, the continuity of $\bar{J}$ implies :

$$\frac{\hbar}{m_A^*} \mathrm{Im}(F_A^*(o) \frac{\partial}{\partial z} F_A(o)) = \frac{\hbar}{m_B^*} \mathrm{Im}(F_B^*(o) \frac{\partial}{\partial z} F_B(o)) \quad (21)$$

It is now apparent that the boundary conditions (17) on the envelope functions imply that the average of the probability current is constant (see Eq. (21)). They are, therefore, meaningful on physical grounds and compatible with the limit behaviour of the effective-mass equation for z-dependent mass and band-edge, and we shall adopt them in our treatment of heterostructures. A further discussion of the boundary conditions is given Section 1.4.

1.3 Coupled bands

The conditions for the validity of the simple-band approach are certainly violated in many situations of interest. This may happen because of various reasons :

a) Band degeneracy near an extremum, as in the case of the valence band maximum at Γ in all cubic semiconductors (Fig. 2).
b) Coupling between bands producing deviations from parabolicity, as in the conduction band of direct gap semiconductors. For narrow-gap materials, like InAs or InSb the non-parabolicity of the conduction band, due to coupling with the valence bands, is quite large for energies very near the band minimum [2], but even in GaAs it has a sizeable effect on levels with energy> 0.1 eV above the band minimum.

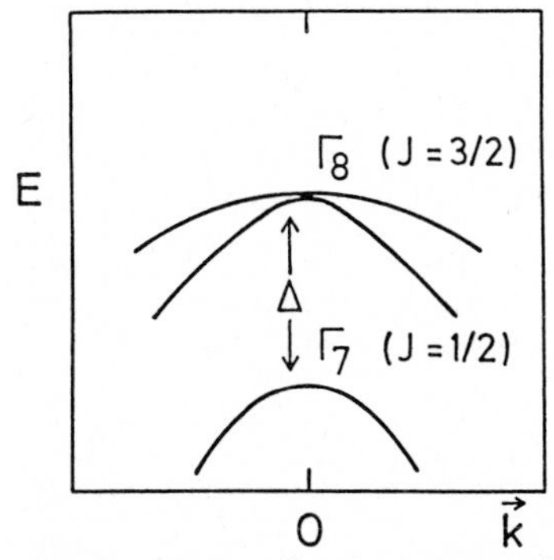

Fig.2 Schematic representation of the top of the valence band of cubic semiconductors. Notations appropriate to zinc-blende semiconductors are used. Δ denotes the spin-orbit splitting.

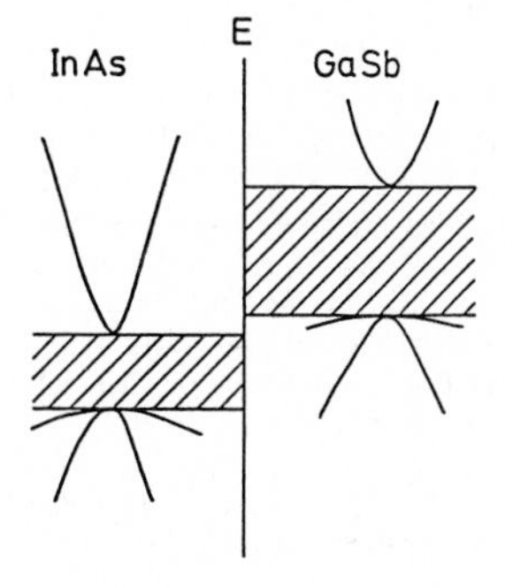

Fig.3 Schematic representation of the energy-gap line-up at the InAs-GaSb interface. The hatched areas denote the energy-gap.

c) There are situations specific to heterostructures in which the single band approach fails; if the two materials have a "staggered" energy gap configuration (see Fig. 3) then, in a large and interesting energy range, the wavefunction has conduction band character on one side of the junction and valence band character on the other. InAs-GaSb superlattices provide an example of this situation.

The simple-band case was treated in a way modeled on the theory of donor impurities; the case in which many bands contribute with comparable weight to the formation of the eigenfunctions is modeled on the theory of acceptors [1]. We start by identifying the bulk bands which are to be treated on the same footing, and describe their bahaviour near $\vec{k} = 0$ by the generalization of Eq. (13), i.e. we write [2]

$$H_{1m}(\vec{k}) = E_1(o)\delta_{1m} + \sum_{\alpha=1}^{3} p_{1m}^{\alpha} k_{\alpha} + \sum_{\alpha,\beta=1}^{3} D_{1m}^{\alpha,\beta} k_{\alpha} k_{\beta} \quad (22)$$

where l, $m = 1,2, \ldots, n$, and α, β run over the x, y and z directions. Given a $\vec{k}$-vector, the n band energies $E_l(\vec{k})$ are given by the eigenvalues of the nxn matrix $H_{lm}(\vec{k})$. The direct $\vec{k}.\vec{p}$ coupling between the n bands is thus retained in the terms $p^{\alpha}_{lm}k_{\alpha}$, where the matrix p^{α} is given by :

$$p^{\alpha}_{lm} = \frac{\hbar}{m_0} < u_l | p^{\alpha} | u_m > \tag{23}$$

The k-quadratic terms proportional to the matrix $D^{\alpha,\beta}$, on the other hand, represent the indirect $\vec{k}.\vec{p}$ coupling between two of the n bands via the other bands (n+1 to ∞) not included in the set. They have indeed an expression very similar to the term in parentheses in the l.h.s. of Eq. (13). To give a specific example, consider the 6x6 matrix given in Table I, which represents the conduction and the upper spin-orbit split component of the valence band. (The matrix is in atomic units, in which $\hbar = m_o = 1$). This is a very good description of these bands for materials with large spin-orbit splittings. We have, therefore, a conduction band with s-like character at Γ and two spin states, s↑ and s↓. The effective-mass m* appearing in the corresponding diagonal terms originates from coupling to the bands not included in the set, and primarily from the split-off valence band [6]. The valence band has p-like character and the upper spin-orbit manifold corresponds to $J = 3/2$ states, classified by the four possible J_z values from -3/2 to 3/2. The parameter P is defined in terms of the interband momentum matrix element, $iP = \langle s | p^x | p_x \rangle$, where p_x indicates the p_x-like valence wavefunction and s the conduction wavefunction at Γ. Then it turns out that $m^{*-1} = 1 + 2P^2/3(E_c-E_v+\Delta)$, where Δ is the valence band spin-orbit splitting. If we wish to consider the energy region close to the valence band top, we can ignore the conduction band altogether, i.e. remove row and columns 1 and 4 (and renormalize the value of γ_1, γ_2 and γ_3). The resulting 4x4 matrix is the Luttinger Hamiltonian [7], which describes the valence band top and the splitting in light and heavy holes bands at $\vec{k} \neq 0$. The parameters γ_1, γ_2 and γ_3 are specific of each material, and they make up the $D^{\alpha,\beta}$ matrices.

We are now ready to generalize Eq. (14), (15) and (17) for the many-band case, in analogy with the many-band effective-mass equation written by Luttinger and Kohn [8] for acceptor impurities. We obtain a system of n differential equations :

$$\sum_{m=1}^{n} [H_{lm}(-i\vec{\nabla}) + U(\vec{r})\delta_{lm}] F_m(\vec{r}) = E F_l(\vec{r}) \tag{24}$$

for the n-component envelope-function $F_l(\vec{r})$, $l = 1,2,\ldots,n$. The total wavefunction, $\psi(\vec{r})$, is expressed, in analogy to Eq. (15), as :

$$\psi(\vec{r}) = \sum_{l=1}^{n} [F_l(\vec{r}) u_{lo}(\vec{r}) + \sum_{m>n} \frac{-i(\vec{\nabla}F_l(\vec{r})).\vec{p}_{lm}}{m_o(E_l(o) - E_m(o))} u_{mo}(\vec{r})] \tag{25}$$

Two facts are worth noticing in Eq. (24). The "kinetic energy" part is just obtained by replacing $\vec{k}$ with $-i\vec{\nabla}$ in the $\vec{k}.\vec{p}$ matrix (22) and the potential energy term is diagonal in the band index. This is a consequence of its slow spatial variation : as it can be taken as a constant in each unit cell, its off-diagonal matrix elements vanish by Bloch function orthogonality. It is also straightforward to generalize the boundary conditions. One finds $F^A_l(-\varepsilon) = F^B_l(+\varepsilon)$ $l = 1, 2, \ldots n$.

$$\sum_{m=1}^{n} [p^z_{lm} - i \sum_{\alpha=1}^{3} (D^{z\alpha}_{lm} + D^{\alpha z}_{lm})\nabla_{\alpha}] F_m \text{ continuous} \tag{26}$$

at $z = 0$ for $l = 1, 2, \ldots, n$

One must, of course, assume Eq. (18) for the n bands of interest. In the second of Eq. (26), notice the terms without derivative coming from P_{lm}, and those involving derivatives with respect to x and y, coming from the mixed terms in k_xk_z and k_yk_z

and $k_y k_z$ in H_{lm}. Let us write down these boundary conditions for an ideal planar interface between lattice matched semiconductors A and B and for the 6-band case of Table I. In this situation the potential $U(\vec{r})$ depends only on z and k_x and k_y are good quantum numbers. That means that we can write

$$F^{A,B}(\vec{r}) = e^{ik_x x} e^{ik_y y} F^{A,B}(z) \tag{27}$$

so that the second of Eqs. (26) reads

$$\sum_{m=1}^{6} [p^z_{lm} + \sum_{\alpha=x,y} (D^{z\alpha}_{lm} + D^{\alpha z}_{lm}) k_\alpha - 2iD^{zz}_{lm} \frac{\partial}{\partial z}] F_m \quad \text{continuous} \tag{28}$$

A further simplification of the boundary conditions occurs by noticing that, if Eq. (18) holds for l=1, ... n, then also the P_{lm} matrix elements between these Bloch functions must be equal, as they are momentum matrix elements connecting the same conduction and valence band functions. It is empirically known [9] that such matrix elements depend essentially only on the lattice constant and, therefore, are indeed about equal in two lattice-matched materials. Thus the continuity of the $p^z_{lm}F_m$ terms follows from the continuity of the F_m alone, and Eq. (28) further simplifies to :

$$\sum_{m=1}^{6} [\sum_{\alpha=x,y} (D^{z\alpha}_{lm} + D^{\alpha z}_{lm}) k_\alpha -2i\, D^{zz}_{lm} \frac{\partial}{\partial z}] F_m \quad \text{continuous} \tag{28'}$$

This is the form of boundary conditions [10,11], used e.g. in the calculations on InAs-GaSb to be discussed soon.

1.4 Discussion of the Envelope-Function Approximation

In deriving the equations for the envelope function approximation,we stated the many, sometimes severe assumptions which are necessary. In the next section it will be apparent that the predictions of the envelope-function calculations are in very good overall agreement with experiments, perhaps more than one would expect. It is worthwhile to consider some of the possible objections to the method, and see if and when they do not influence the accuracy of the calculations very much.

Let us consider a GaAs-AlGaAs superlattice, for the sake of definiteness. One could argue that a low-lying quantum well state in the conduction band has an energy close to the GaAs band minimum, but deep into the gap of AlGaAs. The corresponding wavefunction decays very rapidly into the AlGaAs barrier, and does not satisfy the assumption of slow variation. On the other hand, the amplitude of the wavefunction in AlGaAs is, for the very same reason, quite small, so that an inaccurate treatment of this region has little influence on the calculation.

Another, more serious source of possible uncertainty is in the boundary conditions. The plausibility arguments presented here rely on the assumption of Eq. (18), which can only be approximately right, and further on the neglect of the k.p correction embodied in the second term on the r.h.s. of Eq. (15). Any attempt to improve upon Eq. (18), however, must necessarily introduce some parameters of unknown value into the theory, and as long as no obvious conflict with experimental facts arises, we shall retain the simple assumption. Furthermore, the discussion of boundary conditions is based on the idea of perfect periodicity of both media up to a geometrical plane defining the interface.

This is certainly an idealization and, in the very important case in which one of the components is an alloy (e.g. GaAs-$Al_xGa_{1-x}As$) it is impossible even to define such a plane. It would be more realistic to talk about an interface layer, comprising several atomic planes, separating the two semiconductors, and characterized by given reflection and transmission coefficients. These would be again parameters of unknown value to be used as input. Altogether, the theoretical situation concerning the boundary conditions is very unsatisfactory.

Finally, a word of caution concerning the use of $\vec{k}.\vec{p}$ matrices to describe the bulk band edges was recently put forward by Schuurmans and 't Hooft [11]. They pointed out that a Hamiltonian like Eq. (22) can have unphysical solutions; in the case of the 8x8 description of GaAs including conduction, upper- and split-off valence bands, for example, they find bulk solutions with E in the gap and very large real k-values. Care must be exerted to prevent that such unphysical branches are included in the construction of the superlattice solution.

2. Results for Systems of Interest

2.1 GaAs-$Al_xGa_{1-x}As$ superlattices : non-self-consistent calculations

In this section, application of the envelope-function method to the calculation of electronic states in superlattices and quantum wells will be discussed, with reference to GaAs-$Al_xGa_{1-x}As$ systems. The method described before is applicable to this system in the range [12] of x values ($0 < x < 0.45$) in which $Al_xGa_{1-x}As$ has a direct band gap at the Γ point. For larger x-values, the conduction band wavefunctions do not satisfy Eq. (18); one can, nonetheless, still make calculations for the valence subbands. We shall consider superlattices made out of intrinsic semiconductors, for which no charge accumulation in the GaAs quantum well takes place. In this case the potential $U(\vec{r})$ (see Eq. (24)) can be regarded as vanishing, corresponding to wells with a flat bottom. Situations involving charge rearrangements across the interfaces will be dealt with in the next section.

An extremely important input parameter for the envelope-function calculations is the band-gap discontinuity, i.e. the relative energy position between the valence band edges of the two materials making up the structure. Since the band gaps are known, the relative lineup of the conduction bands is then uniquely determined. Thus, this parameter determines the height of the barriers confining the particles in the quantum wells. In spite of its central role in the physics of heterostructures, this parameter is not accurately known, even for an extensively investigated system like GaAs - $Al_xGa_{1-x}As$. From the experimental point of view, published values of the valence band discontinuity range from the 15% of the gap difference [13] to $\sim$ 40% [14,15], with more recent investigations [16] suggesting 20-25%. Theoretical predictions can hardly be of any help, given the normal uncertainty ($>$ 0.1 eV) of state-of-the-art band theory in bulk materials. In the following, the value of this parameter is specified in each calculation, while the results are very sensitive to it. The problem of the band gap discontinuity is of central importance, and it would be very desirable to have a firm determination of this parameter.

We start by some general considerations. The growth axis of MBE heterostructures is in a <001> direction [17]. In a superlattice, one has, therefore, a Brillouin zone which is very thin in the z direction ($2\pi/d$, where d is the period) and has the usual $\sim 2\pi/a$ size, where a is the lattice constant, in the x and y direction. One has, therefore, a very asymmetric band structure with a bulk-like bandwidth ($\sim$10 eV) in the k_x, k_y plane and narrow ($\sim(d/a)^2.10eV\sim10meV$) bands in the k_z-direction.

The dispersion in the k_z-direction is relatively simple to handle, even in the coupled-band model of Table 1, because most of the off-diagonal terms vanish for k_x, $k_y = 0$. In detail, we see that $J_z = \pm 3/2$ states, corresponding to the heavy holes, completely decouple and behave as simple particles with effective mass $m_{hh} = 1/(\gamma_1 - 2\gamma_2)$. The conduction band states are coupled via the P matrix element to the light holes ($J_z = \pm 1/2$). This 6x6 model is reasonably good for GaAs-AlGaAs, as long as we consider energies differing from the conduction or valence band extrema by less then the spin-orbit splitting of the valence band (340 meV). For the heavy holes we have a Kronig-Penney type of eigenvalue problem, with the boundary condition

$$\left.\frac{1}{m_{hh}^A}\frac{dF^A}{dz}\right)_{o^-} = \left.\frac{1}{m_{hh}^B}\frac{dF^B}{dz}\right)_{o^+}$$

Table I

k.p Hamiltonian describing interaction of the s-like spin-degenerate conduction band with the upper spin-orbit partner ($p_{3/2}$) of the valence band. The four degenerate (at $\vec{k}$ = 0) valence band edge states are labelled by the value of J_z = - 3/2, -1/2, 1/2 or 3/2 (see text). The lower half of the matrix is obtained by hermitian conjugation.

	S↑	3/2	-1/2	S↓	1/2	-3/2
S↑	$E_c \frac{1}{2m^*} k^2$	$\frac{iP}{\sqrt{2}}(k_x+ik_y)$	$\frac{-i}{\sqrt{6}} p(k_x-ik_y)$	0	$-i\sqrt{\frac{2}{3}}\, pk_z$	0
3/2		$E_v-(\frac{\gamma_1}{2} - \gamma_2)k_z^2$ $\frac{-(\gamma_1+\gamma_2)}{2}(k_x^2+k_y^2)$	$\frac{\sqrt{3}}{2}\gamma_2(k_x^2-k_y^2)$ $-i\sqrt{2}\gamma_3 k_x k_y$	0	$\sqrt{3}\gamma_3$ $(k_x-ik_y)k_z$	0
-1/2			$E_v-(\frac{\gamma_1}{2}+\gamma_2)k_z^2$ $-(\frac{\gamma_1-\gamma_2}{2})(k_x^2+k_y^2)$	$i\sqrt{\frac{2}{3}}\, Pk_z$	0	$-\sqrt{3}\gamma_3$ $(k_x-ik_y)k_z$
s↓				$E_c+\frac{1}{2m^*}k^2$	$\frac{i}{\sqrt{6}} P(k_x+k_y)$	$-\frac{i}{\sqrt{2}} P(k_x-ik_y)$
1/2					$E_v-(\frac{\gamma_1}{2}+\gamma_2)k_z^2$ $-(\frac{\gamma_1-\gamma_2}{2})(k_x^2+k_y^2)$	$\frac{\sqrt{3}}{2}\gamma_2(k_x^2-k_y^2)$ $-i\sqrt{3}\gamma_3 k_x k_y$
-3/2						$E_v-(\frac{\gamma_1}{2}-\gamma_2)k_z^2$ $-(\frac{\gamma_1+\gamma_2}{2})(k_x^2+k_y^2)$

As shown e.g. by Bastard [18] one finds analytically the dispersion relation :

$$\cos(k_z d) = \cos(k_{zA} d_A)\cos(k_{zB} d_B) - \frac{1}{2}\left(\frac{m_{hh}^A k_{zB}}{m_{hh}^B k_{zA}} + \frac{m_{hh}^B k_{zA}}{m_{hh}^A k_{zB}}\right)\sin(k_{zA} d_A)\sin(k_{zB} d_B) \quad \text{where} \tag{29}$$

$$k_{zA,B}^2 = 2m_{hh}^{A,B}(E - E_v^{A,B}) \tag{29'}$$

and the superlattice period d is made up of thickness $d_A(d_B)$ of material A(B). As for light-holes and conduction electrons, one has the 2x2 matrix between s↑ and $J_z = + 1/2$, or an equivalent one for s↓ and $J_z = - 1/2$:

$$\begin{matrix} E_c + \frac{1}{2m^*}k_z^2 & -i\sqrt{\frac{2}{3}}\,p\,k_z \\ i\sqrt{\frac{2}{3}}\,p\,k_z & E_v - \frac{1}{2}(\gamma_1 + 2\gamma_2)k_z^2 \end{matrix} \tag{30}$$

One could solve directly the superlattice problem for these coupled bands as discussed previously or, a simplification valid for energies very near the band edges, find the eigenvalues of (30), $\xi_c(k_z)$, $\xi_v(k_z)$ and describe their non-parabolic dispersion via energy-dependent effective masses :

$$\xi_c(k_z) = E_c + \frac{1}{2m^*(E)}k_z^2 \qquad \xi_v(k_z) = E_v - \frac{1}{2m_{1h}(E)}k_z^2 \tag{31}$$

One can then use for each band the dispersion relation (29) with the appropriate energy-dependent mass replacing m_{hh} (also in (29')). For a complete discussion of k_z-dispersion in GaAs-AlGaAs superlattices, we refer the reader to the treatment of Schuurmans and t'Hooft [11].

Consider now the dispersion for k_x, $k_y \neq 0$. This is much more complicated, but also very interesting. The main features of interest can be investigated in a model in which the conduction band is treated as uncoupled from light- and heavy-hole bands, but the hole bands are strongly coupled with one another via the Luttinger Hamiltonian. For the conduction band one has again a simple Kronig-Penney-like problem with possible non-parabolic effects embodied in an energy-dependent effective-mass. An equation like (29) is still valid, but (29') is replaced in both A and B by :

$$k_{zA,B}^2 = 2m^*(E)(E - E_v^{A,B}) - k_x^2 - k_y^2 \tag{29''}$$

Notice that k_x, k_y are good quantum numbers, and they are the same in A and B (translation invariance along the interfaces).

For the valence bands we have to deal with the 4x4 Luttinger Hamiltonian, which we rewrite for convenience in the form :

$$H = \begin{matrix} & 3/2 & -1/2 & 1/2 & -3/2 \\ \frac{3}{2} & a_+ & c & b & 0 \\ -\frac{1}{2} & c^* & a_- & 0 & -b \\ \frac{1}{2} & b^* & 0 & a_- & c \\ -\frac{3}{2} & 0 & -b^* & c^* & a_+ \end{matrix} \tag{32}$$

with :

$$a_{\pm} = E_v - \frac{1}{2}(\gamma_1 \pm \gamma_2)(k_x^2 + k_y^2) - \frac{1}{2}(\gamma_1 \mp 2\gamma_2)k_z^2$$

$$b = \sqrt{3}\,\gamma_3\,(k_x - ik_y)k_z \qquad (32')$$

$$c = \sqrt{\frac{3}{2}}\,(\gamma_2(k_x^2 - k_y^2) - 2i\gamma_3 k_x k_y)$$

The bulk solutions of (32) give the spin-degenerate light- and heavy-hole bands. This degeneracy is actually lifted in non-inversion symmetric zincblende materials, like GaAs, by terms linear in the k-vector [19]. The terms are, however, extremely small and we neglect them here. The valence bands are anisotropic and, therefore, one has to expect that also for superlattices the band dispersion will be different for different directions in the k_x, k_y plane. As we shall see, however, this warping of the bands is small and a good approximation is obtained by the "axial model" in which [20,21] γ_2 and γ_3 in the c matrix elements (see the last of Eq. (32')) are replaced by $\bar{\gamma} = 1/2\ (\gamma_2 + \gamma_3)$. With this replacement, Eq. (32) acquires cylindrical symmetry about the z- axis, and the bands are isotropic in the k_x, k_y plane. The boundary conditions for the four-component envelope-function are given by Eqs. (26), the second of which demands, in this case, that

$$\begin{bmatrix} i(\gamma_1-2\gamma_2)\frac{\partial}{\partial z} & 0 & \sqrt{3}\gamma_3(k_x-ik_y) & 0 \\ 0 & i(\gamma_1+2\gamma_2)\frac{\partial}{\partial z} & 0 & -\sqrt{3}\gamma_3(k_x-ik_y) \\ \sqrt{3}\gamma_3(k_x+ik_y) & 0 & i(\gamma_1+2\gamma_2)\frac{\partial}{\partial z} & 0 \\ 0 & -\sqrt{3}\gamma_3(k_x+ik_y) & 0 & i(\gamma_1-2\gamma_2)\frac{\partial}{\partial z} \end{bmatrix} \begin{bmatrix} F_{3/2}(z) \\ F_{-1/2}(z) \\ F_{1/2}(z) \\ F_{-3/2}(z) \end{bmatrix} \qquad (33)$$

be continuous across the z = 0 interface. The solution of the 4-component envelope-function equation, with the boundary conditions of continuity and Eq. (33) must be pursued numerically. In Fig. 4 results [21] are shown for the valençe subbands of a superlattice consisting of GaAs layers 140Å thick, separated by 200Å thick barriers of $Al_{0.21}Ga_{0.79}As$. The barriers are thick enough to prevent sizeable coupling of different GaAs wells. The k_z -dispersion is, therefore, very flat, and we can regard the subband of Fig. 4 as those of an isolated 140Å GaAs quantum well. Notice the anisotropy of the in-plane dispersion, that increases with k, and the good average represented by the axial approximation. Notice also the scale of k, which is of the order of $\sim 10^6\,cm^{-1}$ that is of $\sim(100Å)^{-1}$, the inverse period being the characteristic scale for this quantity. At $k_x = k_y = 0$, there is no mixing of the light and heavy components. It is, therefore, possible to identify each subband as 100% light, l, or heavy, h. As soon as we move out of the k=0 axis, the mixing grows rapidly, and is the reason for the conspicuous non-parabolicities to be seen in Fig. 4. Notice in particular the light-hole subband, which starts off with a positive (i.e. electron-like) mass and has a maximum for a finite k-value, of the order $\sim 2\pi/d_A$ where d_A is the quantum well thickness. This strong mixing and consequent non-parabolicity were first pointed out by Nedorezov [22], who solved exactly the problem for an infinitely deep well, obtaining results in qualitative agreement with those shown here. Experimentally, hole band mixing is detectable by polarized luminescence experiments [23]. A strong non-parabolicity of the light-hole subband is probably the explanation for the broad line seen in resonant Raman scattering from p-type quantum wells [24]. A strong confirmation of the theoretical picture comes also from magneto-optical experiments, as we shall see soon.

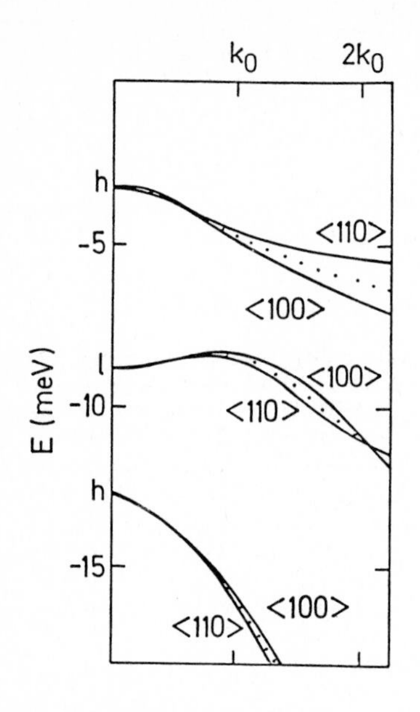

Fig.4 Dispersion of valence subbands of a 140Å GaAs - 200Å $Al_xGa_{1-x}As$ superlattice, with x = 0.21, in the plane perpendicular to the <001> growth axis. Solid lines : dispersion for $\vec{k}$ in the <100> or <110> directions. Dotted line : axial approximation. k_o denotes $\pi/340Å = 9.24.10^5\ cm^{-1}$. h and l denote "heavy" or "light" character at $\vec{k} = 0$. (After Ref. 21).

Finally, a word of caution about the sensitivity of these results to the values of the input parameters. The band parameters [25] of GaAs and AlAs are given in Table II, those of $Al_xGa_{1-x}As$ following by linear interpolation. The energy gap of GaAs at 0°K is taken to be 1.52 eV, that of $Al_xGa_{1-x}As$ (as long as x < 0.4 so that the gap is direct) is given [26] by

$$E_g(x) = E_g(GaAs) + 1.04x + 0.47x^2 \tag{34}$$

Table II Band parameters of GaAs and AlAs. P, m*, γ, γ_2 and γ_3 are appearing in Table I; γ is needed for magnetic field calculations in section 3. They are all in atomic units. The (a) set of parameters corresponds to the full 6x6 matrix of Table I. The (b) set corresponds to removing the explicit coupling of conduction and upper valence band. The direct energy gap, E_g and the valence band spin-orbit splitting are in eV.

	GaAs		AlAs	
	(a)	(b)	(a)	(b)
P	0.65	---	0.65	---
m*	0.195	0.067	0.307	0.124
γ_1	1.80	6.85	1.00	3.45
γ_2	-0.42	2.10	-0.54	0.68
γ_3	0.38	2.90	0.07	1.29
γ	-1.32	1.20	-1.10	0.12
E_g	1.52		3.13	
Δ	0.34		0.275	

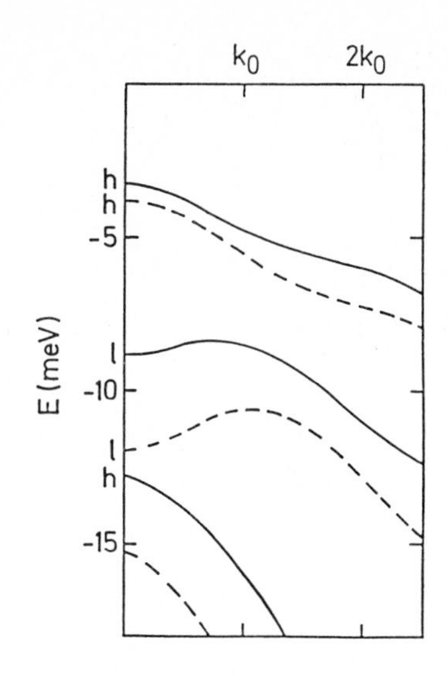

Fig.5 Comparison between the axial model results of Fig.4, obtained following the prescription of Ref. 13 for the valence band discontinuity (solid lines), and those obtained from prescription of Ref. 14 (dashed lines). (After Ref. 21).

As we already mentioned above, this information is relatively well established and the results are not so sensitive to it. The really important question is how the band gap difference from Eq. (34) is distributed between valence and conduction band edges. In Fig. 4 Dingle's 15% prescription was followed. In Fig. 5 we compare the axial model results for the 15% [13] and the 43% [14] prescriptions : the qualitative aspects of the non-parabolic dispersion retain their validity, but the positions of the individual subbands are very strongly affected by the value of the discontinuity parameter.

2.2 Heterojunctions and type-II superlattices : self-consistent calculations

In the example of GaAs-AlGaAs superlattices discussed up to now, one has to deal with systems in which no significant charge rearrangement across or near the interfaces takes place, at least as long as we consider heterostructures involving intrinsic semiconductors. ("Modulation doping" can produce filling of the quantum wells with charge carriers released from impurities located in far away layers).

Whenever such transfers of charge take place, they contribute to the definition of the one-particle potential $U(\vec{r})$ appearing in the effective-mass equation (14) or (24). The potential depends on the charge density corresponding to the wavefunctions of the quantum states filled by the transferred carriers. Since these wavefunctions depend in turn on the potential $U(\vec{r})$ the problem must be solved self-consistently, within some scheme of approximation, e.g. the Hartree approximation, the local-density approximation, etc. These procedures are familiar from the prototype 2-dimensional semiconductor system, the Si-SiO_2 MOS structures [27]. F. Stern's lectures in this school will also discuss self-consistent calculations, so that we shall simply mention some applications to GaAs-AlGaAs heterojunctions and to InAs-GaSb superlattices. The main message is that the envelope-function approximation can handle self-consistent calculations in a natural way, in contrast e.g., to tight-binding methods.

A good example of application of the many-band envelope function in a self-consistent way [28] is provided by the so-called type-II superlattices, exemplified by the InAs-GaSb system [29], which will be discussed in the lectures of L.L. Chang. This system is characterized by the "staggered" band line-up of Fig. 3, in which some GaSb occupied valence states lie higher in energy than the empty conduction band bottom of InAs. It is clear that electrons will tend to transfer to the InAs layers, except in the case in which the layers are so thin that the quantization energy reverses the level order, locating empty InAs quantum well

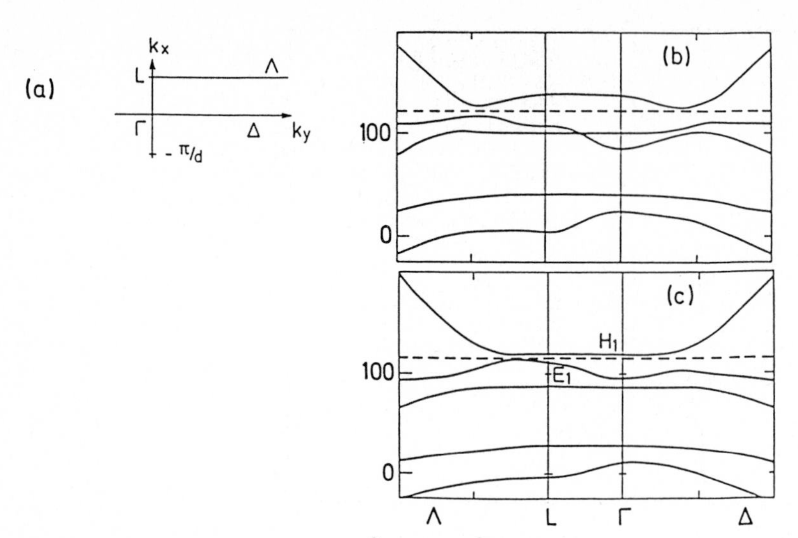

Fig.6 Band structure of a 120Å InAs-80Å GaSb superlattice (b) non self-consistent; (c) in the self-consistent Hartree approximation. The Brillouin-zone nomenclature is shown in (a), where x denotes the growth axis. (After Ref. 28).

conduction states above full GaSb well valence states. Thus, for superlattice periods above a critical value, $d_c \approx 180Å$, self-consistency is necessary. The results of such a calculation, performed in the Hartree approximation, are shown in Fig. 6. Assuming perfectly intrinsic materials, the Fermi energy lies in the small hybridization gap. Notice the sizeable effect of self-consistency, and the strong non-parabolicity of the subbands. In this case, the density of transferred carriers is $\sim 10^{12}$ cm^{-2}. The Hartree approximation is accurate when the dimensionless parameter r_s satisfies $r_s \ll 1$. In this case, this parameter is defined as

$$r_s = r_o/a_B^* \tag{35}$$

where r_o is the mean electron distance expressed in terms of the 3-dimensional density n as

$$r_o = \left(\frac{3}{4\pi n}\right)^{1/3} \tag{35'}$$

and a_B^* is the effective Bohr radius, scaled by the effective mass and the dielectric constant ε :

$$a_B^* = \left(\frac{\hbar^2}{m_0 e^2}\right)\frac{m_0}{m^*}\,\varepsilon \tag{35''}$$

In small gap systems, m^*/m_o is very small ($\sim$0.02 for InAs) and ε large ($\sim$ 16) so that the $r_s < 1$ regime is achieved at moderate electron densities. The Hartree approximation is, therefore, accurate.

Recently, calculations in the Hartree approximation for the 2-dimensional hole gas in the single heterojunction between GaAs and p-type $Al_xGa_{1-x}As$ with $x \sim 0.5$, were reported [20,30-33]. This was stimulated by experiments [34,35] on samples with $\sim 5\times10^{11}$ holes cm^{-2}. Fig. 7 shows [30] the band-bending profile and the roughly triangular potential well binding the holes, and resulting from the ionization of impurities in the depletion layers, from the band discontinuity and from the holes themselves. An interesting feature of this system, in contrast to

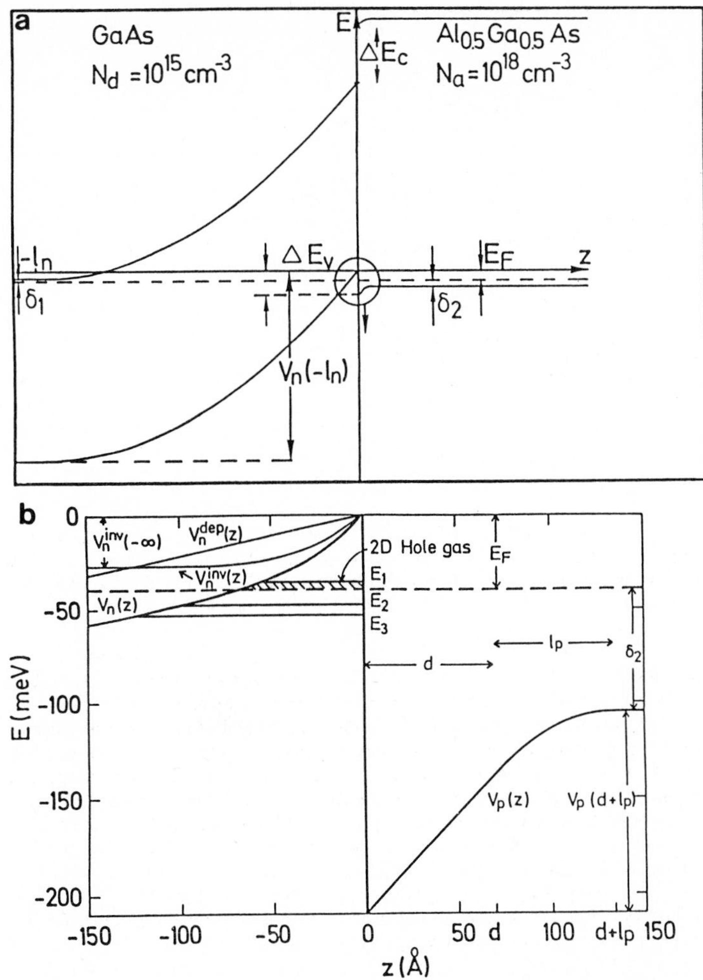

Fig.7 (a) Band diagram for the GaAs-$Al_{0.5}Ga_{0.5}As$ heterojunction. A portion of (a) is magnified in (b) where the energy levels and the potentials due to the inversion layer V_{inv} and to the depletion layers V_n^{dep} and V_p and the total potential in GaAs, $V^n = V_n^{dep} + V_{inv}$ are drawn. (After Ref. 30).

superlattices and quantum wells, is that it lacks inversion symmetry. This leads to a removal of the Kramer's spin degeneracy, much larger than the one present in bulk GaAs because of inversion symmetry breaking due to the difference between Ga and As ionic potentials. Fig. 8 shows the subband dispersion parallel to the interface, in the axial approximation [33]. The twofold degeneracy is present only at $k_x = k_y = 0$ and is removed, with a splitting of bout 2 meV at the Fermi energy. In this heterojunction the Hartree approximation can no longer be justified, because of the large hole mass giving $r_s > 1$. Nonetheless, as will be shown after discussing the Landau levels in Section 3.1, the agreement of the results with experiment is rather good.

3. Effect of External Magnetic Fields

One of the advantages of the envelope-function method is that it is easily extended to include an external field. There are various kinds of such fields which are of great experimental interest. One is a strain field, which is almost invariably

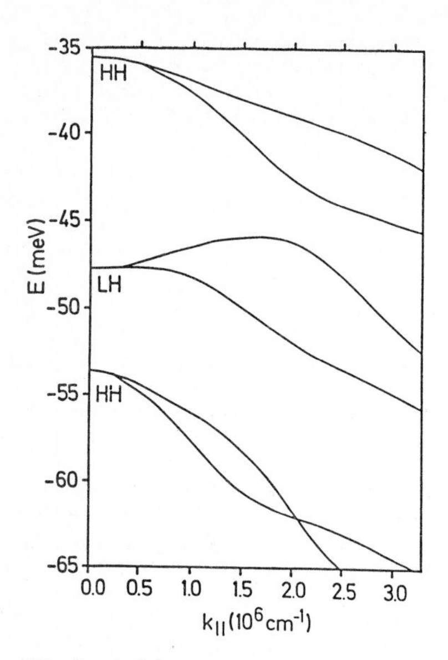

Fig.8 Self-consistent results for the subband dispersion of the heterojunction of Fig. 7 in the axial approximation (after Ekenberg and Altarelli, Ref. 33).

present because lattice match is never perfect. J.Y. Marzin is going to discuss strain effects in his lectures. Another "field" is the potential generated by impurity atoms, which will be discussed in Section 4. Here, the effect of an external magnetic field perpendicular to the layers is discussed. Many of the most revealing experiments concerning 2-dimensional electronic systems are performed with an external magnetic field, which has a dramatic influence on the energy spectrum. In a 2-dimensional system, a perpendicular magnetic field quantizes both available degrees of freedom, producing an entirely discrete spectrum. This leads to an enrichment and sharpening of optical structures, and to striking transport phenomena, like the quantum Hall effect. Our main motivation in describing the Landau level spectrum is the stringent test of electronic calculations provided by the quantitative interpretation of magneto-optical experiments.

3.1 Landau levels in heterostructures

The inclusion of a magnetic field in the many-band envelope-function formalism [36,37] follows the lines of the classical work of Luttinger [7] on the cyclotron resonance of holes in semiconductors. The field $\vec{B} = (0, 0, B)$ is described by the vector potential $\vec{A}$. (It is convenient to choose a gauge with $A_z = 0$). In the $\vec{k}.\vec{p}$ bulk Hamiltonian, Eq. (22), $\vec{k}$ is to be replaced by $\vec{k}' = \vec{k} + (e/c)\vec{A}$. Then it is easy to see that the x and y components of this new operator do not commute, but instead

$$[k'_x, k'_y] = -i(e/c)B \tag{36}$$

also, new diagonal terms arise, representing the direct coupling of the electron and hole spins to the field. The conduction band diagonal elements in the 6x6 Hamiltonian of Table I now become

$$H_{cc} = E_c + \frac{1}{2m^*}(\vec{k} + (e/c)\vec{A})^2 + \frac{e}{2c} g^* s_z B \tag{37}$$

where $\vec{s}$ is the electron spin and the effective g-factor [6] g^* is simply $2/m^*$. For the valence band diagonal elements, one must add the term

$$\frac{e}{c} \kappa J_z B + \frac{e}{c} q J_z^3 B \tag{38}$$

where J_z is the spin 3/2 operator and κ and q are two material parameters [7]. Actually q is very small for the semiconductors of interest here and the second part of (38) can be neglected. It is easy to see that as a consequence of (36) we can define operators a, a^+

$$a = \sqrt{\frac{c}{2eB}}\,(k'_x - ik'_y)$$
$$a^+ = \sqrt{\frac{c}{2eB}}\,(k'_x + ik'_y) \qquad (39)$$

with commutator

$$[a, a^+] = 1 \qquad (39')$$

so that all terms in k_x or k_y in the Hamiltonian can be expressed in terms of these harmonic oscillator raising and lowering operators.

It is not possible to write a general solution of this Hamiltonian in closed form. Fortunately, however, it is possible to do it if we take the axial approximation in the valence-band portion of the 6x6 Hamiltonian, (see the discussion following Eq. (32), (32') in Section 2.1). In this case one can see by inspection that the general solution in each material takes the form :

$$\Psi_n=(c_1(z)\phi_n, c_2(z)\phi_{n-1}, c_3(z)\phi_{n+1}, c_4(z)\phi_{n+1}, c_5(z)\phi_n, c_6(z)\phi_{n+2}) \qquad (40)$$

where the ϕ_n are harmonic oscillator wavefunctions with n = -2, -1, 0, 1, ... and the c coefficients are automatically vanishing for the components with negative oscillator index for n < 1 (e.g. if n = 0, then $c_2(z) = 0$; if n = -2, only $c_6(z) \neq 0$). For the superlattice case all six $c_j(z)$ have the common prefactor exp $(ik_{sz}z)$ where k_{sz} is a vector in the superlattice Brillouin zone, parallel to the field. Acting on a state with the structure of (40), the effective Hamiltonian is a 6x6 matrix with the form :

$$H_{jj'} = (D^{zz}_{jj'}k_z^2+E_j)\delta_{jj'}+(p^z_{jj'}+A_{jj'}(n,B))k_z+C_{jj'}(n,B) \qquad (41)$$

where j, j' run over the 6 basis states, $p^z_{jj'}$ is the coefficient of k_z in the terms present at zero field in the k.p Hamiltonian, and the A and C matrices represent the "new", magnetic-field induced terms. With the replacement $k_z = -i\,\partial/\partial z$, Eq.(41) gives a differential system for the $c_j(z)$ functions. Deriving in the usual way the boundary conditions at the interfaces [36,37], one obtains, besides the continuity of the c_j functions, the following conditions, in analogy to Eq. (28') :

$$-i\,D^{zz}_{jj}\frac{\partial}{\partial z}c_j(z) + \frac{1}{2}\sum_{j'} A_{jj'}c_{j'}(z) \text{ continuous for } j=1,\dots 6 \qquad (42)$$

One can then proceed to the solution of Eq. (41) with the boundary conditions (42), as a function of the external field B and of the superlattice wavevector k_{sz}. For superlattices with thick barriers the latter dependence is very small and can be neglected (quantum well limit).

This procedure was applied to the GaAs-AlGaAs and to the InAs-GaSb system. The computed Landau levels are directly comparable to magneto-optical experiments once the selection rules for optical transitions have been established. In terms of the eigenstates Ψ_n, Eq.(40), the rules for Faraday configuration (light propagation along z, σ^+ or σ^- circular polarization) are [36] $\Delta n = \pm 1$, $\Delta k_{sz} = 0$. Terms breaking the axial symmetry, proportional to $(\gamma_3-\gamma_2)$, also induce $\Delta n = \pm 3$ transitions [38], which are, however, expected to be weaker.

3.2 Comparison with magneto-optical experiments

Magneto-optical experiments in the interband region for GaAs-AlGaAs quantum well systems were recently shown [39,40] to provide a large amount of clearly resolved structure. In Fig. 9 we show the computed [39] Landau levels for a superlattice

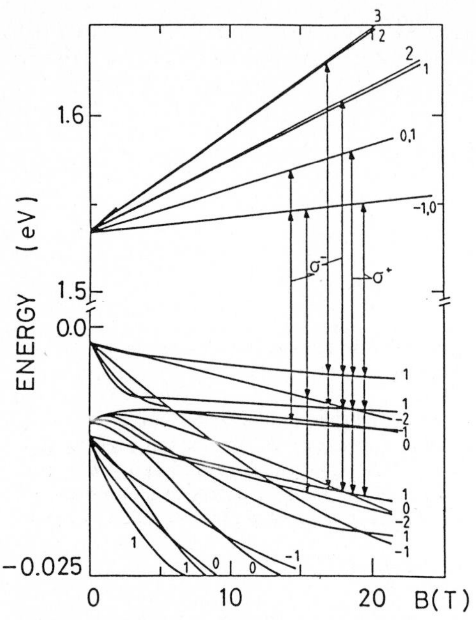

Fig.9 Calculated Landau levels of conduction and valence subbands of a 125Å GaAs - 80A $Al_{0.21}Ga_{0.79}As$ superlattice as a function of the field B. The arrows indicate allowed transitions in the σ^+ and σ^- circular polarizations. The numbers are the quantum number n of Eq. 40.

formed by 125Å GaAs quantum well between 80Å thick barriers of $Al_{0.21}Ga_{0.79}As$, at $k_{sz} = 0$. These results were obtained using the 15% rule for the band discontinuity, and the spherical approximation for the valence bands. The spherical approximation [41] assumes $\gamma_2 = \gamma_3 = \bar{\gamma}$, with $\bar{\gamma} = (2\gamma_2+3\gamma_3)/5$ throughout the valence band part of the Hamiltonian. Notice in this Figure the striking contrast between the upper, conduction-band part of the spectrum and the lower, valence-band part. Conduction band Landau levels are almost linear in the B field, small deviations being related to the slight non-parabolicity induced by the k.p coupling to the valence band. Valence band Landau levels show on the other hand striking deviations from linearity, consistent with the large deviations from parabolicity in the subband structure for B = 0. It is also interesting to notice how some of the levels, more specifically the topmost valence group with $n = 0, \pm 1$ have an extremely weak field dependence over a large range of field values. In the customary language this would correspond to a very large cyclotron mass. In the Figure also the allowed transitions for σ^+ and σ^- polarization are shown. We can then directly compare the theory with the luminescence excitation spectra [39] as shown for a 90Å well in Fig. 10. It is quite apparent that although the correspondence is for most transitions very good, there are some experimental transitions which do not fit at all in the picture. It is believed that such transitions are related to excitons, which are not included in the theoretical picture. The peaks in good agreement with theory can be then associated to interband transitions and one can, by looking at the B = 0 limit in Fig. 10, measure the exciton binding energy directly. It appears that the binding energy is larger than commonly held, for reasons which shall be detailed in the next chapter.

In Fig. 11 we show the result of a Landau level calculation by Ekenberg and Altarelli [33] for the single heterojunction with the band structure of Fig. 8. These calculations were performed in the axial approximation and they are only approximately self-consistent because the potential was calculated only at B = 0, and assumed not to vary with the field. In this case also some Landau levels are very flat, in particular the one cut by the Fermi level in the region $7 \leqslant B < 10$ T.

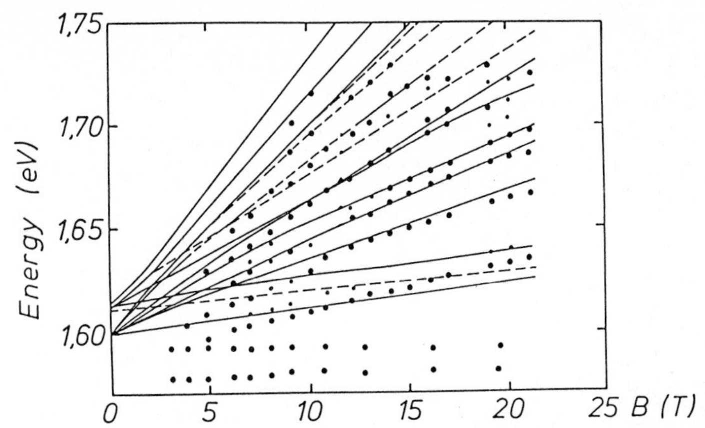

Fig.10 Comparison between the calculated transitions (lines) and maxima in the luminescence excitation spectrum (circles) for σ^+ polarization (after Ref. [39]) of a 90Å GaAs quantum well between $Al_{0.21}Ga_{0.79}As$ barriers. Large circles and solid lines correspond to stronger transitions, small circles and dashed lines to weaker ones.

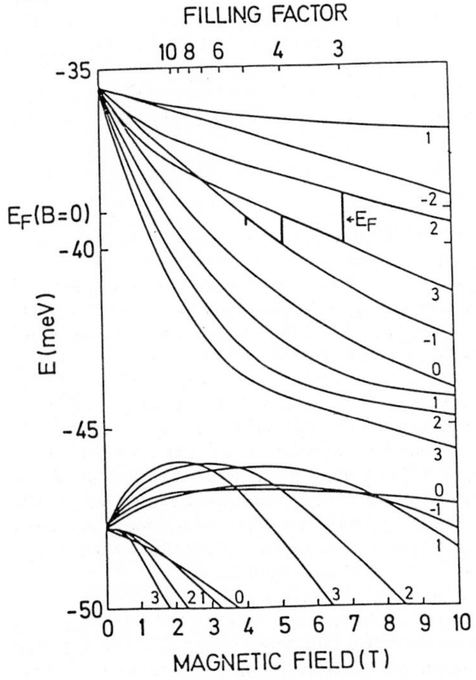

Fig.11 Landau levels of the GaAs-$Al_{0.5}Ga_{0.5}As$ heterojunction of Fig.8 as a function of field B. The Fermi level position for $5 \cdot 10^{11}$ holes cm^{-2} is indicated for higher fields by the thick line (after Ekenberg and Altarelli, ref. 33).

It can be shown that these results compare very favourably with the existing cyclotron resonance experiments [34,35]. It would be interesting to have more results on samples with different parameters, to see if one can really conclude that many-body effects are not as important as expected.

Finally, Fig. 12 shows the comparison between theory [36] and experiment [42] for the magneto-optical transitions in the InAs-GaSb superlattice of Fig.6. In this

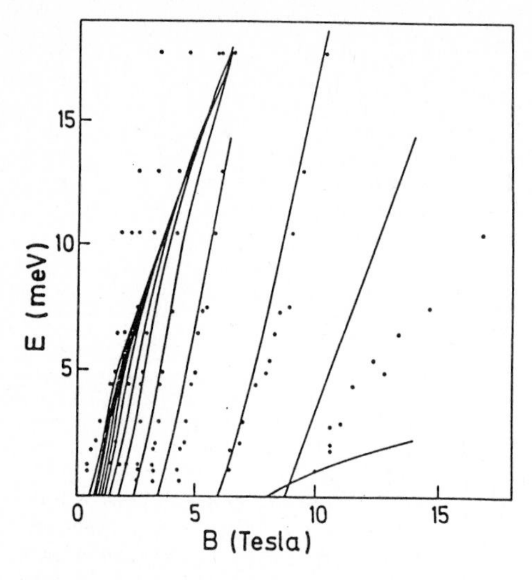

Fig.12 Solid lines : predicted optical transition energies for the InAs-GaSb superlattice of Fig.6 as a function of field (after Ref. 36). Dots : experimental transition energies (from Ref. 42).

case too it appears that, in spite of the complicated Landau level structure, good agreement is obtained.

We, therefore, restate as an overall conclusion that experiments in a magnetic field provide a very strong confirmation of the envelope-function approach to the calculation of electronic states in heterostructures. The Landau level field dependence, however, is so complicated in the valence-band region that interpretation of experiments cannot be based on the $B = 0$ band structure, but requires a detailed calculation including the field.

4. Impurity and Exciton States in Heterostructures

4.1 Impurity states in quantum wells and superlattices

The subject of impurity states is one of the central topics of semiconductor physics, because impurities influence the properties of semiconductors very much. There is no reason to expect that heterostructures will be different in this respect, and in fact many optical and transport measurements point to an important role of impurities in these systems.

Most theoretical work on the subject has been within the framework of the effective-mass approximation. Like in the bulk, this method is of course only valid for shallow impurities, because it is necessary to assume (see section 1.1) that the states are very extended in real space, and that the potential is weak and slowly varying. These conditions are satisfied (except perhaps in the few cells surrounding the impurity atom) by the Coulomb potential, screened by the dielectric function of the superlattice, of simple donor or acceptor impurities.

We shall consider first the problem of donor impurities, which does not suffer from the complications of a degenerate band edge, like acceptor impurities. In bulk GaAs a purely Coulombic donor impurity has a hydrogenic spectrum

$$E_n = E_c - R^*/n^2 \tag{43}$$

with

$$R^* = \frac{m^* e^4}{2\hbar^2 \varepsilon^2} \tag{43'}$$

and a Bohr radius, for the ground state effective-mass wavefunction :

$$a^* = \frac{\hbar^2 \varepsilon}{m^* e^2} \tag{44}$$

Since for GaAs [25] $m^* = 0.067$, $\varepsilon = 12{,}5$, one finds $R^* = 5.8$ meV, $a^* = 98.7$Å. Notice that the energy scale of Coulomb binding is comparable to or smaller than typical quantization energies of quantum wells. For an impurity sitting in a quantum well, there are two more lengths, the well thickness L and the distance of the impurity from the well center, z, which are important.

Some exact results for limiting cases are helpful in this problem. For a strictly 2-dimensional hydrogenic problem [43], the binding energy is $E_B = 4R^*$. On the other hand, Satpathy [44] has found an elegant analytic solution for a 3-dimensional hydrogenic impurity at arbitrary distance z from a hard wall (infinitely high potential barrier). When $z \gg a^*$, then of course $E_B \approx R^*$. When $z \to 0$, the ground state acquires $2p_z$ character, with $E_B = R^*/4$, as found by Levine [45]. Between these two limits there is a smooth behaviour of the binding energy and of the wavefunction character. However, as soon as two barriers, as in a quantum well, are present, no exact solution to the problem is known. Bastard [46] has proposed a simple variational approach, which can be carried through analytically, for the problem of a quantum well bounded by two infinite barriers. As one can expect, confinement increases binding; furthermore, impurities sitting at the center of the well have a larger binding energy then those at the edges (binding energy is always referred to the lowest subband). It is interesting and probably relevant to experiment that the curve of binding energy versus impurity position for $L/a^* \gg 1$ is very flat around $z = 0$ (center of the well). This means that most randomly distributed impurities will have a binding energy close to that of the center position. On the other hand, for $L/a^* < 1$ the curve flattens also at the edges, so that most impurities have either an "on-center" or "on-edge" binding energy.

Various more sophisticated variational calculations have been performed [47-51], which relax the assumption of infinite barriers. The most detailed one is by Mailhiot et al. [47], who include the current-conserving boundary conditions, and the effect of image forces, due to the difference in dielectric constant between GaAs and AlGaAs. When the wavefunction can penetrate the barrier, the $L \to 0$ limit of the binding energy is not $4R^*$, but rather the binding energy of bulk $Al_xGa_{1-x}As$. One finds that as L decreases, the binding energy starts from the GaAs value (for $L \gg a^*$), increases with increasing confinement up to a maximum (which never exceeds $4R^*$, and decreases with decreasing barrier height, i.e. Al percent) and then goes down again to the bulk $Al_xGa_{1-x}As$ value. These trends for the isolated quantum well can become more complicated when also the barrier thickness is varied, and true superlattice effects are taken into account [50].

The $L \to 0$ limit, of course, should not be taken too seriously, because the whole effective-mass formalism collapses in that limit. A more suitable approach proposed by Priester et al. [52] for superlattices with very thin wells is based on the observation that in such case the subband spacing is much larger then the characteristic Coulomb energy R^*. Subband mixing is then irrelevant, and one can write a 2-dimensional effective-mass equation for the impurity levels attached to subband n :

$$\left[E_n(\vec{k}_{\parallel}^{0}) - \frac{\hbar^2}{2m_{\parallel}^*} \nabla_{\parallel}^2 + U_n(\vec{r}_{\parallel})\right] F(\vec{r}_{\parallel}) = E\, F(\vec{r}_{\parallel}) \tag{45}$$

where $E_n(\vec{k}_{\parallel}^{0})$ is the energy of the subband minimum, $m_{\parallel}^*$ the parallel effective-mass describing the curvature of the subband, and $U_n(r_{\parallel})$ is the impurity potential after averaging its z-dependence over the subband wavefunction $\psi_n(z)$. If the n-th subband in question is not the lowest one, the bound states of Eq. (45) will be degenerate with the 2-dimensional continuum of the lower subbands. The mixing and resulting broadening of the discrete bound level, however, appears to be small [53] (a result in agreement with the experimental observation of sharp excitonic lines from higher subbands, and with simplified theoretical models [54]).

We consider now acceptor impurities. In this case, the kinetic energy operator in the effective-mass equation is the Luttinger Hamiltonian describing the upper spin-orbit split valence band. In other words, we obtain the acceptor Hamiltonian in each material by adding the impurity potential, $e^2/\varepsilon r$, on the diagonal elements of Eq. (32). The resulting problem in quite complex also in the bulk case. The non-spherical character of heavy- and light-hole bands leads to an effective-mass wavefunction which is a superposition of different angular momenta (e.g. s, d, g, etc. for the ground state). Variational calculations are accordingly more complex than for donors. The situation in a quantum well is not simplified. One could argue that the quantum well removes the degeneracy of the band edge (Fig. 4). However, this leads to hardly any simplification, because the confinement-induced splitting is comparable or smaller (see e.g. Fig. 4) than the typical acceptor binding energy ($\sim$ 30 meV for GaAs). Therefore, it is to be expected that several subbands are strongly mixed together by the acceptor potential. Only for very thin wells a one-subband formulation as in Eq.(45), may become possible. In spite of these difficulties, a realistic inclusion of the valence band coupling is present in the variational calculation of Masselink et al. [55], who consider a mixing of s and d-waves only for the ground state, and use an approximate way to deal with boundary conditions. This is a remarkable piece of work, providing results that are directly comparable to experiment. A short-range potential is added ad-hoc to simulate chemical shifts between different acceptors. The comparison with experiment is favorable.

It would be interesting to explore if it is possible to find simplifications of the acceptor Hamiltonian in a superlattice which would make more insight possible without too much loss of accuracy. An example of one such simplification for bulk acceptors is the "spherical model" of Baldereschi and Lipari [41]. In the quantum well geometry an "axial model" seems more appropriate, but no attempts in this direction are known to the author.

4.2 Excitons in quantum wells and superlattices.

Excitons are bound electron-hole pairs and are the lowest electronic excited states of non-metallic crystals. They are easily detected in optical spectra, because they give rise to sharp line structures, in contrast to broad continuum transitions. In a semiconductor like GaAs bulk excitons have a binding energy of about 4 meV. The theoretical determination of this binding energy may at first seem an even more challenging problem then the acceptor one. In the bulk, however, an important simplification is possible [56], which makes a very good approximate calculation easy. In fact, the bulk effective mass Hamiltonian for excitons [57] can be written :

$$H_{exc} = H_e(\vec{p}_e) - H_h(\vec{p}_h) - e^2/\varepsilon|\vec{r}_e - \vec{r}_h| \tag{46}$$

Here the first term is simply $p_e^2/2m_e^*$, the second is the Luttinger Hamiltonian for the hole (i.e. equation (32) with $\vec{p}_h = -i\partial/\partial\vec{r}_h$ replacing $\vec{k}$). We can define a relative coordinate $\vec{r} = \vec{r}_e - \vec{r}_h$, its conjugated momentum $\vec{p}$, and a coordinate $\vec{R} = 1/2\,(\vec{r}_e + \vec{r}_h)$, with conjugate momentum $\vec{P} = \vec{p}_e + \vec{p}_h$. $\vec{P}$ is actually a constant of the motion, corresponding to the overall momentum of the pair (we avoid mentioning the "center of mass" because, due to the structure of the valence band edge, there is no center of mass for the electron-hole pair [58]). Optical transitions can only create or destroy $\vec{P} = 0$ excitons, for which the Hamiltonian becomes

$$H_{exc} = H_e(\vec{p}) - H_h(\vec{p}) - e^2/\varepsilon r \tag{47}$$

Notice that this Hamiltonian is a 4x4 matrix. The electron term and the Coulomb term are assumed multiplied by the unit matrix, i.e. they are added to the diagonal terms of the Luttinger Hamiltonian (32). Now, apart from the Coulomb term, there are two kind of terms : the terms proportional to $(\gamma_1 + 1/m_e^*)$, and other terms proportional to γ_2 or γ_3. For GaAs $m_e^* = 0.067$, $\gamma_1 = 6.85$, $\gamma_2 = 2.10$, $\gamma_3 = 2.90$. Therefore, $\gamma_1 + 1/m_e^* \gg \gamma_2, \gamma_3$. Neglecting all γ_2, γ_3 terms in first approximation one finds four degenerate exciton states with binding energy

$$R^* = \frac{e^4}{2(\gamma_1 + 1/m_e^*)\hbar^2 \varepsilon^2} \tag{48}$$

The off-diagonal terms can be included as a second-order perturbation [56], which does not remove the fourfold degeneracy, but provides a correction to the binding energy.

This procedure cannot be applied to acceptor states calculations, because the large value of $1/m_e^*$ ($\sim$ 15) is essential to the argument. Unfortunately it is not easy to carry out this procedure for a quantum well either. First of all, we lose translation invariance in the z direction : we can, therefore, set the component of $\vec{p}$ along the interfaces, $P_{\parallel} = 0$, but p_z is no longer a good quantum number. Also, the boundary conditions depend separately on z_e and z_h, not on $z_e - z_h$. The two-particle nature of the problem actually gives rise to a situation opposite to the bulk case : the exciton problem is more complicated than the acceptor problem, and to the best of the author's knowledge, there is no satisfactory calculation, comparable, say, to the work of Masselink et al. [55] for acceptors.

The existing calculations [59-61] try to exploit the splitting of the valence band top into a heavy hole subband and a light-hole subband in the quantum well (we know from section 2.1 that this nomenclature is strictly correct only for $k_{\parallel} = 0$) to force a diagonalization of the Luttinger Hamiltonian. One simply ignores the off-diagonal terms in Eq. (32), and obtains two heavy holes and two light hole bands, all decoupled. The "heavy holes" have a mass

$$\frac{1}{m_{hh}} = \gamma_1 - 2\gamma_2 \text{ in the z direction} \tag{49}$$

$$\frac{1}{m_{hh}} = \gamma_1 + \gamma_2 \text{ in the x,y directions.}$$

The "light holes" have a mass

$$\frac{1}{m_{lh}} = \gamma_1 + 2\gamma_2 \text{ in the z direction}$$

$$\frac{1}{m_{lh}} = \gamma_1 - \gamma_2 \text{ in the x, y directions.} \tag{50}$$

Then two sets of excitons, light and heavy, are deduced and their binding energy is obtained by a variational trial function, most often of the form

$$\Psi = f_e(z_e) f_h(z_h)\, g(x, y, z) \tag{51}$$

with x, y and z relative coordinates. Many trends in the result are similar to the donor results.

Although this approach receives some support in the experimentally confirmed fact [59] that there are two exciton peaks in the energy region near the transitions from the heavy and from the light-hole subbands to the first conduction subband, it is obvious that it cannot be right. Indeed in the $L \to \infty$ limit, these theories give, of course, two excitons, heavy and light; but in bulk GaAs, we know that the ground exciton state is fourfold degenerate. Similarly, for $L \to 0$, one gets two split excitons for bulk AlGaAs. The quantum well geometry removes the fourfold degeneracy, but the splitting must depend on L in some other way. The two observed excitons must both involve a mixture of light and heavy subbands, a rather natural conclusion if we observe the energy and momentum scales of, e.g. the subband structure in Fig. 4.

The neglect of subband mixing via the off-diagonal Luttinger term is also a restriction on the variational freedom of the wavefunction. Therefore, it is not too surprising that the direct observation of the exciton binding energy in the magneto-optical experiments of Ref. 39-40 (see section 3.2) provides values considerably larger then the theoretical predictions.

Acknowledgement

It is a pleasure to thank my co-workers Ulf Ekenberg, A. Fasolino and J.C. Maan for allowing me to quote results prior to publication.

References

1. W. Kohn : in "Solid State Physics", F. Seitz and D. Turnbull ed.s, (Academic, New York, 1957), v. 5, p. 257
2. E.O. Kane : in "Semiconductors and Semimetals", R.K. Willardson and A.C. Beer ed. s, (Academic, New York, 1966) v. 1, p. 75
3. W.A. Harrison : Phys. Rev. 123, 85 (1961)
4. D.J. Ben Daniel and C.B. Duke : 152, 683 (1966)
5. This is not a universal result, but can be shown to hold for the valence and conduction band edges of direct-gap III-V compounds, see Ref. 2.
6. L.M. Roth, B. Lax and S. Zwerdling : Phys. Rev. 114, 90 (1959) ; L.M. Roth in "Handbook of Semiconductors", W. Paul ed. (North-Holland, Amsterdam, 1982) vol. 1
7. J.M. Luttinger : Phys. Rev. 102, 1030 (1956)
8. J.M. Luttinger and W. Kohn : Phys. Rev. 97, 869 (1955)
9. M. Cardona : in "Atomic Structure and Properties of Solids", E. Burstein ed. (Academic, New York, 1972) p. 514
10. M. Altarelli : in "Applications of High Magnetic Fields in Semiconductor Physics", G. Landwehr, ed. (Springer, Berlin 1983) p. 174
11. M.F.H. Schuurmans and G.W. 't Hooft : to be published
12. H.C. Casey and M.B. Panish : "Heterostructure Lasers" (Academic, New York, 1978) part A
13. R. Dingle : in "Festkörperprobleme", H.J. Queisser, ed. (Vieweg, Braunschweig, 1975) vol. 15, p. 21
14. R.C. Miller, D.A. Kleinman and A.C. Gossard : Phys. Rev. B29, 7085 (1984)
15. W.I. Wang, E.E. Mendez and F. Stern : Appl. Phys. Lett. 45, 639 (1984)
16. Y. Horikoshi, A. Fischer and K. Ploog : private communication
17. See e.g. K. Ploog : Ann. Rev. Mat. Science 12, 123 (1982)
18. G. Bastard : Phys. Rev. B24, 5693 (1981)
19. E.O. Kane : in "Handbook of Semiconductors", W. Paul ed. (Academic, New York, 1982) vol. 1, p. 193
20. D.A. Broido and L.J. Sham : in "Proc. of the 17th Int. Conf. on the Physics of Semiconductors, San Francisco 1984", in press
21. M. Altarelli, U. Ekenberg and A. Fasolino : to be published
22. S.S. Nedorezov : Soviet Phys. Sol. State 12, 1814 (1971), (Fiz. Tverd Tela 12, 2269 (1970))
23. R. Sooryakumar, D.S. Chemla, A. Pinczuk, A. Gossard, W. Wiegmann and L.J. Sham : in "Proceedings of The 17th Int. Conf. on the Physics of Semiconductors, San Francisco, 1984", in press
24. A. Pinczuk, H.L. Störmer. A.C. Gossard and W. Wiegmann : in "Proceedings of the 17th Int. Conf. on the Physics of Semiconductors, San Francisco, 1984", in press
25. "Landölt-Börnstein : Numerical Data and Functional Relationship in Science and Technology" O. Madelung, ed. Group III, Vol. 17, (Springer, Berlin, 1982)
26. A. Onton : in "Festkörperprobleme", H.J. Queisser ed., (Vieweg, Braunschweig, 1973)) vol. 13, p. 59
27. For a comprehensive review see T. Ando, A.B. Fowler and F. Stern : Rev. Mod. Phys. 54, 437 (1982)
28. M. Altarelli : Phys. Rev. 28, 842 (1983)
29. See e.g. L.L. Chang : J. Phys. Soc. Japan 49, Suppl. 1, 997 (1980) and references therein

30. U. Ekenberg and M. Altarelli : Phys. Rev. B30, 3369 (1984)
31. E. Bangert and G. Landwehr : Superlattices and Microstructures, (1985), in press
32. A. Broido and L.J. Sham : Phys. Rev. B31, 888 (1985)
33. U. Ekenberg and M. Altarelli : to be published
34. H.L. Störmer, Z. Schlesinger, A. Chang, D.C. Tsui, A.C. Gossard and W. Wiegmann : Phys. Rev. Lett. 51, 126 (1983)
35. J.P. Eisenstein, H.L. Störmer, V. Narayanamurti, A. C. Gossard and W. Wiegmann : Phys. Rev. Lett. 53, 2579 (1984)
36. A. Fasolino and M. Altarelli : Surf. Science 142, 322 (1984)
37. A. Fasolino and M. Altarelli : in "Two-dimensional Systems, Heterostructures and Superlattices", G. Bauer, F. Kuchar and H. Heinriched.s, : (Springer, Berlin, 1984) p. 176
38. H.R. Trebin, U. Rössler and R. Ranvaud : Phys. Rev. B20, 686 (1979)
39. J.C. Maan, G. Belle, A. Fasolino, M. Altarelli and K. Ploog : Phys. Rev. B30, 2253 (1984)
40. N. Miura, Y. Iwasa, S. Tarucha and H. Okamoto : in "Proc. of the 17th Int. Conf. on the Physics of Semiconductors, San Francisco 1984", in press
41. A. Baldereschi and N.O. Lipari : Phys. Rev. B8, 2697 (1973)
42. J.C. Maan, Y. Guldner, J.P. Vieren, P. Voisin, M. Voos, L.L. Chang and L. Esaki : Solid State Commun., 39, 683 (1981)
43. L. Pauling and E.B. Wilson : "Introduction to Quantum Mechanics", (Mc Graw-Hill, New York, 1935)
44. S. Satpathy : Phys. Rev. B28, 4584 (1983)
45. J.D. Levine : Phys. Rev. 140, A586 (1965)
46. G. Bastard : Phys. Rev. B24, 4714 (1981)
47. C. Mailhiot, Y.C. Chang and T.C. Mc Gill : Phys. Rev. B26, 4449 (1982)
48. R.L. Greene and K. Bajaj : Solid State Comm. 45, 825 (1983)
49. K. Tanaka, M. Nagaska, T. Yamabe : Phys. Rev. B 28, 7068 (1983)
50. S. Chaudhuri : Phys. Rev. B 28, 4480 (1983)
51. S. Chaudhuri and K.K. Bajaj : Phys. Rev. B29, 1803 (1984)
52. C. Priester, G. Allan and M. Lannoo : Phys. Rev. B 28, 7194 (1983)
53. C. Priester, G. Allan and M. Lannoo : Phys. Rev. B 29 3408 (1984)
54. S. Satpathy and M. Altarelli : Phys. Rev. B23, 2977 (1981)
55. W.T. Masselink, Y.C. Chang and H. Morkoc : Phys. Rev. B28, 7373 (1984); J. Vac. Sci. Tech. B2, 376 (1984)
56. A. Baldereschi and N.O. Lipari : Phys. Rev. B3, 439 (1971)
57. See e.g. G. Dresselhaus : J. Phys. Chem. Solids 1, 14 (1956)
58. M. Altarelli and N.O. Lipari : Phys. Rev. B 15, 4898 (1977)
59. R.C. Miller, D.A. Kleinmann, W.T. Tsang and A.C. Gossard : Phys. Rev. B 24, 1134 (1981)
60. G. Bastard, E.E. Mendez, L.L. Chang and L. Esaki : Phys. Rev. B 26, 1974 (1982)
61. R.L. Greene, K.K. Bajaj and D.E. Phelps : Phys. Rev. B 29, 1807 (1984)

Electrons in Heterojunctions

F. Stern

IBM Thomas J. Watson Research Center, Yorktown Heights, NY 10598, USA

Properties of electrons - and, in a few cases, of holes - in heterojunctions are described, with emphasis on simple treatments of energy levels and low-temperature transport properties in $Al_xGa_{1-x}As/GaAs$ heterojunctions. A few related aspects of electrons in quantum wells are also described.

1. Introduction

Although heterojunctions have been structures of basic interest and of device potential for many years, [1] the rapid developments in recent years have been spurred by the availability of growth techniques for making nearly ideal structures and by the superlattice proposal of Esaki and Tsu, subjects covered in other lectures at this School. The two-dimensional aspects of the properties of electrons in heterojunctions had as a forerunner the related properties of electrons in semiconductor inversion and accumulation layers, a field that was in some sense opened by the experiment of Fowler, Fang, Howard, and Stiles [2], although there was a considerable body of related work on that system and on thin films even earlier. For the purpose of these lectures I will not discuss these earlier developments, some of which are described in the review article by Ando et al.[3]. The reader should recognize that much of the elementary physics of electrons in heterojunctions is taken bodily, or in some cases adapted, from the earlier literature.

First, a few basic results. The density of states in a two-dimensional electrons gas is given by :

$$\rho_0(E) = m/\pi\hbar^2 \tag{1}$$

where m is the density-of-states effective mass for motion along the interface and where a spin degeneracy of 2 and a valley degeneracy of 1 have been assumed. When, as described below, there is a ladder of subbands i, with energies beginning at E_i, then the total density of electrons is given, with Fermi-Dirac statistics, by

$$N_s = \frac{mk_BT}{\pi\hbar^2} \sum_i \log\left[1 + \exp\left(\frac{E_F - E_i}{k_BT}\right)\right] \tag{2}$$

where the density-of-states mass has been taken to be a constant, an approximation which ignores the nonparabolicity of the conduction band of GaAs. The simple notion of a constant, isotropic effective mass fails altogether for holes, as will be discussed in the lectures by Altarelli [4].

2. Heterojunction in Equilibrium

The simple, ungated heterojunction considered here is shown in Fig. 1. This figure omits the band bending associated with a gate or with Fermi level pinning at the outer surface of a real device, and therefore applies to structures in which the $Al_xGa_{1-x}As$ layer is sufficiently thick that these influences are absent. The additional effects induced by a gate will be discussed in the lectures by Vinter [5].

In most high-mobility heterojunctions, electrons are supplied to the GaAs channel by donors in the $Al_xGa_{1-x}As$ [6,7]. In a heterojunction in equilibrium, the Fermi level must be constant across the junction, leading to the condition

$$V_b = E_{Db} + V_1 + V_{sp} + E_o + (E_F - E_o) \tag{3}$$

where V_b is the conduction band offset at the interface, E_{Db} is the donor binding energy in the $Al_xGa_{1-x}As$, V_1 and V_{sp} are the potential energy drops across the ionized part of the doped layer and across the undoped spacer layer, respectively, E_o is the lowest quantum level in the GaAs channel, and E_F-E_o is the Fermi energy in the GaAs channel measured from the bottom of the lowest subband.

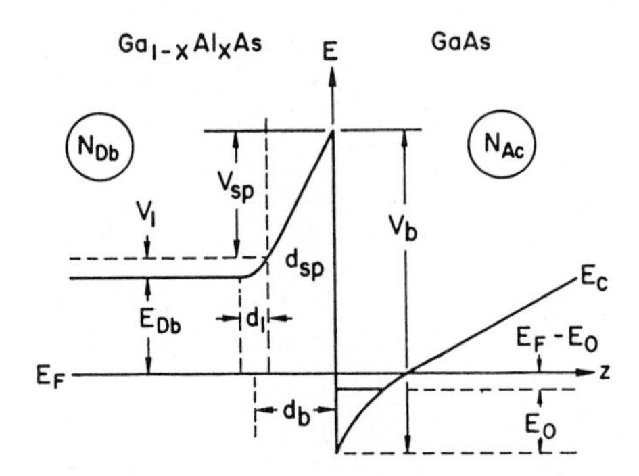

Fig.1 Schematic conduction band diagram for $Al_xGa_{1-x}As$/GaAs heterojunction at low temperature. The doping levels in the barrier region and in the GaAs are N_{Db} and N_{Ac}, respectively, and the thickness of the nominally undoped spacer is d_{sp}. The Fermi level in the barrier is assumed to be fixed by the donor binding energy E_{Db}.

Although the considerations leading to Eq. (3) are well known [8,9], this equation will be discussed here briefly, because it has turned out to be one way to determine band offsets, as applied to holes in $Al_xGa_{1-x}As$/GaAs heterojunctions by Wang et al. [10,11] and to electrons in GaInAs/AlInAs heterojunctions by Chaudhuri et al. [12]. The case shown in Fig. 1 assumes that the donor binding energy in $Al_xGa_{1-x}As$ is deep enough for the Fermi level to be pinned at the impurity level at low temperatures. That implies that any shallow donor levels, if present, are outnumbered by the residual acceptors often found in $Al_xGa_{1-x}As$. If the AlAs mole fraction x is small, or in the presence of light, Eq. (3) must be modified to account for a Fermi level above the conduction band edge in the $Al_xGa_{1-x}As$ [13].

The potential energy changes V_1 and V_{sp} of Fig. 1 are given by

$$V_1 = e(N_d + N_s)^2/2\varepsilon_b N_{Db}, \tag{4}$$

$$V_{sp} = e d_{sp}(N_d + N_s)/\varepsilon_b, \tag{5}$$

where ε_b is the permittivity of the $Al_xGa_{1-x}As$, N_{Db} is the net donor impurity concentration in the $Al_xGa_{1-x}As$, N_s is the channel electron density, and N_d is the density of fixed charges in the GaAs, given approximately by

$$N_d = (2\phi_d \varepsilon_c N_{Ac}/e)^{1/2}, \tag{6}$$

where ε_c is the permittivity of GaAs, N_{Ac} is the net acceptor concentration in the GaAs, and $e\phi_d$ is the energy difference between the conduction band edge in the bulk of the GaAs and the Fermi level (or the quasi-Fermi-level in the bulk, if there is a substrate bias or back gate bias). A more complete discussion is given, for example, in Sec. III.A.1 of Ref. 3.

The last term in Eq. (3) is simply given from the channel density N_s and the density of states, Eq. (1) or, for temperatures above absolute zero, from Eq. (2). The quantum energy E_o is found from methods like those outlined in the following section. The channel density itself is determined from a Hall effect measurement or from analysis of the magnetoconductance oscillations. If Eq. (3) is to be used to determine the band offset V_b, the remaining unknowns are the donor binding energy and N_d, the density of fixed charges. N_d is difficult to determine accurately because the impurity concentration in the GaAs may not be well known, especially for relatively pure samples. Fortunately, in those cases N_d is often small compared to N_s. It may be possible to measure N_d using back gate bias [14] or to infer it from measured subband splittings in conjunction with theoretical calculations [15,16].

The donor binding energy in $Al_xGa_{1-x}As$ has been known for many years to have a strong dependence on composition [17,24] and is difficult to measure accurately. Wang et al. [10,11] used p-type samples to determine the valence band offset from the valence band analog of Eq. (3). This has advantages over the corresponding procedure for n-type samples, because the acceptor binding energy in $Al_xGa_{1-x}As$ varies more gradually with composition than the donor binding energy, and because p-type samples are relatively free of the persistent photoconductance [25,32] that complicates many measurements on n-type samples. Wang et al. [10,11] studied p-$Al_xGa_{1-x}As$/GaAs heterojunctions with x = 0.5 and x = 1. They found that the valence band offset is given approximately by 0.45x (in eV) although other measurements, some of which are noted in Ref. 11, give values closer to 0.5 x. Most of the recent values (see Ref. 11 for a partial compilation) are considerably larger than the valence band offsets ($\sim$ 0.2 x, in eV) originally found by Dingle et al. [33] from analysis of quantum well optical absorption spectra, which were generally accepted for many years.

This abbreviated discussion does not cover the possible sources of error, both theoretical and experimental, that may affect this so-called charge transfer method of determining band offsets, nor does it describe the other methods that have been used. It shows, however, that very simple physics can sometimes lead to unexpected and useful results.

3. Subband Structure

The energy level structure of quasi-two-dimensional systems is easily calculated within the framework of effective mass approximation, provided that the complications associated with valley degeneracy, effective mass anisotropy, and with the complex valence band structure of most III-V and group IV semiconductors are put aside. We assume that the electron states (the simple considerations that follow are generally not valid for holes) have effective mass m and are derived from a single conduction band minimum. Then the one-electron wave function can be written as a product

$$\psi(x,y,z) = \zeta_i(z)\, u(r) \exp(iK.R), \tag{7}$$

where $\zeta_i(z)$ is the envelope wave function for the ith subband, u(r) is the periodic Bloch function associated with the bottom of the conduction band, and K and R are vectors in two-dimensional wave-vector and configuration space, respectively.

Within the spirit of effective mass approximation, the Schrödinger equation for the envelope function is

$$\frac{-\hbar^2}{2m_3}\frac{d^2\xi_i}{dz^2} + [V(z) - E_i]\,\xi_i(z) = 0 \tag{8}$$

where m_3 is the effective mass for motion normal to the interface and V(z) is the effective potential, which can contain both electrostatic contributions and contributions due to band structure variations. The energy levels measured from the bottom of the conduction band are

$$E = E_i + \hbar^2K^2/2m, \tag{9}$$

where m is the effective mass, assumed isotropic, for motion parallel to the interface.

For silicon inversion layers, where V(z) is very large (≈3eV) in the oxide, one can usually assume that only the semiconductor region, with $z > 0$, enters in the Schrödinger equation, and use the boundary condition $\zeta_i(0) = 0$. In heterojunctions and quantum wells, where the interface barrier height is smaller and wave functions can spread into regions with varying effective mass, a modified Schrödinger equation [34]

$$\frac{-\hbar^2}{2}\frac{d}{dz}\left(\frac{1}{m_3}\frac{d\xi_i}{dz}\right) + [V(z) - E_i]\,\xi_i(z) = 0 \qquad (10)$$

can be used provided there is no change in conduction band character across the interface. This form conserves probability current.

It is clear that some kind of grading is present in an electronic sense even for crystallographically ideal surfaces, as indicated schematically in Fig. 2. There must be at least one layer of atoms with bonding - and therefore also other electronic properties such as energy levels and polarizability - different from that of either of the adjoining materials. Equation (10) is an approximate way of dealing with such a transition layer. Other methods have been discussed, for example, by Kroemer and Zhu [35] and by White et al [36]. Stern and Schulman [37] used both the effective mass approach of Eq. (10) and a tight-binding approach to study energy levels in quantum wells, and found that a modest amount of grading has a relatively small effect on the energy levels.

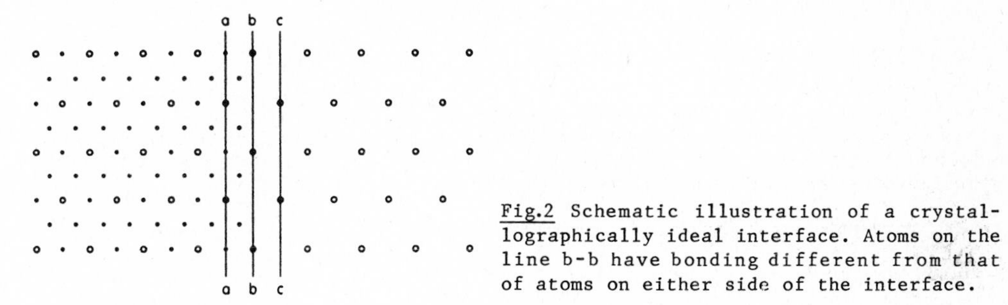

Fig.2 Schematic illustration of a crystallographically ideal interface. Atoms on the line b-b have bonding different from that of atoms on either side of the interface.

Some interesting formal consequences of a modified Schrödinger equation with a rapidly but continuously varying effective mass in a graded transition layer have been discussed by Price and Stern [38]. Interface boundary conditions enter in a crucial way in the coupling of the two valleys for a (001) Si surface, as reviewed by Ando [39]. Interface boundary conditions and general aspects of the effective mass approximation will be discussed in the lectures by Altarelli [4].

Solutions of the Schrödinger equation within the simple effective mass approximation have been given by many authors. The potential energy V(z) includes a term which depends on the eigenfunctions $\zeta_i(z)$, and the Schrödinger equation and Poisson's equation must therefore be solved self-consistently. The Hartree approximation, in which many-body effects are ignored, has been used by Stern and Howard [40] for silicon inversion layers and similar systems, including the effects of effective mass anisotropy, and detailed results have been given by Stern [41]. Many-body effects must, however, be included to give agreement with experiment. These effects have been treated by many authors, including Vinter [42], Jonson [43], and Ando [44]. A discussion of many-body effects in relation to the temperature dependence of energy levels has been given by Kalia et al [45,46]. Experimental tests of theory are provided by optical absorption measurements (see, for example, [3], but the absorption peaks are not expected to coincide with

Vinter p. 238.

$-\frac{\hbar^2}{2m}\frac{d^2\zeta_n}{dz^2} + V(z)\zeta_n(z) = E_n\zeta_n(z)$ $\quad n_{el}(z) = \sum_n |\zeta_n(z)|^2 N_n$

$N_n = \frac{k_B T m}{\pi\hbar^2}\ln\left[1+\exp\left(\frac{E_F - E_n}{k_B T}\right)\right]$

$\frac{d^2 V}{dz^2} = \frac{e^2}{\epsilon}\left[N_D^+(z) - n_{el}(z)\right]$

the calculated energy differences. The necessary corrections have been reviewed by Ando [47]. Results for (001) Si inversion layers show good agreement with the theory, but recent measurements [48] on other Si surfaces are in less satisfactory agreement with the theory [49]. Subband separations can also be found from Raman scattering measurements, a subject covered in the lectures by Abstreiter [50].

Self-consistent solutions of the Schrödinger equation and Poisson's equation are required unless there is negligible charge in the channel. The band bending and the envelope wave function for the lowest subband in an $Al_xGa_{1-x}As/GaAs$ heterojunction are shown in Fig. 3, from a self-consistent calculation by Ando [9].

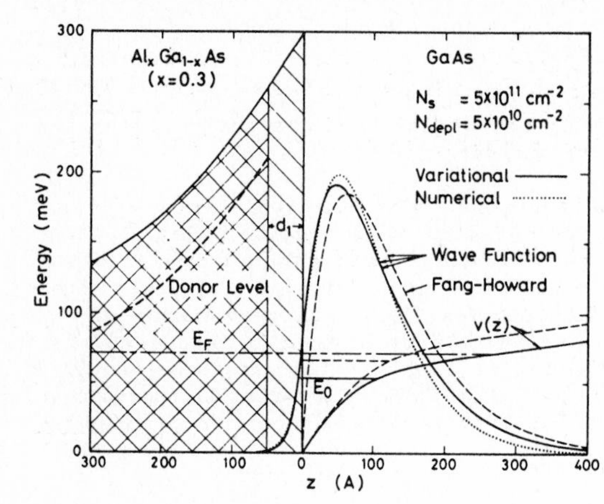

Fig.3 Self-consistent results for an $Al_xGa_{1-x}As/GaAs$ heterojunction calculated using both variational (full and dashed curves) and numerical (dotted curve) envelope functions (after Ando, Ref. 9)

Self-consistency may also be important in a quantum well if carriers are introduced by adding impurities in the barrier. In relatively wide wells and for sufficiently high carrier densities, the quantum well then approaches the character of two isolated heterojunctions, and the lowest energies occur in pairs, with symmetric and antisymmetric envelope functions. The evolution of the eigenvalue spectrum for such a situation is shown in Fig. 4.

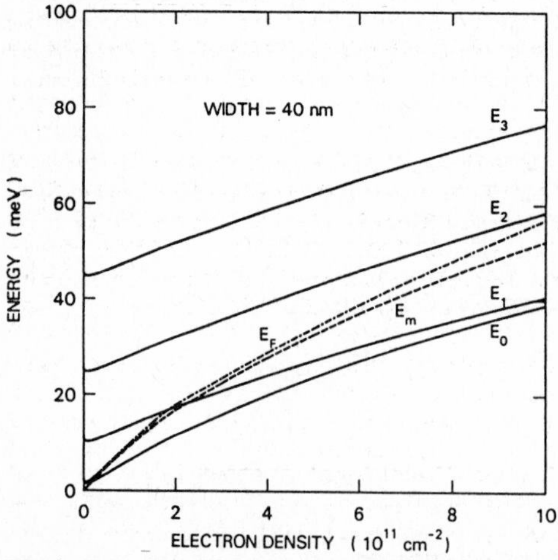

Fig.4 The four lowest calculated energy levels in an $Al_{0.4}Ga_{0.6}As$-GaAs-$Al_{0.4}Ga_{0.6}As$ quantum well 40 nm wide versus the density of electrons in the well. Also shown are the Fermi energy E_F and the maximum energy E_m of the self-consistent potential in the well. All energies are measured from the bottom of the empty well. Many-body effects, which have been included via a local-density scheme as in Ref. 16, account for the dip at low electron densities (after Stern and Schulman, Ref. 37).

4. Static Screening

A system with mobile charges will respond to changes in the energies of electronic states in such a way as to screen out fluctuations. The conventional treatment of such systems assumes that the fluctuations are sufficiently small that Poisson's equation can be linearized. Let us assume that we are dealing with an electron system that is two-dimensional in a quantum sense, so that at low enough temperatures only one subband is occupied, and that the charge distribution of electrons is given by

$$g(z) = \zeta_0^2(z), \tag{11}$$

where $\zeta_0(z)$ is the normalized envelope wave function for the lowest subband. Suppose that some perturbation, for example a charged impurity, induces a weak perturbing potential $\psi(r)$. Then the physically relevant potential to be screened by the two-dimensional electron system is the average potential in the electron layer,

$$\phi_{av}(R) = \int \phi(r)g(z)dz,$$

where $R \equiv (x,y)$. Poisson's equation for the perturbing potential is [40]

$$\nabla^2\phi(r) - 2q_s\phi_{av}(R)g_0(z) = -p_{ext}(r)/\varepsilon, \tag{12}$$

where the screening parameter q_s is

$$q_s = \frac{me^2}{2\pi\varepsilon\hbar^2\ [1 + \exp\ (-y)]}\ , \tag{13}$$

and $y = (E_F-E_0)/k_BT$. (As before, we assume a valley degeneracy of 1 here). At low temperatures, the bracketed expression in Eq. (13) approaches unity. At high temperatures, Eq. (13) gives $q_s = N_s e^2/2\varepsilon k_B T$, but this result and the simplified one-subband picture used here become invalid when carriers in higher subbands contribute to the screening.

Equation (13) gives the two-dimensional screening effect for static potentials and long-wavelength perturbations. The linear response under more general conditions, the two-dimensional analog of the Lindhard [51] dielectric response of a free electron gas, has been derived by Stern [52]. In the limit of long wavelengths and zero frequency, his result is equivalent to the long-wavelength treatment given above, but for wave vectors greater than $2k_F$ it leads to a rapid decrease in the screening parameter with increasing q [52]. Screening, like other aspects of two-dimensional physics, sometimes has a different character from that which might be expected from intuition based on physics in three dimensions. The wave-vector independence of screening for $q < 2k_F$ is one example. Another is the fact that the screening does not depend on electron density, at least for an ideal two-dimensional system at absolute zero. A treatment of the dielectric response of a two-dimensional electron system, including its nonlocal character, was given by Dahl and Sham [53].

5. Scattering Mechanisms

Although many of the aspects of transport in two-dimensional systems are worthy of detailed examination, the discussion here concentrates on the simplest case, the Coulomb scattering that limits the mobility in GaAs-based heterojunctions at low temperatures, and refers only briefly to other mechanisms.

Scattering by a charge outside a sheet of electrons is most simply described in Born approximation, which gives the cross-section for scattering through an angle θ as [40]

$$\sigma(\theta) = (me^2/2\pi\hbar^3 v)|(\psi_f|\phi(r)|\psi_i)|^2, \tag{14}$$

where v is the carrier velocity. The perturbing potential $-e\phi(r)$ reflects both the external charge and the screening effect of the carrier in the two-dimensional layer. If we simplify the problem further by assuming that the electrons lie in a sheet of zero thickness in the plane $z = 0$ and that the surrounding media have permittivity ε, then the screened Coulomb potential in the electron plane associated with a charge Ze located a distance d away is

$$\phi(R) = (Ze/4\pi\varepsilon)\int_0^\infty q(q+q_s)^{-1} J_0(qR)\exp(-qd)dq, \tag{15}$$

where J_0 is the Bessel function of order zero, R is the distance from the point in the layer closest to the charge, and q_s is the screening parameter given in Eq. (13). A more complete expression that takes both the finite thickness of the electron layer and the dielectric discontinuity into account was first given by Stern and Howard [40] using the Fang-Howard [54] variational envelope function. They showed that, at least for silicon inversion layers, the Born approximation is not seriously in error. To the best of my knowledge, a similar analysis of the validity of the Born approximation has not been carried out for GaAs-based heterojunctions. The most detailed calculation of Coulomb scattering is that of Vinter [55], who went beyond the linear screening approximation and found somewhat smaller mobilities for silicon inversion layers than those deduced from the model indicated here, based on linear screening and the Born approximation.

The dominant Coulomb scatterers in silicon inversion layers are thought to be residual charged impurities at the $Si\text{-}SiO_2$ interface, with impurities in the Si itself making a negligible contribution in the samples typically used. In GaAs heterojunctions, on the other hand, the Coulomb scatterers arise both from the impurities that donate electrons to the channel and from residual impurities in the GaAs itself.

The effect of an interface on the conductance of a bulk semiconductor has long been discussed in terms of specular and diffuse scattering by the boundary. For a dynamically two-dimensional system, the notion of specularity has no meaning because the carriers have no component of velocity in the direction normal to the interface. Instead, the scattering must be treated microscopically. This subject has been discussed in some detail in Ref. 3, primarily with silicon inversion layers in mind. For GaAs-based heterojunctions and quantum wells, there are several experiments showing that the interface is quite smooth, with lateral correlation lengths greater than 10 nm and vertical steps of atomic dimensions. Because interface scattering increases strongly with electron density, and the densities in these structures are generally smaller than in silicon inversion layers, interface scattering is not believed to play an important role in limiting the mobility in heterojunction devices [9], as opposed to the strong role played by this mechanism in silicon inversion layers at high electron densities. Alloy scattering connected with penetration of the wave function into the $Al_xGa_{1-x}As$ is also weak [9], but may be important in other systems in which the material on the channel side of the heterojunction is an alloy, as discussed, for example, by Bastard [56].

In GaAs-based heterojunctions, where interface scattering is believed to be relatively unimportant in limiting the mobility, the principal low-temperature scattering mechanism in good samples arises from the ionized impurities in the barrier that contribute charges to the channel and from residual impurities in the GaAs itself. A calculation of the combined effect of these two contributions [57], shown in Fig. 5, gives a peak mobility of order 10^6 $cm^2V^{-1}s^{-1}$, close to the highest values that have been measured in $Al_xGa_{1-x}As/GaAs$ heterojunctions in the dark (see, for example, [58]). The predicted peak mobility depends mainly on the residual ionized impurity concentration in the GaAs, and the measured values indicate that this is of order 10^{14} cm^{-3} in the best samples. Further improvement in the mobility of ungated heterojunctions would appear to require even purer GaAs.

When carriers occupy more than one subband, the factors that influence the mobility change in at least two ways. First, the screening is changed by the presence of carriers in the second subband. Here the wave vector dependence of

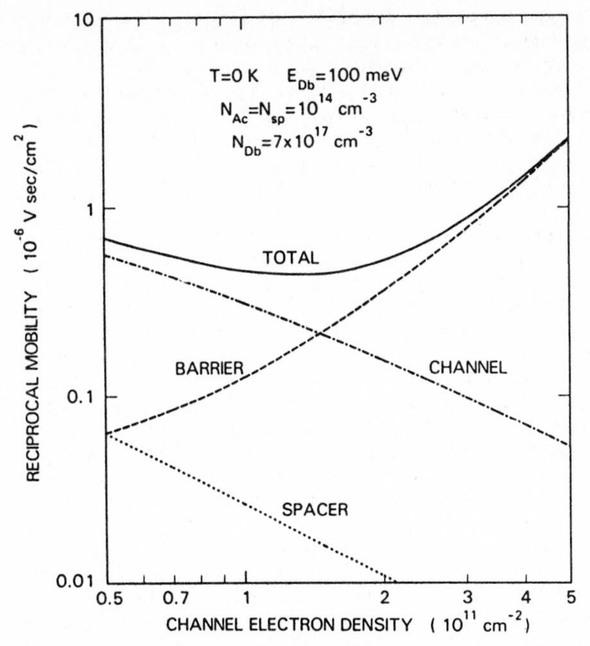

Fig.5 Calculated low-temperature mobility of elecrons in $Al_xGa_{1-x}As/GaAs$ heterojunctions as limited by Coulomb scattering (after Stern, [57])

screening [52], noted briefly above, becomes nontrivial because the Fermi wave vectors in the two subbands are generally different. Second, the presence of the second subband may open a new scattering channel. The first of these effects appears to increase the mobility [59], while the second reduces it [60]. For GaAs-based heterojunctions, mobility reduction is expected [9] and observed [61].

Although temperature dependence of mobility in semiconductors outside the range of strong or weak localization is usually attributed to phonon scattering, the temperature dependence of screening can lead to a decrease of mobility with increasing temperature. This effect was first pointed our for silicon inversion layers by Stern [62]. His calculation gave a linear dependence on temperature, consistent with experiments of Cham and Wheeler [63]. Kawaguchi et al. [64], however, found a more nearly quadratic dependence. Theoretical attempts to resolve this are still inconclusive [65].

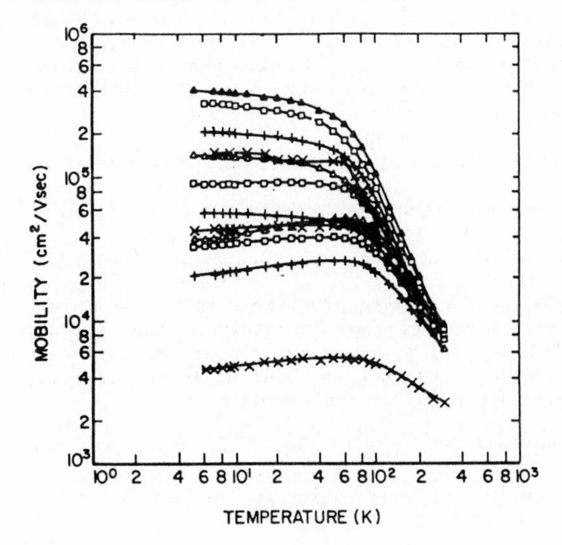

Fig.6 Temperature dependence of electron mobility in a series of $Al_xGa_{1-x}As$ / GaAs heterojunction samples (after Lin, [69])

For electrons in GaAs-based heterojunctions the situation is quite different. The Coulomb scattering is primarily forward scattering because the screening is relatively weak (in part because of the small electron effective mass). Thus the mechanism described above is not effective. Instead, in very high mobility samples the mobility decreases with increasing temperature, [66-68] whereas in low mobility samples it first increases [67,68]. These effects are illustrated in Fig. 6, from the Ph.D. thesis of Lin [69]. The former effect is attributed to phonon scattering, including a piezoelectric contribution [66]. The mobility increase is thought to result from averaging of the relaxation time over the thermal distribution of carrier energies, [65,68,69]. At temperatures above $\sim$10 K the mobility in the best sample is dominated by phonon scattering, including polar optical phonons. The theory of phonon scattering has been discussed, for example, by Price [70], and will be covered in part in the lectures by Vinter [5].

Acknowledgments

I am indebted to the organizers of the Winter School on Heterojunctions and Superlattices in Semiconductors for the opportunity to participate and thank B.J.F. Lin for permission to include the as yet unpublished results of Fig. 6. I also thank many of my colleagues for useful discussions, particularly P.M. Mooney, M.I. Nathan, T.N. Theis, and D.J. Wolford for discussions of deep donors and persistent photoconductance.

References

1. A.G. Milnes, D.L. Feucht: Heterojunctions and Metal-Semiconductor Junctions (Academic, New York, 1972)
2. A.B. Fowler, F.F. Fang, W.E. Howard, P.J. Stiles: Phys. Rev. Lett. 16, 901 (1966)
3. T. Ando, A.B. Fowler, F. Stern: Rev. Mod. Phys. 54, 437 (1982)
4. M. Altarelli: these lectures
5. B. Vinter: these lectures
6. L. Esaki, R. Tsu: IBM Research Report RC 2418 (1969, unpublished)
7. R. Dingle, H.L. Störmer, A.C. Gossard, W. Wiegmann: Appl. Phys. Lett. 33, 665 (1978)
8. D. Delagebeaudeuf, N.T. Linh: IEEE ED-29, 955, (1982)
9. T. Ando: J. Phys. Soc. Jpn 51, 3900 (1982)
10. W.I. Wang, E.E. Mendez, F. Stern: Appl. Phys. Lett. 45, 639 (1984)
11. W.I. Wang, F. Stern: Conference on Physics and Chemistry of Semiconductor Interfaces, Tempe, Arizona, 1985 ; to be published in J. Vac. Sci. Technol. B 3 (1985).
12. D. Ray Chaudhuri, J.B. Roy, P.K. Basu: Phys. Status Solidi (a) 86, K79 (1984)
13. F. Stern: Bull. Am. Phys. Soc. 30, 208 (1985)
14. See, for example, Z. Schlesinger, J.C.M. Hwang, S.J. Allen, Jr.: Phys. Rev. Lett. 50, 2098 (1983)
15. T. Ando: J. Phys. Soc. Jpn 51, 3893 (1982)
16. F. Stern, S. Das Sarma: Phys. Rev. B 30, 840 (1984)
17. A.J. SpringThorpe, F.D. King, A. Becke: J. Electron. Mater. 4, 101 (1975)
18. A.K. Saxena: Appl. Phys. Lett. 36, 79 (1980)
19. T. Ishikawa, J. Saito, S. Sasa, S. Hiyamizu: Jpn. J. Appl. Phys. 21, L675 (1982)
20. H. Künzel, K. Ploog, K. Wunstel, B.L. Zhou: J. Electron. Mater. 13, 281 (1984)
21. L.G. Salmon, I.J. D'Haenens: J. Vac. Sci. Technol. B 2, 197 (1984)
22. N. Chand, T. Henderson, J. Klem, W.T. Masselink, R. Fischer, Y.C. Chang, H.Morkoç: Phys. Rev. B 30, 4481 (1984)
23. E.F. Schubert, K. Ploog: Phys. Rev. B 30, 7021 (1984)
24. J.C.M. Henning, J.P.M. Ansems, A.G.M. de Nijs: J. Phys. C 17, L915 (1984)
25. H.L. Störmer, A.C. Gossard, W. Wiegmann, K. Baldwin: Appl. Phys. Lett. 39, 912 (1981)
26. D.M. Collins, D.E. Mars, B. Fischer, C. Kocot: J. Appl. Phys. 54, 857 (1983)
27. M.I. Nathan, T.N. Jackson, P.D. Kirchner, E.E. Mendez, G.D. Pettit, J.M. Woodall: J. Electron. Mater. 12, 719 (1983)

28. J. Klem, W.T. Masselink, D. Arnold, R. Fischer, T.J. Drummond, H. Morkoç, K. Lee, M.S. Shur: J. Appl. Phys. 54, 5214 (1983)
29. H. Künzel, A. Fischer, J. Knecht, K. Ploog: Appl. Phys. A 32, 69 (1983)
30. J.M. Mercy, C. Bousquet, J.L. Robert, A. Raymond, G. Gregoris, J. Beerens, J.C. Portal, P.M. Frijlink, P. Delescluse, J. Chevrier, N.T. Linh: Surf. Sci. 142, 298 (1984)
31. M.I. Nathan, M. Heiblum, J. Klem, H. Morkoç, J. Vac. Sci. Technol. B 2, 167 (1984)
32. A. Kastalksky J.C.M. Hwang: Solid State Commun. 51, 317 (1984) ; Appl. Phys. Lett. 44, 333 (1984)
33. R. Dingle, W. Wiegmann, C.H. Henry: Phys. Rev. Lett. 33, 827 (1974) ; R. Dingle: in Festkörperprobleme/Advances in Solid State Physics, ed. by H.J. Queisser (Pergamon/Vieweg, Braunschweig 1975), Vol. XV, p. 21
34. D.J. BenDaniel, C.B. Duke: Phys. Rev. 152, 683 (1966)
35. H. Kroemer, Q.G. Zhu: J. Vac. Sci. Technol. 21, 551 (1982)
36. S.R. White, G.E. Marques, L.J. Sham: J. Vac. Sci. Technol. 21, 544 (1982)
37. F. Stern J.N. Schulman: International Conference on Superlattices, Microstructures, and Microdevices, Champaign, Illinois, 1984 ; Superlattices and Microstructures (to be published)
38. P.J. Price, F. Stern: Surf. Sci. 132, 577 (1983)
39. T. Ando: Surf. Sci. 98, 327 (1980)
40. F. Stern, W.E. Howard: Phys. Rev. 163, 816 (1967)
41. F. Stern: Phys. Rev. B 5, 4891 (1972)
42. B. Vinter: Phys. Rev. B 13, 4447 (1976) ; 15, 3947 (1977)
43. M. Jonson: J. Phys. C 9, 3055 (1976)
44. T. Ando: Surf. Sci. 58, 128 (1976) ; Phys. Rev. B 13, 3468 (1976)
45. R.K. Kalia, S. Das Sarma, M. Nakayama, J.J. Quinn: Phys. Rev. B 18, 5564 (1978)
46. S. Das Sarma, R.K. Kalia, M. Nakayama, J.J. Quinn: Phys. Rev. B 19, 6397 (1979)
47. T. Ando: Surf. Sci. 73, 1 (1978)
48. T. Cole, B.D. McCombe: Phys. Rev. B 29, 3180 (1984)
49. S. Das Sarma, B. Vinter: Phys. Rev. B 28, 3639 (1983)
50. G. Abstreiter: these lectures
51. J. Lindhard: K. Dan. Vidensk. Selsk. Mat.-Fys. Medd. 28, (8) 1 (1954)
52. F. Stern: Phys. Rev. Lett. 18, 546 (1967)
53. D.A. Dahl and L.J. Sham: Phys. Rev. B16, 651 (1977)
54. F.F. Fang and W.E. Howard: Phys. Rev. Lett. 16, 797 (1966)
55. B. Vinter: Phys. Rev. B26, 6808 (1982)
56. G. Bastard: Surf. Sci. 142, 284 (1984)
57. F. Stern: Appl. Phys. Lett. 43, 974 (1983)
58. G. Weimann, W. Schlapp: Appl. Phys. Lett. 46, 411 (1985)
59. F. Stern: Surf. Sci. 73, 197 (1978)
60. S. Mori, T. Ando: Phys. Rev. B19, 6433 (1979)
61. H.L. Störmer, A.C. Gossard, W. Wiegmann: Solid State Commun. 41, 707 (1981)
62. F. Stern: Phys. Rev. Lett. 44, 1469 (1980)
63. K.M. Cham, R.G. Wheeler: Phys. Rev. Lett. 44, 1472 (1980)
64. Y. Kawaguchi, T. Suzuki, S. Kawaji: Solid State Commun. 36, 257 (1980) ; Surf. Sci. 113, 218 (1982).
65. F. Stern, S. Das Sarma: Solid-State Electronics 28 (1985)
66. E.E. Mendez, P.J. Price, M. Heiblum: Appl. Phys. Lett. 45, 294 (1984)
67. M.A. Paalanen, D.C. Tsui, A.C. Gossard, J.C.M. Hwang: Phys. Rev. B29, 6003 (1984)
68. B.J.F. Lin, D.C. Tsui, M.A. Paalanen, A.C. Gossard: Appl. Phys. Lett. 45, 695 (1984)
69. B.J.F. Lin: Ph. D. thesis, Electrical Engineering Department, Princeton University (1985)
70. P.J. Price: Surf. Sci. 113, 199 (1982) ; and in Proceedings of the 17th International Conference on the Physics of Semiconductors, San Francisco, 1984 (to be published)

Part III

Experimental Studies

Pictures of the Fractional Quantized Hall Effect

H.L. Störmer

AT & T Bell Laboratories, Murray Hill, NJ 07974, USA

These lecture notes present the basic experimental facts of the integral and fractional quantized Hall effect. By means of a phenomenological comparison between integral and fractional quantized Hall effect, it is concluded that the fractional quantized Hall effect is of many-particle origin. A simplified approach to the presently prevailing theoretical model of the electronic groundstate and its quasi-particle excitations is presented, together with a set of illustrations, in the hope of providing some physical insight into this novel many-particle state.

INTRODUCTION

The essence of the experimental observations, termed the quantized Hall effect, is quickly stated. At low temperatures and in high perpendicular magnetic fields, the Hall resistance ρ_{xy} of a two-dimensional (2D) electron gas is quantized to $\rho_{xy} = h/\nu e^2$ to an accuracy as high as $\sim$ 1 part in 10^7. The quantum number ν can be either an integer (integral quantized Hall effect, IQHE) [1] or a rational fraction with exlusively odd denominator (fractional quantized Hall effect, FQHE) [2]. Concomitant with the quantization of ρ_{xy}, the normal resistivity ρ_{xx} seems to vanish as the temperature tends towards T = 0.

Experimental observations of this nature must be surprising even to the most experienced physicists. Traditionally, solid state physics is not the discipline for high-precision measurements of quantum numbers, and yet this phenomenon shows an accuracy which rivals the accuracy of experimental data achieved in atomic physics. (Interestingly enough, an ideal 2D system would not even show the effect; a small amount of randomness is essential). Fractional quantum numbers generally belong to the realm of elementary particle physics, and yet the fractional quantized Hall effect exhibits quantum numbers such as ν = 1/3 and ν = 2/5. Vanishing resistivity, the absence of dissipation, is interesting in itself and promises some kind of ideality of the underlying electronic state.

All these features have a fundamental ring to them and the appearance of Planck's constant h and the electronic change e make the quantized Hall effect a macroscopic quantum phenomenon.

These lecture notes restate the basic observations of the IQHE and the FQHE. The IQHE is being derived in terms of the single-particle density of states of a 2D electron system in a magnetic field. From a phenomenological comparison between the IQHE and the FQHE it is concluded that the FQHE is of many-particle origin. These parts of the lecture notes are a reiteration of an earlier review [3]. In a final chapter, the presently prevailing theoretical model for the electronic groundstate and its quasi-particle excitations are presented. Though the approach is not rigorous, it is hoped that this train of thought and the accompanying illustrations might help the reader to develop an intuitive understanding of this novel, highly correlated electronic state. For these lecture notes no attempt is being made to present a complete listing of citations. For this purpose we refer the reader to a few recent review articles and lecture notes listed at the end.

1. Two-Dimensional Systems.

Two-dimensionality of the electronic system is an important prerequisite for the observation of the QHE. This first chapter briefly reviews the classical device structures in which these conditions can be achieved [4].

In a three-dimensional world two-dimensional (2D) systems are necessarily associated with a surface or an interface. The motion of a particle along the plane (x,y) is essentially free, while its normal motion is restricted. For quantum mechanical objects these restrictions can be made so severe that the particle loses any degree of freedom in the z-direction, and hence assumes a truly 2D character. This situation is achieved by confining the particle to a narrow potential well in the z-direction. Quantum mechanically, the bound states of such a potential well are discrete, with a typical energy spacing of

$$\Delta E \sim \frac{\hbar^2}{2m^\bullet}\frac{1}{d^2} \sim 10\ \mathrm{meV} \tag{1}$$

(for a well width $d \sim 50\text{Å}$ and $m^\bullet \sim 1/10m_\bullet$ typical for electrons in semiconductors). At low temperatures ($kT \ll \Delta E$) carriers trapped in the lowest bound state of such a well lack any degree of freedom in the z-direction, while being able to move freely along the interface following the free-particle dispersion relation

$$E = \frac{h^2 k_{xy}^2}{2m^\bullet} \tag{2}$$

Two devices are presently preferred to generate 2D carrier systems. In a Si-MOSFET, carriers are confined to the interface between Si and SiO_2 via a strong electric field establishing a quasi-triangular potential well [4] (Fig. 1).

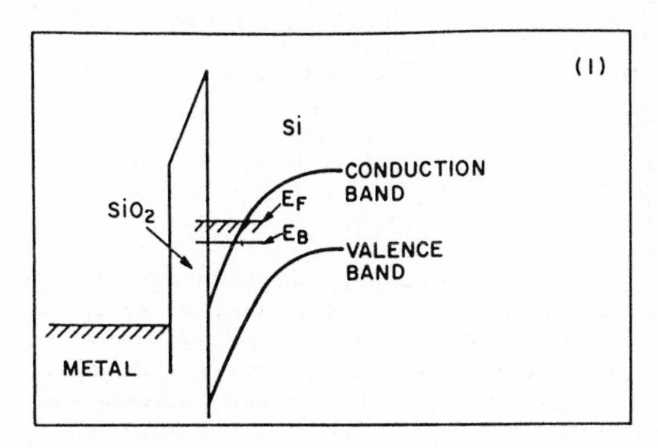

Fig.1 Carrier confinement in a Si MOSFET. A quasi-triangular potential well is established via the Si-SiO_2 interface and an external voltage which is applied to the gate metal.

The second structure is termed modulation-doped GaAs-(AlGa)As heterostructure, in which the equivalent situation is achieved by an abrupt doping profile, resulting in carrier confinement at the GaAs-(AlGa)As interface [3] (Fig. 2). The Si-MOSFET has the advantage over GaAs-(AlGa)As that the 2D carrier concentration can readily be varied by the application of an external gate-voltage. Conversely, the mobility μ of carriers in the GaAs-(AlGa)As ($\mu \sim 10^6\ \mathrm{cm^2/Vsec}$) far exceeds those of carriers in a Si-MOSFET ($\mu \sim 10^4\ \mathrm{cm^2/Vsec}$). Since a low carrier scattering rate is crucial for the observation of the FQHE the GaAs-(AlGa)As system is presently preferred for studies of this kind.

2. Integral Quantum Hall Effect (IQHE) in Modulation-Doped Systems

This chapter describes some aspects of the integral quantum Hall effect (IQHE) as it is observed in modulation-doped GaAs-(AlGa)As and thereby introduces some of the notation used in the next chapter. Figure 3 shows one of the most distinct

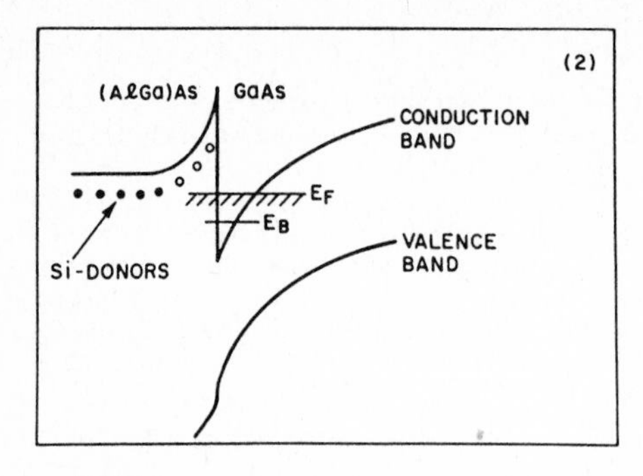

Fig.2 Carrier confinement in a modulation-doped GaAs-(AlGa)As heterojunction. A quasi-triangular potential well is established via the GaAs-(AlGa)As interface and a strong electric field generated by an abrupt doping profile and subsequent charge-transfer.

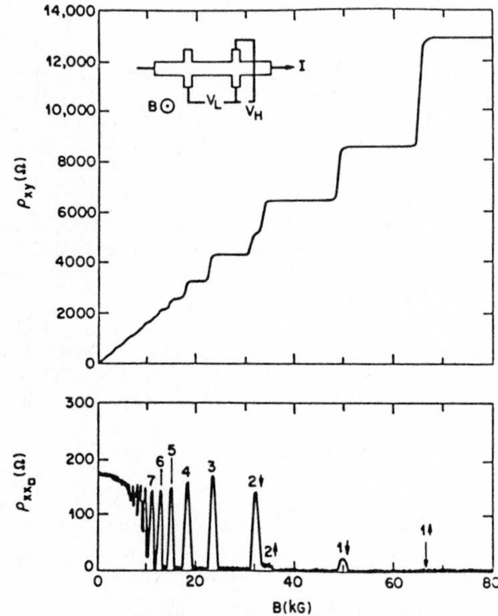

Fig.3 The integral quantum Hall effect IQHE. Low-temperature Hall resistance ($\rho_{xy} = V_H/I$) and magneto-resistance ($\rho_{xx} = V_L/I$) of a modulation-doped GaAs-(AlGa)As sample with a density of $n = 4.0\times10^{11}\ cm^{-2}$ and $\mu = 8.6\times10^{4}\ cm^2/Vsec$. Insert shows sample configuration [5].

manifestations of the IQHE in a GaAs-(AlGa)As heterostructure at 50 mK [5]. The sample configuration is shown as an insert. Two characteristic voltages V_H and V_L are measured as a current I is imposed onto the 2D system and the perpendicular magnetic field B is varied. V_H, the Hall voltage across the current path, and V_L, the longitudinal voltage along the current path, are normalized to I. This yields the components of the resistivity tensor $\hat{\rho}$ with $\rho_{xy} = V_H/I$ and $\rho_{xx} = gV_L/I$ where g is a geometry factor typically of the order of 1. The relations between the components of the resistivity tensor $\hat{\rho}$ with $\rho_{xy} = V_H/I$ and $\rho_{xx} = gV_L/I$ where g is a geometry factor typically of the order of 1. The relations between the components of the resistivity tensor $\hat{\rho}$ and the components of the conductivity tensor $\hat{\sigma} = \hat{\rho}^{-1}$ are :

$$\sigma_{xx} = \frac{\rho_{xx}}{\rho_{xx}^2 + \rho_{xy}^2}, \quad \sigma_{xy} = \frac{-\rho_{xy}}{\rho^2_{xx} + \rho^2_{xy}}, \quad \sigma_{yy} = \sigma_{xx}, \quad \sigma_{yx} = -\sigma_{xy} \tag{3}$$

$$\rho_{xx} = \frac{\sigma_{xx}}{\sigma_{xx}^2 + \sigma_{xy}^2}, \quad \rho_{xy} = \frac{-\sigma_{xy}}{\sigma^2_{xx} + \sigma^2_{xy}}, \quad \rho_{yy} = \rho_{xx}, \quad \rho_{yx} = -\rho_{xy} \tag{4}$$

At first glance these equations seem to be surprising, since vanishing resistivity ($\rho_{xx} \to 0$) implies also vanishing conductivity ($\sigma_{xx} \to 0$), while in the absence of a magnetic field one quantity is the inverse of the other. Figure 4 illustrates this

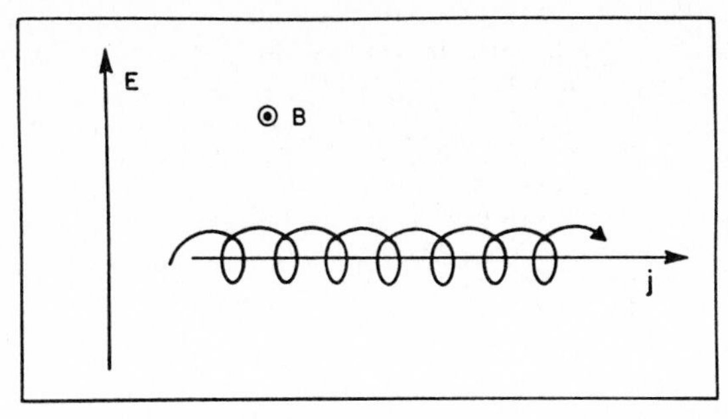

Fig.4 Cycloid motion of a carrier in crossed electric (E) and magnetic fields (B).

counterintuitive relationship for carriers in a crossed magnetic (B) and electric field (E). Under such conditions carriers move on a cycloid orbit in the direction perpendicular to E and B. Since there is no electric current along E the conductivity $\sigma = i/E = 0$. Conversely, the absence of any electric field component along the current direction implies for the resistivity $\rho = E/j = 0$.

From purely classical considerations, i.e., balancing the Lorentz force acting on a carrier moving in a magnetic field against the force from the electric Hall field, it is expected that ρ_{xy} shows a linear magnetic field dependence

$$\rho_{xy} = B/en \tag{5}$$

with n being the 2D electron density. Rather than this linear B-dependence, Fig. 3 shows a Hall resistivity ρ_{xy} which assumes a staircase-like structure with plateaus quantized to

$$\rho_{xy} = h/ie^2 \ , \ i = 1,2,3... \tag{6}$$

The accuracy of this quantization has been verified to approximately 1 part in 10^7. Concomitant with the appearance of plateaus in ρ_{xy}, the diagonal resistivity ρ_{xx} seems to vanish over large portions of B. Resistivities as low as $\rho_{xx} < 10^{-10}\ \Omega/$ equivalent to roughly $10^{-16}\ \Omega$ cm, have recently been established.

Following the present understanding of the IQHE,the formation of plateaus in ρ_{xy} and vanishing values of ρ_{xx} are directly related to the singularities in the density of states (DOS) of a 2D system in a strong perpendicular magnetic field (Fig. 5a). The DOS of an ideal two-dimensional system consists of spin-split Landau levels with energies

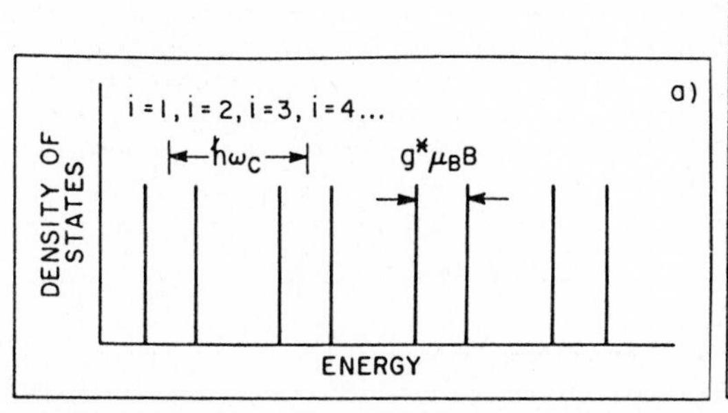

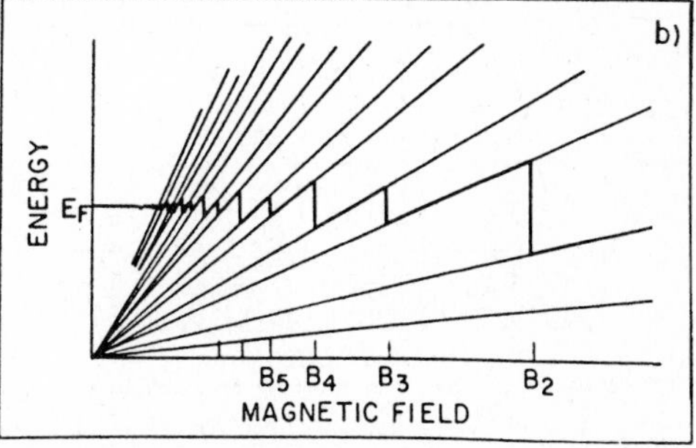

Fig.5a Density of states of an ideal 2D system. $\hbar\omega_c$ = Landau splitting, $g^*\mu_B B$ = spin splitting. The degeneracy of each singularity is d = eB/h.
Fig.5b Landau fan and position of Fermi energy E_F as a function of field for an ideal 2D system at low temperature.

$$E = (j + 1/2)\hbar\omega_c \pm Sg^{\cdot}\mu_B B, \quad j = 0,1,2... \tag{7}$$

$\hbar\omega_c = \hbar eB/m^{\cdot} = 0.17$ [meV/kG]xB is the Landau level splitting for an effective mass of GaAs of $m^{\cdot} = 0.067\ m_o$. S is the spin of the carriers, $g^{\bullet}$ their effective g-factor and $\mu_B = eh/2m_o$ is Bohr's magneton. For our purposes it is not important to discern between Landau level splitting and spin-splitting. We only retain a sequence of singularities (from now on called magnetic levels or levels) numbered by i = 1,2,3 ... starting with i = 1 at the lowest energy. The number of states per level for any 2D-system is

$$d = eB/h = B/\Phi_o = 2.42 \times 10^9\ cm^{-2}\ kG^{-1} \times B \tag{8}$$

where $\Phi_o = h/e$ is the magnetic flux quantum. This value is independent of any material parameter. Through it one can define a filling factor

$$\nu = n/d. \tag{9}$$

At low temperatures($kT \ll \hbar\omega_c$, $g^{\cdot}\mu_B B$) and at any given field, ν indicates the number of populated levels. Noting that n is the number of carriers per unit area and d is the number of flux quantum per unit area, ν is also a measure of the number of flux quantum associated with each electron. For a system with fixed carrier density, the filling factor decreases as B is raised. The variation of the Fermi level E_F is periodically abrupt due to the strongly singular DOS (Fig. 5b). At any given field, E_F resides in the close vicinity of level i = int (ν) + 1. However, an exceptional situation arises at fields

$$B_1 = n\ \Phi_o/i \tag{10}$$

where an exact multiple i of levels is filled. Then E_F is intermittent and lies in the gap region between level i and level i + 1.

The value of ρ_{xy} and the vanishing of ρ_{xx} can then be derived apparently in the following way : The diagonal conductivity σ_{xx} is entirely dependent on the DOS at the position of E_F. Since the DOS vanishes in the gap region, σ_{xx} vanishes as well and with Eq. 4 we derive $\rho_{xx} = 0$ as long as $\rho_{xy} \neq$. The classical expression ρ_{xy} = B/en holds also for quantum mechanical free electrons. Hence, at a sequence of singular points on the field axis $B_1 = n\Phi_o/i$ where ρ_{xx} vanishes, the Hall resistance is $\rho_{xy} = \Phi_o/ie = h/ie^2$.

Such a derivation of ρ_{xx} and ρ_{xy} neither accounts for the finite width of the plateaus nor for the width of the zero resistance regions,both of which are the truly outstanding features of the IQHE. Explanation of a finite width of these

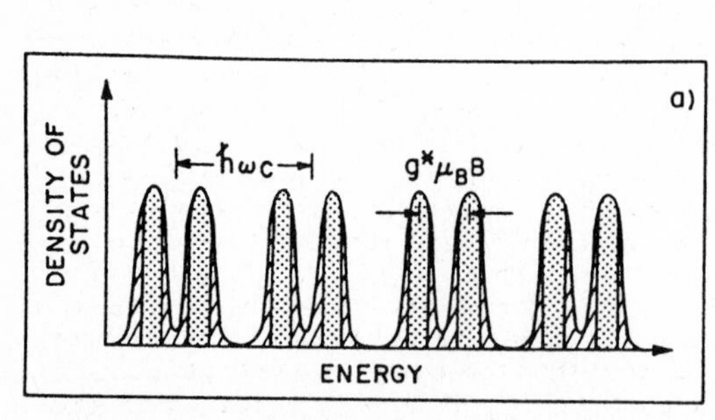

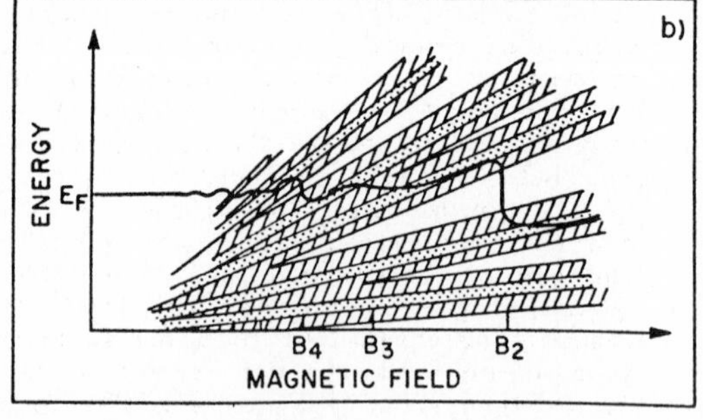

Fig.6a Density of states of a real 2D system. Disorder broadens the singularities (Fig.5a) into bands. The central part of each band contains delocalized states while the states in the flanks are localized, not participating in transport.
Fig.6b Landau fan and position of Fermi energy E_F as a function of field for a real 2D system at low temperature.

features requires the existence of localized states. Localized states are expected to be present in real two-dimensional systems, due to disorder as a result of random distribution of impurities or random interface steps. They lift the degeneracies of the magnetic levels and broaden them (Fig. 6a). States at the center of this distribution will be extended, while those in its tails will be localized, not participating in electronic transport. This broadening of the magnetic levels moderates the abrupt jumps of E_F from one level to the next as B is varied about the crucial values B_1 (Fig. 6b). Hence, for finite ranges of field, E_F moves through regions of localized states between magnetic levels. In an elegant gedanken experiment [6] LAUGHLIN has shown that under these conditions ρ_{xy} remains quantized and ρ_{xx} tends towards zero in spite of the disorder. He demonstrates quite generally that independent of the strength of the disorder

$$\sigma_{xx} = 0 \text{ and } \sigma_{xy} = ie/\Phi_o = ie^2/h, \quad i = 0, \pm 1, \pm 2... \tag{11}$$

whenever E_F lies within localized states (mobility gap) or within the region of a true gap. The value of i may be zero, which describes the case of an insulator. Except for this degenerate case i is finite and, hence, with Eq. 4

$$\rho_{xx} = 0 \text{ and } \rho_{xy} = h/ie^2, \; i = 0, \pm 1, \pm 2... \tag{12}$$

which are the quantities one generally obtains in transport measurements. The quantized region may be wide (as wide as 95% plateaus and 5% transitions) indicating that the major part of the DOS consists of localized states and still Eq. 12 holds. LAUGHLIN's gedanken experiment does not specify the value of i. For weak disorder i is expected to coincide with the value obtained from the ideal case, in agreement with the experiment.

We summarize this chapter stating that the IQHE is understood in terms of gaps in the single particle DOS of a 2D electron system in a strong perpendicular magnetic field. Disorder leads to the formation of localized states in the gap region between magnetic levels. Whenever the Fermi energy lies in this range $\rho_{xx} = 0$ and $\rho_{xy} = h/ie^2$ (i = 1,2,3...) excluding the degenerate case i = 0 of an insulator. A single-electron picture is sufficient for a description of the IQHE.

3. The Fractional Quantized Hall Effect (FQHE)

Experiment

The availability of low-density 2D modulation-doped heterostructures with unprecedently high mobilities ($\mu \approx 10^6$ $cm^2/Vsec$) led to magneto-transport studies on GaAs-(AlGa)As structures at low temperatures in extremely high magnetic fields (up to 280 kG) where the extreme quantum limit could be reached. In terms of the filling factor $\nu = n/d$ the extreme quantum limit, where only the lowest magnetic level is populated, is characterized by $\nu \leq 1$. Figure 7 shows ρ_{xy} data from a sample of constant electron density $n = 2.13 \times 10^{11}$ cm^{-2} and mobility $\mu \approx 10^6$ $cm^2/Vsec$.[7]. For this low concentration, exact multiples ($\nu = i$) of magnetic levels are filled at $B = nh/ie \approx 88$ kG, 44 kG, 22 kG... In the vicinity of these field positions one expects the appearance of the IQHE. The last of these plateaus and concomitant resistance minimum are indeed observed at $B_1 \approx 88$ kG. Data below 70 kG were not accessible, due to the use of a fixed base field from a superconducting magnet. Measurement in the absence of this base field (not shown) reveals clearly the higher orders (ν = 2,3...) of the IQHE. The majority of Fig. 7 covers the region of the extreme quantum limit where $\nu < 1$. Contrary to expectation, a rich sequence of structures is observed, reminiscent of the IQHE. However, the deduced quantum numbers are not integers but rational fractions with exclusively odd denominators. At the present time, structures in ρ_{xx} have been observed in the vicinity of filling factors

$$\begin{aligned} \nu &= 1/3,\ 2/3,\ 4/3,\ 5/3,\ 8/3 \\ \nu &= 1/5,\ 2/5,\ 3/5,\ 4/5,\ 6/5,\ 7/5,\ 8/5 \\ \nu &= 2/7,\ 3/7,\ 4/7 \\ \nu &= 4/9,\ 5/9 \end{aligned} \tag{13}$$

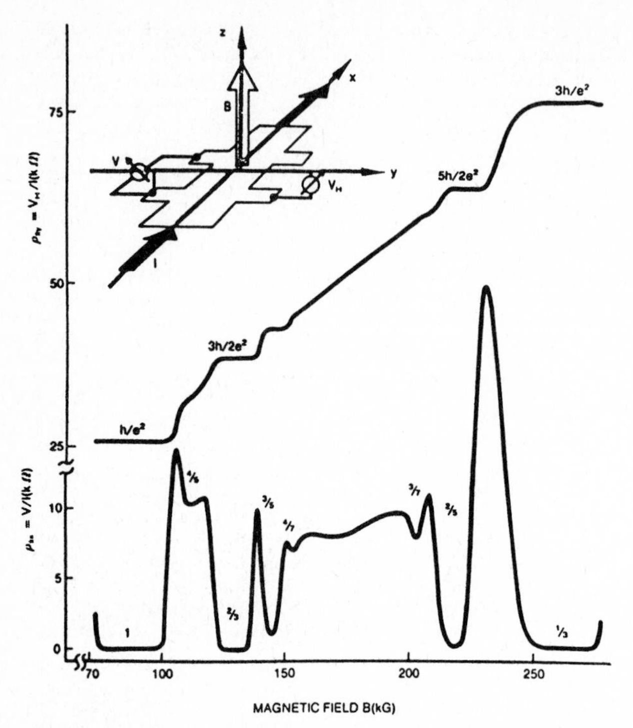

Fig.7 The fractional quantum Hall effect FQHE. Low-temperature Hall resistance ρ_{xy} and magneto-resistance ρ_{xx} of a very high-mobility modulation-doped GaAs-(AlGa)As sample with density $n = 2.13 \times 10^{11}$ cm^{-2} and mobility $\mu \sim 10^6$ $cm^2/Vsec$ in the extreme quantum limit ($\nu < 1$). Structures resembling the IQHE (Fig. 3) occur at fractional filling factor ν. [7].

The concomitant Hall plateaus of the more prominent of these structures have been determined to be quantized to $\rho_{xy} = h/\nu e^2$ to an accuracy as high as 3 parts in 10^5 (limited by the equipment). While the existence of plateaus in ρ_{xy} and vanishing resistance in ρ_{xx} at integer filling factor ν are well accounted for by the IQHE, the appearance of similar phenomena at fractional ν is inconsistent with such an interpretation. Not only do these structures appear at fractional occupation of a magnetic level but, moreover, ρ_{xy} is quantized to $\rho_{xy} = h/\nu e^2$ with ν being an exact rational fraction and not an integer.

Since phenomenologically these new features resemble those of the IQHE, this phenomenon is termed the fractional quantum Hall effect (FQHE), though both must be of different origin. While the IQHE can be explained in terms of non-interacting 2D electrons in a high magnetic field, no such interpretation seems to be possible for the FQHE.

4. Phenomenological interpretation of the FQHE

In order to assess the possible origin of the FQHE, we return to Laughlin's gedanken experiment described in section 2. As an example, we choose the $\nu = 1/3$ state. The other fractions can be discussed in an analogous way. From $\rho_{xx} - 0$ and $\rho_{xy} \neq 0$ at $\nu \approx 1/3$, we can deduce $\sigma_{xx} \to 0$. Hence, the DOS at the position of E_F for partial filling of the lowest magnetic level is vanishingly low, being either zero and forming a true gap, or finite but localized, forming a mobility gap.

The appearance of such gaps in the single particle DOS of 2D electrons in the extreme quantum limit is totally unexpected. A description of the minima in ρ_{xx}

cannot be given in terms of a non-interacting particle DOS. One has to involve electron interaction for their explanation.

The formation of the long predicted electron Wigner solid, where a finite gap separates the condensed state from the single particle excitations, initially seems to provide a basis for the observed anomalies. This would require such an electron solid to form preferentially around given filling factors, e.g.ν = 1/3. Numerical studies on the groundstate energy of a Wigner solid and the related CDW in a 2D system in the extreme quantum limit indicate no preference for any given fractional ν and, hence, call in question any interpretation of the FQHE in terms of A Wigner solid. Experimental data also dismiss such an interpretation. At low temperature, and in the presence of disorder, a Wigner lattice is pinned to potential fluctuation and a non-linear current/voltage characteristic is expected to occur as the solid becomes depinned at small electric fields. Measurements at ν = 1/3 down to electric fields as low as 10 μV/cm did not produce any such non-linearities.

Since a Wigner solid does not seem to explain the experimental results, we must look beyond such an interpretation. For this we return to the earlier gedanken experiment, which requires $\rho_{xy} = h/ie^2$, $i = \pm 1, \pm 2, \ldots$ whenever E_F lies in a gap region (excluding here the trivial case i = 0). The experimental result $\rho_{xy} = h/1/3\ e^2$ is clearly in conflict with such a conclusion, indicating that the assumptions under which the statement was derived do not hold for the electronic state responsible for the FQHE. However, with an ad hoc assumption, Eq. 12 can be reconciled. This will shed some light on the possible nature of the underlying electronic state.

Laughlin's gedanken experiment [6] relies on gauge invariance of the vector potential (by which the flux quantum Φ_o enters the derivation of ρ_{xy}), and on the quantization of the electric charge, e. The final result is actually stated as a ratio of these quantities $\rho_{xy} = \Phi_o/ie = h/ie^2$. The experimentally observed value of $\rho_{xy} = \Phi_o/\ 1/3\ e$ in the FQHE can be regained if we assume the formation of carriers with effective fractional charge $e^* = 1/3\ e$.

Fractionally charged quasi-particles as current-carrying units, and the existence of a gap at E_F for $\nu = p/q$ do provide a phenomenological explanation of the FQHE with $\rho_{xy} = \Phi_o/e^*$. The above deduction is by no means rigorous. This picture is rather brought forward here, guided by recent theoretical studies on the groundstate of 2D systems in the extreme quantum limit, which suggest the formation of a novel electron liquid with fractionally charged quasi-particles of fraction ν.

5. Present Understanding of the FQHE

This chapter retraces the lines of thought which led to the presently prevailing theoretical model for the electronic state underlying the FQHE.

The discovery of the FQHE has initiated a reexamination of the groundstate of a 2D electron system in the extreme quantum limit. A numerical calculation by YOSHIOKA, HALPERIN and LEE [8] for a finite size system of 4, 5, and 6 electrons in a rectangular box with periodic boundary conditions in a high magnetic field, yielded three important results :

1. Over a wide range of ν, the groundstate of the collection of electrons is significantly lower than that of a Wigner solid.
2. At ν = 1/3 (and possibly at ν = 2/5, but also at ν = 1/2), the groundstate energy, as a function of ν, develops a downward cusp, indicating a commensurate energy at these filling factors.
3. The pair correlation function of the groundstate differs considerably from that of a Wigner crystal.

All these results indicate that the Wigner crystal is not the groundstate for this finite system. While extrapolation to many electrons is unreliable, these numerical data, nevertheless, are suggestive for the groudstate of a real system.

An analytic expression for the groundstate of a 2D system in the extreme quantum limit at rational filling factor was recently proposed by LAUGHLIN [9]. This many-particle wavefunction, with built-in-pair correlation, presently forms the basis for most theoretical models of the FQHE.

LAUGHLIN's wavefunction has the following properties :

1. It describes a state which only has a filling factor $\nu = 1/m$, where m is an integer. Assuming electron/hole symmetry, a case can also be made for $\nu = 1 - 1/m$.
2. It is antisymmetric only for odd m, hence, only odd denominators are allowed.
3. Its pair correlation function suggests it to be a novel quantum-fluid rather than a Wigner solid for $m \lesssim 10$.
4. The elementary excitations are separated from the groundstate by a finite gap.
5. These quasi-particle excitations have fractional charge $e^{*} = e/m$.
6. The quantum-fluid is incompressible and has no low-lying excitations. Hence, it flows resistance-less at $T = 0$.
7. For $m \gtrsim 10$, the quantum liquid is expected to crystalline into a Wigner solid.

For a rigorous derivation of these properties I refer the reader to the original literature and to some recent review articles on the subject.

5.1 Illustration of the Wavefunction

In the remainder of the paper I would like to present a greatly simplified approach to Laughlin's wavefunction. Though it lacks rigor it might assist the reader in developing a physical understanding of the electronic state at fractional filling of a Landau level. The lines of thought follow closely a suggestion by Halperin [10].

The wavefunction proposed by LAUGHLIN to describe the state at filling factor $\nu = 1/m$ is :

$$\Psi(Z_1, Z_2 \ldots Z_N) = \prod_{i<j}^{N} (Z_i - Z_j)^m \exp\left[-\frac{1}{4} \sum_k |Z_k|^2 \right] \tag{14}$$

The square of this N-particle wavefunction describes the probability to find the N participating electrons at positions Z_1, Z_2 ... Z_N.

The complex plane has been chosen to represent the 2D plane. Such a choice is a matter of mathematical convenience since it simplifies Eq. 14 considerably. A particle at (x,y) in the real 2D plane is described by a single complex number $z = (x - iy)l_o$ where $l_o = \sqrt{\hbar/Be}$ is the magnetic length. Apart from the scale factor l_o and an inversion of the y-axis, the real 2D plane and the complex plane are equivalent.

Many particle wavefunctions are difficult to visualize. In order to simplify this task we will focus on the motion of one prototype-electron (Z_N) in the presence of all other electrons fixed at positions Z'_1, Z'_2, $Z'_3 \ldots Z'_{N-1}$. The wavefunction for this single particle is then :

$$\Psi(Z_N) = Z_o \exp\left[-\frac{1}{4} |Z_N|^2 \right] \prod_{1}^{N-1} (Z_N - Z'_i)^m \tag{15}$$

where the products over all fixed pairs, $[Z'_i - Z'_j)^m$ and their exponential are collected into Z_o. Equation (15) describes a particle which moves through a set of fixed points Z'_i like a ball through a pin-ball machine trying to stay away from the fixed particles. In the vicinity of each of the fixed electrons its wavefunction decays rapidly with a power m.

It is instructive to try to develop an intuitive understanding of the particular form of Eq. 15. For this we start with a single electron in the lowest Landau level (spin neglected) on an infinite 2D plane in a normal field B restricted to the lowest Landau level. Using a symmetric gauge, its wavefunction can be written as

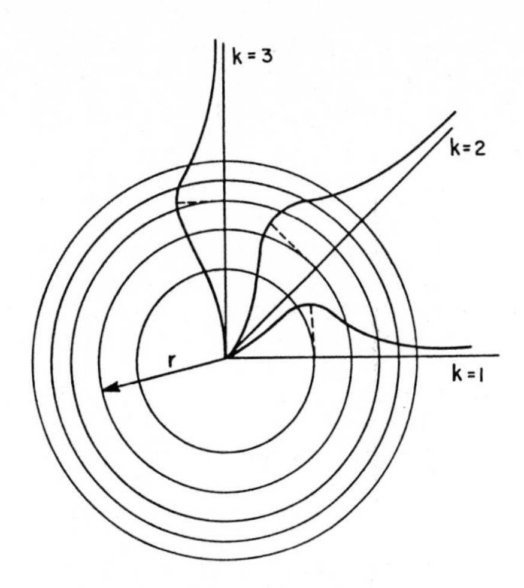

Fig.8 Illustration of the basis functions for a 2D-carrier in the lowest Landau level using a symmetric gauge for the vector potential. The orbits are racetracks about the origin with increasing angular momentum k and increasing radius $r = \sqrt{2k}\, l_o, l_o = \sqrt{\hbar/Be}$.

$$\phi\ (Z) = P\ Z^k \exp\ [\ - \frac{1}{4}\ |Z|^2\ ] \tag{16}$$

where P is a normalization factor and the same complex notation is used. These wavefunctions are racetracks around the origin with angular momentum k and orbital radius $r = \sqrt{2k}$ in units of l_o (see Fig. 8). The general case of a wavefunction for one electron in the lowest Landau level can then be written as a linear combination of these basic functions

$$\Psi(Z) = P'\exp\ [\ - \frac{1}{4}\ |Z|^2\ ]\ \sum_{k-1} a_k Z^k \tag{17}$$

with expansion coefficients a_k.

If we confine the system to a large dis of radius R (in units of l_o) the basic functions remain approximately valid, but the expansion has to be cut off for orbitals bigger than R. The limits k to $k_{max} = S = R^2/2$.

$$\Psi(Z) = P'\ \sum_{k-1}^{s}\ a_k Z^k \exp\ [\ - \frac{1}{4}\ |Z|^2\ ] \tag{18}$$

Since the exponential in Eq. 18 is always a positive real number $\Psi(Z)$ has s roots $(Z_1', Z_2', Z_3' \ldots Z_s')$ in the complex plane. Then $\Psi(Z)$ can obviously be expressed in terms of its roots

$$\Psi(Z) = P'\ \exp\ [\ - \frac{1}{4}\ |Z|^2\ ]\ \prod_{k=1}^{s}\ (z-z'_k) \tag{19}$$

Though in principle some roots might be degenerate (same position), for a general case they are roughly uniformly distributed over the disc (Fig. 9). On a small loop around each root the phase of the wavefunction changes exactly by 2π. We might call these points vortices due to their formal analogy to vortices in superconductors. Their extent is $\sim \sqrt{2}l_o$ and their density in the plane is $\eta = s/\pi R^2 l_o^2 = Be/h = B/\Phi_o$ which coincides with the density of flux quantum Φ_o due to the magnetic field B.

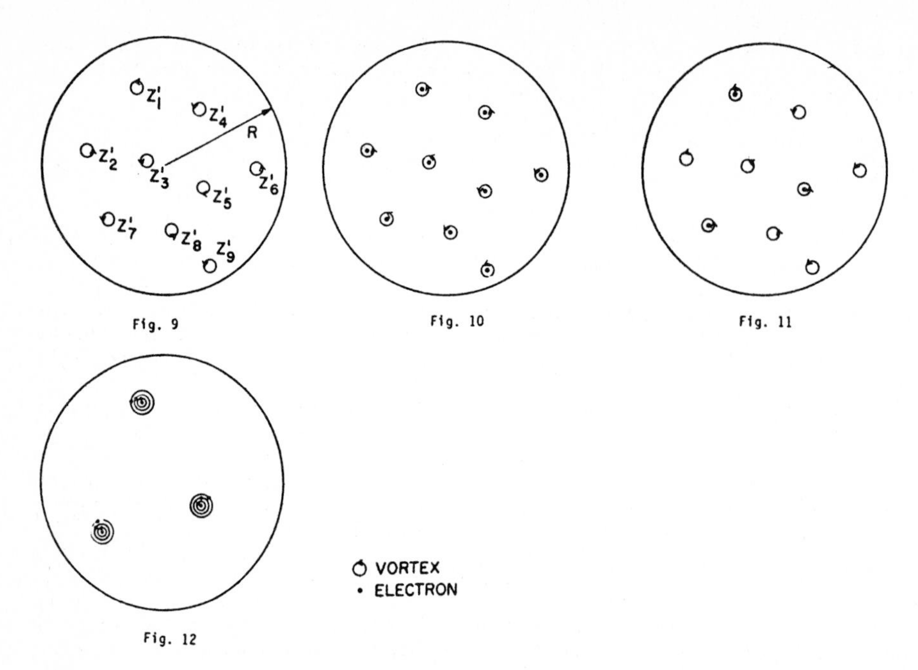

Fig.9 Illustration of a general wavefunction for a single 2D-carrier in the lowest Landau level confined to a large disc of radius R. The z_i' indicate the positions of "vortices" i.e., positions where the wavefunction vanishes and its phase changes by 2π around it. Their density is the same as the magnetic flux density $\eta = B/\Phi_o$ of the magnetic field B.
Fig.10 Additional (fixed) electrons can be positioned only at the vortices of the prototype-electron ($\psi = 0$) to obey Pauli's principle. At maximum filling the electron density equals the magnetic flux density and hence $\nu = 1$, i.e., the Landau level is completely filled.
Fig.11 At $\nu = 1/3$ the electrons occupy only 1/3 of the vortices. The existence of the other vortices is required by the strengh of B but there is no compelling reason for their actual position.
Fig.12 The $\nu = 1/3$ system can considerably reduce its potential energy by placing a three-fold vortex at the position of each electron. This reduces Coulomb interaction since the wavefunction now vanishes like the 3rd power (three vortices) rather than linearly (one vortex).

So far we have considered only a single electron. Were we to add more electrons to the system, we would have to take products over their wavefunctions, antisymmetrize the product and the problem would become quickly unmanageable. However, we are satisfied with observing the motion of a prototype-electron among a set of other fixed electrons. Such additional (fixed) electrons in Fig. 9 can only be positioned at the location of the vortices of the prototype in order to obey Pauli's principle. Only in the center of the vortices does the wavefunction of the prototype vanish. As we keep adding carriers we fill up all vortices until an electron density η is reached. This is the maximum number of electrons which fit into the lowest Landau level (Fig. 10). Since the electron density equals the magnetic flux density, the filling factor is exactly $\nu = 1$, as required. Hence, within the limits of our model, which keeps S - 1 particles fixed, Eq. 19 describes the state of a noninteracting electron gas at $\nu = 1$. The prototype-electron produces a vortex at the position of all other electrons.

If we were to release the fixed electrons, simple illustrations like Fig. 10 would be impossible. However, one can imagine snapshots where at any time each electron generates a vortex at the position of each other electron to satisfy Pauli's principle. Then, by induction from Eq. 19, one might suggest the following wavefunction for this many-particle state.

$$\Psi(Z_1,Z_2,Z_3 \ldots Z_3) = P^H \prod_{|<|}^{s} (z_i-z_j) \exp \sum_{k}^{s} \left[-\frac{1}{4} |z_k|^2 \right] \tag{20}$$

Indeed, Eq. 20 is the totally antisymmetrized solution for the $\nu = 1$ state in very high magnetic fields neglecting electron-electron interaction.

Our aim, however, is to find an intuitive solution to the $\nu = 1/m$ state. As a concrete example we chose $\nu = 1/3$ and return to Fig. 9. At $\nu = 1/3$ the lowest Landau level is only filled to 1/3 capacity i.e., the fixed electrons occupy only 1/3 of the vortices leaving 2/3 of the vortices unoccupied (Fig. 11). Vortices at the position of the fixed electrons are required by the Pauli principle. There are no compelling reasons for vortices at other positions, except that the total vortex-density has to remain η. In an interacting system such unoccupied vortices are actually wasteful, since the prototype electron avoids certain points in the plane without gaining energy. A much more favorable solution is to generate three-fold vortices at the position of each fixed electron (Fig. 12). This keeps the prototype further away from the fixed electrons and, hence, reduces considerably the Coulomb energy of the system. Since each fixed electron is located at a three-fold root of ψ this state is just the illustration of Eq. 15 for $m = 1/\nu = 3$. Through induction we regain Eq. 14 which is Laughlin's wavefunction for the electronic state underlying the FQHE. In this state each electron generates an m-fold vortex around each other's electron. Because the wavefunction drops off like the distance between carrier pairs to the m-th power, such a configuration considerably reduces the Coulomb energy of the 2D system. This is the reason why this highly correlated motion of the carriers is energetically so favorable and is believed to form the groundstate of a 2D electron system in a magnetic field.

5.2 Illustration of the Quasi-particles

Excitation from the groundstate of Eq. 15 forms quasi-particles with fractional charge $e^* = e/m$. This section proposes a gedanken experiment to develop an intuitive picture of such an e/3 quasi-particle.

Figure 13 shows again the $\nu = 1/3$ electronic state. There are four fixed electrons each accompanied by a three-fold vortex of the prototype-electron. The extent of each three-fold vortex is $\sim \sqrt{2m} l_0 = \sqrt{6} l_0$ which coincides with their average spacing. In this sense the vortices are dense in the 2D plane. The probability of finding the prototype-electron is shown as contour lines tending towards zero in the vicinity of the fixed carriers.

At fixed carrier density we slightly raise the magnetic field so that exactly one more flux quantum Φ_0 enters the system. This requires one more single vortex in the wavefunction of the prototype which we might place in the center of Fig. 13. Such an additional vortex requires the wavefunction to vanish at a given point, introduces considerable distortion and raises the total energy of the system. Predominantly this energy increase is caused by the increased cyclotron energy ($1/2\ \hbar\omega_c$). However, a small fraction of it is due to the close proximity of the additional vortex to the neighboring three-fold vortices. It is energetically advantageous for the system to open the cage surrounding the single vortex at the cost of lowering the average distance between all fixed particles and their accompanying three-fold vortices.

In a gedanken experiment we can perform this flux quantum addition three times, creating a three-fold vortex in the center and successively displacing the surrounding carriers. Finally we take an electron from outside the system and place it at the position of the vortices. The resulting state is again a $\nu = 1/3$ state where each fixed electron is associated with a three-fold vortex. This state is

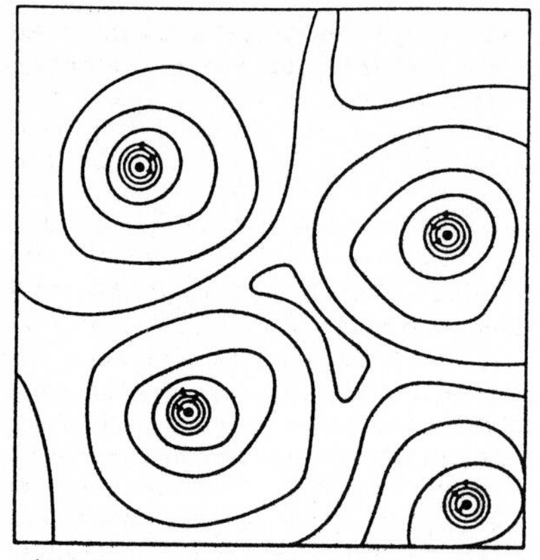
Fig.13

Fig.14

Fig.13 Schematic illustration of $\nu = 1/3$ state. Four electrons are fixed. One electron moves and generates three vortices at the position of each of the fixed electron. Its probability distribution is shown as contour lines tending towards zero near the fixed electrons.

Fig.14 Introduction of one additional vortex (slight increase of magnetic field) into the state of Fig. 13. The vortex considerably perturbs the system. It regains equilibrium by a slight displacement of the fixed carriers. The vortex represents a quasi-particle with charge $e^+/3$.

electrically neutral, since the total charge of the electrons is compensated by the total charge of ionized impurities from which the electrons emerged. Therefore, removing again the additional electron creates locally an apparent positive charge e^+ at the position of the three-fold vortex in the center. With further removal of two of its three vortices, an apparent positive charge of roughly $e^+/3$ will remain i.e., each single vortex in the system appears to be associated with a charge $e^* = e^+/3$. The vortex in the center of Fig. 14 therefore represents a quasi-particle of charge $e^+/3$. They are stable objects and like real carriers these quasi-particles can move through the system carrying a fraction of a charge from one place to another and, hence, give rise to an electrical current. A rigorous calculation shows that their charge is exactly $e^* = e/m$.

Conclusions

These lecture notes have considered only a few selected aspects of the experiments and theoretical models of the IQHE and the FQHE. The form of presentation might mislead the reader to assume that these phenomena are well understood. This is far from being true, beginning with theoretical models for the FQHE which are radically different from the model presented here. And even within this model important questions remain to be answered : What is the groundstate for $\nu = p/q$ where $p \neq 1$ and $p \neq q = 1$? Hierarchical models have been suggested, where e.g., the quasi-particle of the $\nu = 1/3$ state performs a correlated motion and generates the $\nu = 2/7$ and $\nu = 3/5$ state. A wavefunction as aesthetically appealing as Eq. 14 has not been found yet. Maybe a simple expression does not exist. How do the plateaus in ρ_{xy} and minima in ρ_{xx} in the FQHE come about? In analogy to the IQHE, localization of particles (quasi-particles in this case) is probably involved. But other scenarios, like the formation of Wigner lattices of quasi-particles, are also being cited. What is the effect of localization on the size of the quasi-particle energy gap? What is the dispersion relation for quasi-particle?... to pose but a few questions.

The experimental data on the new groundstate are also rather rudimentary. Only electrical transport measurements have so far been performed, and the information gained does not go much beyond what can be read off from Fig. 7. The field is wide open for ingenious, though probably difficult, experiments to probe the nature of the electronic state underlying the FQHE and in general to investigate the rich pattern of behavior of a 2D electron system in the presence and absence of a magnetic field over a wide range of carrier densities.

Acknowledgement

Most of the experimental work described in this paper results from a collaboration with D.C. TSUI, A.M. CHANG, P. BERGLUND, G.S. BOEBINGER, J.C.M. HWANG, A.C. GOSSARD, M.A. PAALANEN, J.S. BROOKS and M.J. NAUGHTON, whom I would like to thank for their cooperation and many stimulating discussions. I also benefited immensely from discussions with R.B. LAUGHLIN, B.I. HALPERIN, M. SCHLUTER and P. LITTLEWOOD. I would like to thank K. BALDWIN, W. WIEGMANN and T. BREMAN for excellent technical support.

References

For these lecture notes no attempt is being made to present a complete listing of citations. We would rather refer the reader to several recent reviews and lectures listed at the end.

1. K. von Klitzing, G. Dorda, M. Pepper : Phys. Rev. B28, 4886 (1983)
2. D.C. Tsui, H.L. Störmer, A.C. Gossard : Phys. Rev. Lett. 48, 1559 (1982)
3. H.L. Störmer : Festkörperprobleme XXIV - Advances in Solid State Physics, ed. by P. Grosse (Vieweg, Braunschweig 1984) p.25
4. T. Ando, A. Fowler, Stern : Rev. Mod. Phys. 54, 437 (1982)
5. M.A. Paalanen, D.C. Tsui, A.C. Gossard : Phys. Rev. B25, 5566 (1982)
6. R.B. Laughlin, Phys. Rev. B23, 5632 (1981)
7. A.M. Chang, P. Berglund, D.C. Tsui, H.L. Störmer, J.C.M. Hwang : Phys. Rev. Lett. 53, 997 (1984)
8. D. Yoshioka, B.I. Halperin, P.A. Lee : Phys. Rev. Lett. 50, 1219 (1983)
9. R.B. Laughlin : Phys. Rev. Lett. 50, 1395 (1983)
10. B.I. Halperin : Helv. Phys. Acta 56, 75 (1983)

Reviews and Lectures

Refs. [3] and [10] are reviews.

R.B. Laughlin : In "Two-Dimensional Systems, Heterostructures and Superlattices", Springer Ser. Solid-State Sci. Vol. 53, ed. by G. Bauer, F. Kuchar, H. Heinrich, pp. 272 and 279.

D.C. Tsui : Proc. of the 17th Intl. Conf. on the Physics of Semiconductors, S. Francisco 1984, to be published.

Photoluminescence in Semiconductor Quantum Wells

C. Delalande

Laboratoire de Physique de l'Ecole Normale Supérieure, 24, rue Lhomond, F-75005 Paris, France

Radiative recombination in semiconductor quantum wells (QW) may occur after the system has been put in an excited electronic state. It is called photoluminescence when this excitation is obtained by light absorption. As described in the first part of this lecture by P. Voisin, an absorption spectrum provides informations essentially on the densities of states in valence and conduction bands and looks like a continuous spectrum above the 2D system energy gap. In sharp contrast, photoluminescence exhibits lines of various widths, often only one line, whose energies are near, even smaller than the energy gap. That is due to the rapid relaxation of photoexcited carriers towards the bottom of the bands or towards energy levels close to these bands. Thus, photoluminescence involves levels which remain populated long enough during the carrier recombination process and for which radiative recombination is effective enough with respect to non-radiative processes (see fig. 1).

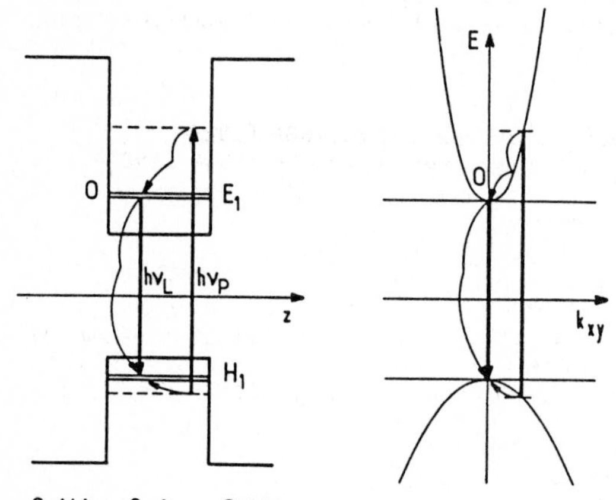

Fig.1 The left part shows the confined electron and hole levels in the z growth direction, and the right part the dispersion relation in the k_{xy} layer plane. The figure is limited to one electron subband and one hole subband and explains schematically the principles of a photoluminescence experiment. A pump photon $h\nu_P$ puts the system in an electronic excited state. A very efficient non-radiative relaxation process occurs and lets the electron (the hole) more to the bottom (the top) of the conduction (valence) band where non-radiative or radiative (photon $h\nu_L$) recombination takes place.

Because of the high sensitivity of the experimental techniques and because of the high radiative quantum efficiency in III-V quantum wells, results can be obtained in a wide range of excitation power even in a single quantum well with a width as small as 25 Å. That has induced the development of excitation spectroscopy experiments which will be discussed in §1.

Most experiments reported here are performed at low temperature, in order to resolve energy details as small as a part of a meV (1 meV = 12°K). In the low excitation regime, the geometry is usually that where the exciting light and the luminescence signal both propagate along the growth axis of the sample, in contrast with the configuration used for lasers of these structures. In CW experiments, the out-of-equilibrium steady state, which is built as a result of the pump and relaxation processes, induces the spectral shape of the luminescence. In a time-resolved set-up, the luminescence signal decay measures the lifetime of the excited state of the QW, which may be radiative or non-radiative. Due to the high quality of the samples which can be obtained by molecular beam epitaxy, the GaAs-Ga(Al)As QW system is clearly the most investigated, and will be the main topic of this lecture.

1. Excitation Spectroscopy

1.1. In an excitation spectroscopy experiment, the variation of the intensity of a given luminescence line is measured versus the wavelength of the exciting light source. Excitation spectroscopy gives, with a higher sensitivity and without the substrate etching problems [1], informations about the structures of the absorption spectrum. It enables also to determine, in a complicated luminescence spectrum, what is due to the quantum well and what is due to other parts of the investigated structure. But excitation spectroscopy involves also the intra-band relaxation process which occurs between the absorption-induced hot electron creation and the luminescence process at the bottom of the bands. That gives rise, in bulk GaAs, to high disparities between absorption [2] and excitation spectroscopy [3].

We will present in this section some results obtained by excitation spectroscopy on the Ga(As)-Ga(Al)As system. Due to the limited wavelength range of dye lasers, this technique has also been used only in the GaAs-In(Ga)As strained system [4]. Very recently, the 1.5 μm InP-$In_{0.53}(Ga)_{0.47}As$ system has been investigated using a classical light source [5].

1.2. Some results of excitation spectroscopy on GaAs-Ga(Al)As QW's

GaAs and $Ga_{1-x}(Al)_xAs$ are two direct gap semiconductors (for $x < 0.435$) whose energy gap is well known [6].

$$E_G[Ga_{1-x}(Al)_xAs] = 1.519 + 1.247x (eV)$$

at low temperature. Excitation spectroscopy, like absorption spectroscopy, is a unique way for the determination of the conduction band $\Delta E_C(x)$ and valence band $\Delta E_V(x) = 1.247x - \Delta E_C(x)$ offsets. The comparison between the optical H_nE_n and L_nH_n absorption transitions (H_n, L_n n^{th} subband of heavy (light) holes, E_n n^{th} subband of conduction electrons) and the results of a simple [7] or more sophisticated theory [8] has given the $\Delta E_C = 0.85\ \Delta E_G$ well-known value. But a careful analysis of recent excitation spectroscopy experiments on parabolic [9] or square [10] multi-quantum-wells gives a value of $\Delta E_C = 0.60\ \Delta E_G$ which has been corroborated by results obtained in Separated Confinement Heterostructures [11] : it has been shown in each case that only some E_nH_m ($m \neq n$) optical transitions are sensitive to the offset ratio, and that the exciton binding energy has to be taken into account. Other values have also been found [12], leaving this problem a subject of discussion.

The energies of bound excitonic states are theoretically known in a perfect 2D system

$$E_{Xn}^{2d} = - R_X/(n-\tfrac{1}{2})^2 \quad \text{where} \quad R_X = e^4\mu/E^2\hbar^2$$ is the 3D Rydberg (μ effective mass), ∿ 4.2 meV in bulk GaAs. That gives a value of $4R_X$ for the ground state binding energy. The finite width of the well [13], the finite barrier height and the problem of the hole mass in the layer plane [14] induce some difficulties in the theoretical considerations. The rough results of these calculations is that the confinement of electrons and holes in the QW produces an increase of the binding energy and a decrease of the transverse size of the exciton. By looking at the 1S and 2S excitonic transitions in the excitonic spectrum of high quality QW, Miller

[15] has measured the heavy-hole and light-hole exciton binding energies ; 9 meV (slightly more for the light hole exciton) is a good estimate for a 50-100 Å well and a concentration in Aluminium near 30 %.

Finally, the width of the excitonic peak in excitation spectroscopy can be correlated to thickness fluctuations of the well of one or two monolayers [16,17], provided their lateral size is larger than the exciton diameter (∿ 250 Å). That is due to the variation of the confinement energy of the electron and hole states with the well width L. As it varies roughly as L^{-3}, low-width wells exhibit larger excitonic peaks (few meV for L ∿ 50 Å) than the wider ones (a part of meV for L > 100 Å).

These features are illustrated in figure 2 on an excitation spectrum obtained in a 50 Å QW grown by molecular beam epitaxy.

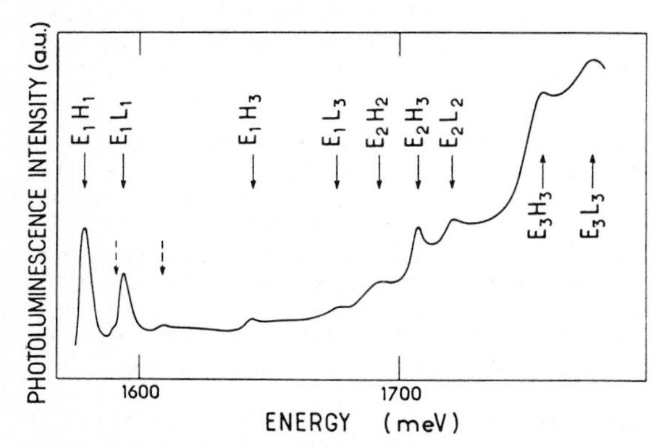

Fig.2 Excitation spectrum of a 50 Å GaAs-$Ga_{0.87}Al_{0.13}As$ quantum well (ref.11). The energy position of the E_1H_3 and E_1L_3 excitonic transitions appear to be highly sensitive to the conduction and valence band offsets. The dashed arrows indicate the position of the transitions involving the 2S state of the E_1H_1 and E_1L_1 excitons. The width of the 1S excitonic peaks, due to interface defects, is about 3 meV in this sample. An E_2H_3 transition occurs because of the asymmetry of the GaAlAs barriers.

2. Quantum Wells Photoluminescence

2.1. Excitonic luminescence

The luminescence of bulk GaAs (and other III-V compounds) is dominated by impurity effects , even for impurity concentrations below $10^{15}cm^{-3}$ [18]. The free exciton luminescence is also complicated by the polariton problem and by reabsorption. In sharp contrast, the luminescence from undoped GaAs-Ga(Al)As QW's presents one dominating line which is attributed to intrinsic heavy-hole free-exciton recombination [1]. The comparison with the absorption spectrum, the strong polarization of the luminescence line, and the fact that the same behaviour is observed in p-type MBE and n-type MOCVD structures strongly supports this interpretation.

In lower quality samples, the width of the excitonic luminescence line increases (typically from 1 meV up to 5 to 10 meV) and a few meV Stokes shift between the Heavy-hole excitonic peak in the excitation spectrum and the luminescence peak appears, as shown in figure 3 [19,20].

This behaviour can be interpreted in terms of exciton trapping on interface defects : considering an isolated interface protuberance of GaAs (typically 300 Å

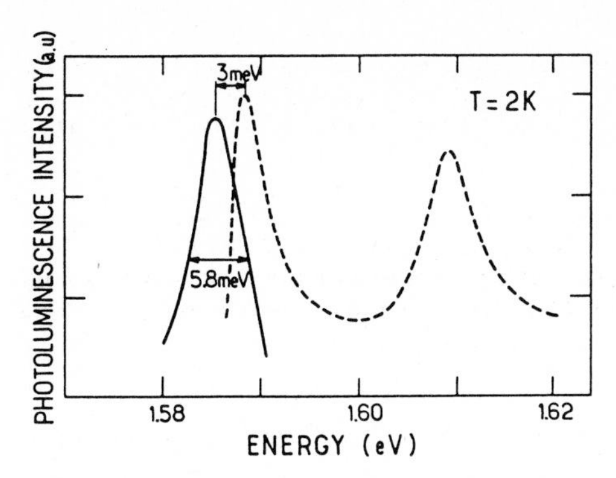

Fig.3 Excitation spectrum (dashed curves) and luminescence spectrum (full lines) of a 70 Å GaAs-$Ga_{0.48}Al_{0.52}As$ MOCVD quantum well, showing the linewidths and the Stokes shift between the free exciton heavy-hole excitation peak and the luminescence peak involving excitons bound to interface defects [20].

diameter and 2 monolayers depth), Bastard et al [20] have shown that the in-plane center of mass motion of the exciton presents bound levels localized on this defect and that the binding energy can be as high as 5 meV. It has been also found that, from a reasonable defect density of $10^{10}cm^{-2}$, an exciton trapped on a defect has not enough time to jump from one defect to another. These results are consistent with the experimental Stokes shift and luminescence width. When the temperature is raised, a thermal detrapping of excitons occurs ; the Stokes shift between excitation and luminescence spectra becomes very low : the luminescence line of these moderate interface quality samples involves then free excitons [21].

As a matter of fact, the interface defects nature changes from one sample to another, depending on the respective sizes of the exciton, the defect, the distance between defects, the exciton diffusion length. Several peaks may occur, each of them corresponding to a discrete well width [22]. A real description seems to be in terms of a continuous spectrum of localized and delocalized exciton states separated by a sharp boundary called the mobility edge [23]. That is consistent with hole-burning [24], Resonant Rayleigh scattering [19] and Transient grating [23] time-resolved experiments. The bound exciton problem in 2D structures appears to be a case where technological improvement of interface quality leads to fundamental problems in theoretical physics.

The direct measurement of the recombination lifetime of quasi 2D excitons has been performed [25] using picosecond light source excitation and streak camera detection with a resolution of 20 ps. This experiment and other results [26] show that the measured lifetime $\tau^{-1} = \tau^{-1}_{rad} + \tau^{-1}_{non\ rad}$, which takes into account both radiative and non-radiative recombination channels, decreases from 1 ns to 200 ps when the well width falls down to 30 Å. A decrease of the luminescence decay time with decreasing well width is indeed expected for both non-radiative interface and radiative recombinations. Free exciton recombination in QW is expected to be enhanced because of an increasing overlap of electron and hole wave functions in quasi 2D structures. The fact that the data for many (but selected) samples fall on to the same curve suggests that, at low temperature, the lifetime is governed by radiative excitonic recombination. The temporal variation of the average energy of the excitonic luminescence has been measured [27] ; it has been found that, as in [20], the results are consistent with a model of phonon-assisted hopping of excitons from one interface defect to another.

In conclusion, steady-state and time-resolved experiments have shown that intrinsic excitonic luminescence is the main recombination channel in GaAs-Ga(Al)As QW's at low temperature, and that these excitons may be bound on interface defects.

2.2. Conduction to valence band luminescence

When the lattice temperature is increased, the pseudo equilibrium of excited carriers induces the ionization of excitons :

$$X \rightleftarrows e + h$$

The electron density N_e, hole density N_h which is equal to N_e because of the neutrality condition, and the exciton density N_x satisfy the two-dimensional mass action law [28]

$$\frac{N_e N_h}{N_x} = \frac{m_e^* kT}{\pi \hbar^2} \exp\left(-\frac{E_x}{kT}\right)$$

At 300°K, in GaAs QW's, the minimum ionization ratio $N_e/(N_e+N_x)$ is about 0.5 [28], supposing a maximum electron density of 1 in an exciton area. At low usual excited carrier concentrations, this calculation indicates that the room temperature luminescence does involve band-to-band recombination, even if the absorption exhibits peaks due to excitons, which are ionized before recombination.

In contrast with the exciton case, where the width of the luminescence line is related to unknown interface defects, the line shape of a band-to-band transition can be calculated in a 2D system. Considering only one electron subband n and one hole subband m, the Fermi golden rule and the $\Delta\vec{k} = 0$ optical selection rule gives :

$$g(h\nu) \propto \iint d^2k^e\, d^2k^h\, M_{nm}(k^e,k^h)\, f_e(k^e)\, f_h(k^h)\, \delta^2(\vec{k}^e-\vec{k}^h)\, [\delta(\varepsilon^e + \varepsilon^h + E_n + H_m + E_G - h\nu)]$$

In this formula E_G is the GaAs bandgap, $E_n(H_m)$ the confinement energy of the electron (hole) subband, $h\nu$ the photon energy, and ε^e (ε^h) the in-plane kinetic energy which is given in the parabolic approximation by $\hbar^2 k^{e2}/2m_e^*$ ($\hbar^2 k^{h2}/2m_h^*$). $f_e(k^e)$ and $f_h(k^h)$ are the Fermi distribution of electrons and holes resulting from the pseudoequilibrium of excited carriers in the steady state situation.

It is easily found, taking a constant matrix element $M(k_e,k_h)$ that

$$g(y) = Y(y)\, \frac{1}{1 + \exp\dfrac{\mu^*/m_e^* y - \mu_e^*}{kT_e^*}} \cdot \frac{1}{1 + \exp\dfrac{\mu^*/m_h^* y - \mu_h}{kT_h^*}}$$

where $y = h\nu - (E_g + E_n + H_m)$; $Y(y)$ is the step function, $\mu_e(\mu_h)$ the Fermi level of electrons (holes), with the energy origin taken at the bottom of the conduction subband n (at the top of the valence subband m), μ^* the reduced mass, $T_e^*(T_h^*)$ the temperature of electrons (holes) in the pseudoequilibrium. When the phonon electron coupling is sufficiently high $T_e^* = T_h^* = T$, the lattice temperature.

In the case of non-degenerate electrons and holes is found a high energy exponential tail $\exp(-y/kT_\mu^*)$, with

$$T_\mu^{*-1} = m_e^*/\mu\, T_e^{*-1} + m_h^*/\mu^*\, T_h^{*-1}$$

In fact, to our knowledge, no fit of 300°K band-to-band luminescence has been made on GaAs QW's. On the contrary Bimberg [29], by following the position of the luminescence line versus the gap of GaAs, and Dawson [30] by comparing the exciton peak in the excitation spectrum and the luminescence peak, claim that the room temperature luminescence is excitonic. Finally, recent time-resolved experiments [31] and high excitation studies in the laser geometry [32] have reported a bimolecular desexcitation consistent with electron-hole plasma recombination. This problem is still a subject for discussion.

The case of a degenerate electron gas and a non-degenerate hole gas has been studied in modulation doped GaAs-Ga(Al)As quantum well heterostructures [33].

Knowing the Fermi energy μ_e (by transport measurements), a Stokes shift $\mu_e[1+(m_e^*/m_K^*)]$ corresponding to the filling of the conduction band has been observed at low temperature between excitation and luminescence. It has also been possible to fit the luminescence line shape with the formula written above, but by considering two transitions involved in luminescence and by adding an inhomogeneous broadening parameter.

In InAs-GaSb superlattices, electrons and holes are not confined in the same material and the binding energy of the exciton is then very low. Band-to-band luminescence occurs rather than exciton recombination, as observed in experimental data from 2°K to 300°K [34].

2.3. Extrinsic luminescence

The best way to characterize extrinsic luminescence is by its appearance in an intentionally doped material. The binding energy E of the impurity, in the case of $A^o e$ or $D^o h$ luminescence, has to be measured with respect to the extrema of the bands. As absorption (and luminescence) exhibits excitonic lines, the binding energy E_X of the exciton has to be taken into account, and the Stokes shift between the excitonic line and the impurity line is then $E-E_X$. Another character of extrinsic luminescence is the possibility of its saturation at high excitation level, as shown in figure 4.

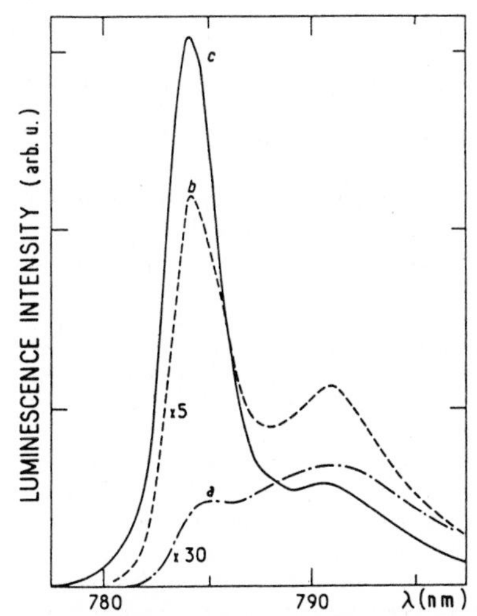

Fig.4 Photoluminescence of a 50 Å GaAs-$Ga_{0.85}Al_{0.15}As$ QW at 3 excitation levels from the lower one (a) to the higher one (c). The impurity line is due to acceptor (Carbon)-electron recombination. The shift with respect to the exciton peak (left side of the figure) is about 15 meV. The binding energy of this acceptor, which is in this case located on the edge of the well, may be obtained by adding the binding energy of the exciton. The saturation of the extrinsic line when the excitation level is increased is also shown here.

The problem of donors and acceptors in QW's is complicated by the fact that the binding energy varies with respect to the position of the impurity in the well. It has been shown [35,36,37] that the binding energy of hydrogenic impurities increases as the well width decreases, until a maximum is reached for some low well width ($L < 30$ Å), and then decreases because of the penetration of the wavefunctions in the barriers. It has also been shown that this binding energy is larger when the impurity is located at the center of the well,and is roughly two times lower when it is located at the edge of the well. So the position of the maximum of the extrinsic luminescence and its shape depend on the distribution of impurities in and near the well.Because of their higher binding energy, the well-center impurities are more populated in the electron (hole) relaxation process and their relative role in the luminescence process is more important. Nevertheless, in non-intentionally doped GaAs MBE structures, it has been shown [38,39] that carbon which is accumulated on the Al(Ga)As-vacuum interface during growth is deposited in the first few monolayers of GaAs and then gives rise to an electron-acceptor luminescence mainly characteristic of impurities at the edge of the well (cf. Fig.4). It has also been shown that the growth of Al(Ga)As induces rough interfaces which segregate these C impurities. Like the free exciton line width, the relative strength of impurity (ie C^oe) and free exciton luminescence lines appears as a test of the interface quality and of the sample purity in single QW's. Inserting a low width GaAs-Ga(Al)As superlattice just before the QW traps the impurities induces a decrease of the extrinsic luminescence of the QW and an improvement of interface quality [40-41]. That explains also why extrinsic luminescence is lower in a multi QW than a single QW.

By selectively doping the center of the well, it has been possible to measure the well width variation of the binding energy of Si donors [42] ($E_B \sim 5.8$ meV in bulk GaAs) and of Be acceptors [43] ($E_B \sim 28$ meV in bulk GaAs). One has also reported recombination involving excitons bound to Si^o donors [44], to Be^o acceptors [45] and other lines [46]. This subject clearly requires further experimental and theoretical investigations.

2.4. Results on various III-V QW's

The understanding of the nature of low-temperature luminescence requires additional absorption or excitation spectroscopy data. Apart from the extensively investigated GaAs-Ga(Al)As system, results are available on the following structure :

InAs-GaSb type II superlattice [34]. The $\sim$ 250 meV luminescence involves mainly band-to-band recombination, with some additional evidence of an extrinsic line at low temperature.

GaSb-AlSb strained type I QW. The lower energy gap is that of GaSb (810 meV). The non-intentionally doped GaSb is p-type and the luminescence seems to involve acceptors [47]. When the well width is decreased, due to the 2D confinement of electrons the Γ valley is above the L and X valleys ; when $L < 45$ A, a decrease of the luminescence yield due to indirect recombination has been observed [48].

$In_xGa_{1-x}As$-GaAs (low gap $In_xGa_{1-x}As \sim 1.3$ eV for $x \sim 0.15$). This strained superlattice will be discussed by J.Y. Marzin in another lecture of this school. The Stokes shift is low, indicating an intrinsic luminescence. [49].

$In_{0.53}Ga_{0.47}As$-InP (low gap $In_{0.53}Ga_{0.47}As \sim 810$ meV). The n-residual impurity content in InGaAs and the Stokes shift may indicate that D^oh recombination is involved [50-51-5].

The $In_{0.53}Ga_{0.47}As/In_{0.48}Al_{0.52}As$ system lattice matched on InP presents an extrinsic luminescence at low temperature [52,53]. In the case of modulation doped superlattice, a band-to-band recombination has been found above 70°K.

This enumeration shows the great diversity of results concerning various materials. Many more investigations are needed for a good knowledge of photoluminescence in these QW structures. It will be partly solved by an improved control of the relevant growth technique.

3. Special Topics

This short lesson does not allow an extensive development of topics close to low excitation and low temperature photoluminescence. We will only give some references on various problems.

When an electron (hole) is photocreated high in the conduction (valence) band, it relaxes its excess kinetic energy before its recombination on the bottom of the bands. The electron-optical phonon polar coupling is specially efficient for electrons having more than 36 meV of excess kinetic energy ; it is known to give oscillations in excitation spectroscopy results in bulk Ga(As). The point is that oscillations do not appear in QW's, indicating a possible relative weakening of this process , as observed and discussed in other experiments by several authors [28,54,55]. At high excitation level, electron-electron interaction give rise to electronic and hole temperatures T_e^*, T_h^* which may be higher than the lattice temperature T_L and which may also populate QW subbands of higher energy [56] ; the appearance of Δn odd forbidden transitions due to many-body interactions [57] has also been reported.

Observation of low-temperature in-plane luminescence after optical carrier injection has been less investigated. At high excitation level and using cleaved edges structures, the laser effect occurs [58]. Apart from the polarization effects (the same as in absorption), the alloy clustering and optical phonon-assisted Stokes shift which have been observed [59] appear to be due to epitaxial causes or to reabsorption problems [6[].

Some last few references for those who like interband luminescence under magnetic field [61,62] or electric field [63,55], and one will understand that photoluminescence in quantum wells is an actively and investigated field whose implications are both fundamental and technological and which could need, for a full understanding, many other studies.

1. C. Weisbuch, R.C. Miller, R. Dingle, A.C. Gossard and W. Wiegmann: Solid State Commun., 37, 219 (1981)
2. M.D. Sturge: Phys. Rev., 127, 768 (1962)
3. C. Weisbuch: Thèse d'Etat, Orsay (1977)
4. J.Y. Marzin: this book
5. C. Weisbuch: 11th Conference on GaAs and related compounds
6. H.C. Casey and M.B. Panish: in Heterostructure Lasers (Academic, New York 1978)
7. R. Dingle: in Festkorperprobleme XV, H.J. Queisser ed (Vieweg) (1975).
8. S.R. White and L.J. Sham: Phys. Rev. Lett., 47, 879 (1981)
9. R.C. Miller, A.C. Gossard, D.A. Kleinmann and O. Munteanu: Phys. Rev. B, 29, 3740 (1984)
10. R.C. Miller, D.A. Kleinmann and A.C. Gossard: Phys. Rev. B, 29, 7085 (1984)
11. M.H. Meynadier, C. Delalande, G. Bastard, M. Voos, F. Alexandre and J.L. Lievin: Phys. Rev. B 31, 5539 (1985)
12. P. Dawson, G. Duggan, H.I. Ralph, K. Woodbridge: Proceedings of the First International Conference on Superlattices, Microstructures and Microdevices, Champaign-Urbana, 1984 (in press)
13. G. Bastard, E.E. Mendez, L.L. Chang and L. Esaki: Phys. Rev. B, 26, 1974 (1982)
14. R.L. Greene and K.K. Bajaj: Solid State Commun., 45, 831 (1983)
15. R.C. Miller, D.A. Kleinmann, W.T. Tsang and A.C. Gossard: Phys. Rev. B, 26, 1974 (1982)
16. C. Weisbuch, R. Dingle, A.C. Gossard and W. Wiegmann: J. Vac. Sci. Technol., 17, 1128 (1980)
17. P.M. Petroff: J. Vac. Sci. Technol., 14, 973 (1977)
18. M. Heiblum, E.E. Mendez and L. Osterling: J. Appl. Phys., 54, 6982 (1983)
19. J. Hegarty, M.D. Sturge, C. Weisbuch, A.C. Gossard and W. Wiegmann: Phys. Rev. Lett., 49, 930 (1982)
20. G. Bastard, C. Delalande, M.H. Meynadier, P.M. Frijlink and M. Voos: Phys. Rev. B, 29, 7042 (1984)
21. C. Delalande, M.H. Meynardier, M. Voos: Phys. Rev. B 31, 2497 (1985).
22. B. Deveaud, J.Y. Emery, A. Chomette, B. Lambert and M. Baudet: Appl. Phys. Lett., 45, 1078 (1984)
23. J. Hegarty, L. Goldner and M.D. Sturge: Phys. Rev. B, 30, 7346 (1984).
24. J. Hegarty: Phys. Rev. B, 25, 4324 (1982)
25. E.O. Gobel, H. Jung, J. Kuhl and K. Ploog: Phys. Rev. Lett., 51, 1588 (1983)

26. R. Hoger, E.O. Gobel, J. Kuhl, K. Ploog and G. Weimann: 17th International Conference on the Physics of Semiconductors (1984)
27. J. Masumoto, S. Shionoya and H. Kawaguchi: Phys. Rev. B, 29, 2324 (1984)
28. D.S. Chemla, D.A.B. Miller and P.W. Smith: in Non-linear optical properties of multiple quantum well structures for optical signal processing, to be published (Academic Press)
29. D. Bimberg, J. Christen, A. Steckenborn, G. Weimann, W. Schlapp: in IUPAP Semiconductor Symposium: "High Excitation and Short Pulse Phenomena", Trieste (1984). To be published in J. of Luminescence
30. P. Dawson, G. Duggan, H.I. Ralph and K. Woodbridge: Phys. Rev. B, 28, 7381 (1983)
31. J.E. Fouquet and A.E. Siegman: Appl. Phys. Lett., 46, 280 (1985).
32. S. Tanaka, M. Kuno, A. Yamamoto, H. Kobayashi, M. Mizuta, H. Kukimoto and H. Saito: Jap. Jour. of Appl. Phys., 23, L427 (1984)
33. A. Pinczuk, J. Shah, R.C. Miller, A.C. Gossard and W. Wiegmann: Solid State Commun., 50, 735 (1984)
34. P. Voisin, G. Bastard, C.E.T. Goncalves da Silva, M. Voos, L.L. Chang and L. Esaki: Solid State Commun., 39, 79 (1981)
35. G. Bastard: Phys. Rev. B, 24, 4714 (1981)
36. R.L. Greene and K.K. Bajaj: Solid State Commun., 45, 852 (1983)
37. W.T. Masselink, Y.C. Chang and H. Morkoç: J. Vac. Sci. Technol. B 2, 376 (1984)
38. R.C. Miller, W.T. Tsang and O. Munteanu: Appl. Phys. Lett., 41, 374 (1982)
39. P.M. Petroff, R.C. Miller, A.C. Gossard and W. Wiegmann: Appl. Phys. Lett., 44, 217 (1984)
40. W.T. Masselink, Y.L. Sun, R. Fisher, T.J. Drummond, Y.C. Chang, M.V. Klein and H. Morkoç: J. Vac. Sci. Tech. B, 2, 117 (1984)
41. R. Fisher, W.T. Masselink, Y.L. Sun, T.J. Drummond, Y.C. Chang, M.V. Klein and H. Morkoç: J. Vac. Sci. Tech. B, 2, 170 (1984)
42. B.V. Shanabrook and J. Gomas: Surf. Science, 142, 504 (1984)
43. R.C. Miller: J. Appl. Phys., 56, 1136 (1984)
44. R.C. Miller and A.C. Gossard: Phys. Rev. B, 28, 3645 (1983)
45. R.C. Miller, A.C. Gossard, W.T. Tsang and O. Munteanu: Solid State Commun., 43, 519 (1982)
46. D.C. Reynolds, K.K. Bajaj, C.W. Litton, P.W. Yu, W.T. Masselink, R. Fisher and H. Morkoç: Phys. Rev. B, 29, 7038 (1984)
47. P. Voisin, C. Delalande, M. Voos, L.L. Chang, A. Segmuller, C.A. Chang and L. Esaki: Phys. Rev. B, 30, 2276 (1983)
48. G. Griffiths, K. Mohammed, S. Subbana, H. Kroemer and J.L. Merz: Appl. Phys. Lett., 43, 1059 (1983)
49. J.Y. Marzin and E.V.K. Rao: Appl. Phys. Lett., 43, 560 (1983).
50. M. Razeghi, J.P. Hirtz, U.O. Ziemelis, C. Delalande, B. Etienne and M. Voos: Appl. Phys. Lett., 43, 583 (1983)
51. P. Voisin: Private Communication
52. D.F. Welch, D.W. Wicko and L.F. Eastman: Appl. Phys. Lett., 43, 762 (1983)
53. A.F.S. Penna, J. Shah, A. Pinczuk, D. Sirico and A.Y. Cho: Appl. Phys. Lett., 46, 184 (1985)
54. J.F. Ryan, R.A. Taylor, A.J. Tuberfield, A. Maciel, J.M. Worlock, A.C. Gossard and W. Wiegman: Phys. Rev. Lett., 53, 1841 (1984)
55. J. Shah, A. Pinczuk, H.L. Stormer, A.C. Gossard and W. Wiegmann: Appl. Phys. Lett., 44, 322 (1984)
56. Z.Y. Xu, V.G. Kreismanis and C.L. Tang: Appl. Phys. Lett., 43, 415 (1983)
57. R.C. Miller, D.A. Kleinman, O. Munteanu and W.T. Tsang: Appl. Phys. Lett., 39, 1 (1981)
58. N. Holonyak Jr, R.M. Kolbas, R.D. Dupuis and P.D. Dapkus: IEEE J. of Quant. Elect. QE 16, 170 (1980)
59. J.J. Coleman,P.D. Dapkus, M.D. Camras, N. Holonyak Jr, W.D. Laidog, T.S. Low, M.S. Burroughs and K. Hess: J. Appl. Phys., 52, 7291 (1981)
60. H.Iwomura, T.Saku, H.Koboyashi Y. Korikoshi: J. Appl. Phys., 54, 2692 (1983)
61. J.C. Maan, G. Belle, A. Fasolino, M. Altarelli and K. Ploog: Phys. Rev. B, 30, 2253 (1984)
62. E.E. Mendez, G. Bastard, L.L. Chang, L. Esaki, H. Morkoç and R. Fisher: Physica 117B and 118B, 711 (1983)

Optical and Magnetooptical Absorption in Quantum Wells and Superlattices

P. Voisin

Groupe de Physique des Solides de l'Ecole Normale Supérieure,
24, rue Lhomond, F-75005 Paris, France

This lecture is devoted to the very basic optical property of superlattices (SL) or Quantum well (QW) structures, namely the way they absorb light. The importance of optical absorption lies in the fact that it reveals essentially intrinsic properties, that is electronic states with large densities of states. As such, optical absorption is an ideal tool for the characterization of band structure properties.

Starting with a simple one-electron model of the absorption process, we shall first derive the general formula of the absorption coefficient. Then, within the framework of the envelope function formalism [1-3], we shall examine the selection rules which arise from the symmetry properties of cell periodic and envelope parts of the wavefunctions [4]. The latter are quite different in type I and type II systems. The shape and the magnitude of the absorption coefficient will be discussed within a simplified model, with special attention to the SL case.

In type I QW structures, the excitonic interaction is enhanced by the two dimensional (2D) character, and it considerably affects the shape of the absorption edges. The theory of excitonic absorption is beyond the scope of this lecture, and only the major theoretical results and their incidence on the absorption profile will be described.

When a magnetic field is applied perpendicular to the layer plane, the in-plane motion is quantized and the density of states becomes quasi-discrete. Interband magnetooptical absorption brings rich experimental informations, which we shall describe qualitatively.

1. Theory of the absorption coefficient

We consider an electromagnetic (em) plane wave propagating along a direction $\vec{k}$ in a weakly absorbing medium. It is characterized by a vector potential $\vec{A}$ or, equivalently, by the associated electric and magnetic fields $\vec{E}$ and $\vec{B}$. We have :

$$\vec{A} = -\vec{\varepsilon}\; E/\omega \;\mathrm{Im}\,(\exp i(\omega t - \vec{k}.\vec{r}))$$

$$\vec{E} = -\frac{\partial \vec{A}}{\partial t} \quad \text{and} \quad \vec{B} = \vec{\nabla}\wedge\vec{A} \tag{1}$$

$\vec{\varepsilon}$ is the polarization vector. The energy density of the em wave is the time average of $\varepsilon_r\varepsilon_0\vec{E}^2$:

$$u = \frac{1}{2}\,\varepsilon_r\varepsilon_0\; E^2 \tag{2}$$

The em energy conservation law writes :

$$\frac{\partial u}{\partial t} = -\,\vec{\nabla}.\vec{\pi} \tag{3}$$

where $\vec{\pi}$ is the time average of Poynting vector $\vec{E}\,\vec{B}/\mu_0$:

$$\vec{\pi} = \frac{1}{2}\,\varepsilon_0 \varepsilon_r^{1/2} c\, E^2\, \vec{k}/k \tag{4}$$

u decreases exponentially with the distance, as

$$u = u_0\, e^{-\alpha z} \tag{5}$$

To calculate the absorption coefficient $\alpha(\omega)$, we first evaluate the rate $P(\omega)$ at which photons disappear, and then we relate $P(\omega)$ to $\alpha(\omega)$. In presence of the em wave, the hamiltonian becomes, to the first order in $\vec{A}$:

$$H = H_0 + \frac{e}{2m_0}\,(\vec{p}.\vec{A} + \vec{A}.\vec{p}) \tag{6}$$

The time-dependent perturbation $H-H_0$ induces transitions between electronic states $|i\rangle$ and $|f\rangle$ of energies $E_i < E_f$. These occur at a rate given by the Fermi golden rule :

$$\tau_{if}^{-1} = \frac{2\pi}{\hbar}\left|\langle f|V|i\rangle\right|^2 \delta(E_f - E_i - \hbar\omega) \tag{7}$$

where V is the time-independent part of $H-H_0$. In the electric-dipole approximation, we neglect the spatial variations of the em wave and write :

$$V = e\, E/2m_0\omega\, \vec{\varepsilon}.\vec{p} \tag{8}$$

The absorption is possible only if $|i\rangle$ is occupied and $|f\rangle$ empty, which adds a factor $f(E_i)(1-f(E_f))$ where $f(E)$ is the Fermi distribution. On the other hand, the em wave also induces $|f\rangle$ to $|i\rangle$ transitions, which correspond to a photon emision rate $\tau_{if}^{-1} f(E_f)(1-f(E_i))$, so that we finally get the rate $P(\omega) = -d/dt \int_\Omega u\, d^3r$ at which the em energy decreases :

$$P(\omega) = \sum_{\substack{i,f \\ (E_f > E_i)}} \tau_{if}^{-1}\,(f(E_i) - f(E_f)) \tag{9}$$

Using Equations 2 to 5, we readily get

$$P(\omega) = \frac{1}{2}\, S\, n\, \varepsilon_0\, c\, E^2(0)\,(1 - e^{-\alpha L_z}) \tag{10}$$

which, in the weak absorption regime, gives :

$$a(\omega) = \frac{\pi e^2}{n\varepsilon_0 c m^2 \Omega\omega} \sum_{i,f} |\langle f|\vec{\varepsilon}.\vec{p}|i\rangle|\quad \delta(E_f - E_i - \hbar\omega)\,(f(E_i) - f(E_f)) \tag{11}$$

$n = \varepsilon_r^{1/2}$ is the refractive index of the medium and $\Omega = S \times L_z$ the volume of the sample, or, more specifically, the volume in which states $|i\rangle$ and $|f\rangle$ are normalized. We now have to estimate the optimal matrix elements $\langle f|\vec{\varepsilon}.\vec{p}|i\rangle$ and perform the summation over the relevant states. In the envelope function formalism, restricting ourselves to the Γ_6, Γ_8, Γ_7 subspace, the wave-function of a state $|i\rangle$ in a two-dimensional system is written

$$\psi^i_{k_\perp n}(\vec{r}) = \sum_{\nu=1}^{8} u_\nu(\vec{r})\,\frac{1}{\sqrt{S}}\,e^{i\vec{k}_\perp\vec{r}}\, f^i_{\nu k_\perp}(z) \tag{12}$$

where the u_ν's are the cell periodic parts of the Bloch wave-functions and the f_ν's the associated envelope function. $\vec{k}_\perp$ and $r_\perp$ are the in-plane wave vector and position. The basic assumption inherent to the effective mass treatment, is that the f_ν's are slowly varying at the scale of a bulk lattice parameter, so that we may write :

$$<f^f_\nu u_\nu|\vec{p}|f^i_\mu u_\mu> \sim <u_\nu|\vec{p}|u_\mu><f^f_\nu/f^i_\mu> + \delta_{\nu\mu} <f_\nu|\vec{p}|f^i_\mu> \quad (13)$$

As long as we are considering transitions between the few low-lying subbands, only one index ν will contribute significantly for each band. In the next section, we examine the case of an idealized type I quantum well.

2. Interband absorption in a schematized quantum well

We consider a quantum well structure built from large band gap materials, and we neglect the difficulties arising from the complex Γ_8-band in-layer kinematics. We suppose that the conduction band is purely S-type and that the valence band consists of uncoupled heavy hole, light hole and split-off subbands. We assume the simple parabolic subband dispersion relations :

$$E_n(k_\perp) = E_n(0) + \frac{\hbar^2}{2m^*_c} k^2_\perp$$
$$H_n(k_\perp) = H_n(0) - \frac{\hbar^2}{2m^*_H} k^2_\perp \quad (14)$$

where H stands for any of the valence bands. By examining the cell-periodic parts of the Bloch functions, we easily obtain the selection rule associated with the polarization, which is indicated in table I, with the notation $P = \langle S|p_x|X\rangle$.

It is seen that in the usual configuration $\vec{k}// z$, the three transitions are allowed. On the contrary, for an em wave propagating in the layer plane, the transitions $HH_n \rightarrow E_m$ become forbidden in the ε_z polarization while the $LH_n \rightarrow E_m$ transitions remain allowed. This cicumstance has been exploited by Marzin [5] to determine the nature of the transitions observed in $In_xGa_{1-x}As$-GaAs superlattices.

The overlap of the envelope functions reads :

$$\frac{1}{S} \int f^{h*}_m(z) e^{-i\vec{k}^h_\perp \vec{r}_\perp} f^e_n(z) e^{i\vec{k}^e_\perp \vec{r}_\perp} d^2r \, dz \quad (15)$$

It leads to two selection rules : the first one, $\vec{k}^h_\perp = \vec{k}^e_\perp$, is the direct consequence of the translational invariance in the layer plane. The second acts on the subband index. In type I systems (Fig. 1a), the conduction and valence wavefunctions are localized in the same QW layer. They have a definite parity with respect to the center of this layer, which leads to the selection rule n-m even. If the quantum wells are infinitely deep, these envelope functions are :

$$f^{e,h}_n(z) = \sqrt{\frac{2}{L_z}} \sin n\frac{\pi z}{L_z} \quad (16)$$

and their overlap is simply δ_{nm}. With finite depth QW's the wave functions have evanescent wings in the adjacent barrier layers. Conduction and valence wave functions are not identical, and transitions with Δn = 2, 4, ... become allowed, though they remain essentially weak. In type II systems, (Fig. 1b), the overlap is always small, as it is due to the leakage of the wavefunctions outside of their confinement layers. For a given conduction well width L_A, it goes to zero as $1/\sqrt{L_B}$. All the states still have a definite parity with respect to the center of the A layers, but even and odd valence states are degenerate, so that there is no

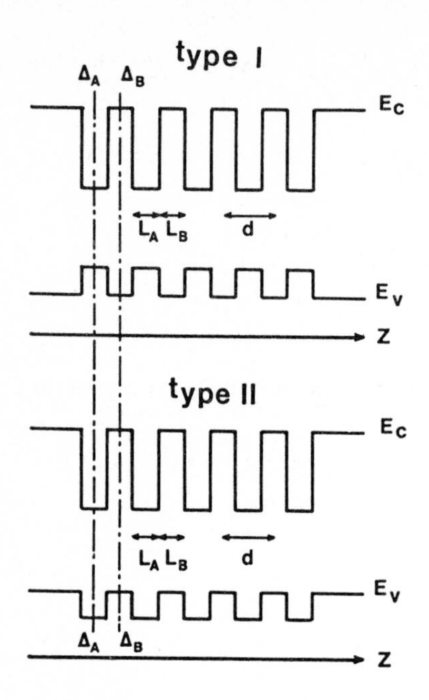

Fig.1 Modulation of the host's band edges along the growth axis for type I and type II heterostructures.

selection rule on the subband index. Transitions with $\Delta n = 0$ or $\Delta n \neq 0$, Δn even or odd have a-priori the same order of magnitude.

We now examine the order of magnitude of the absorption coefficient. To be specific, we consider the $HH_1 \rightarrow E_1$ transition in a type I QW, for which the envelope function overlap is nearly 1, so that the optical matrix element is $P/\sqrt{2}$; we have:

$$\alpha(\omega) = \frac{2\pi e^2}{n\varepsilon_0 c m_0 \Omega\omega} \frac{P^2}{2m_0} \sum_k \delta\left(E_g + HH_1 + E_1 + \frac{\hbar^2}{2}\left(\frac{1}{m_c^*} + \frac{1}{m_{HH}^{\perp}}\right) k_{\perp}^2 - \hbar\omega\right) \tag{17}$$

or :

$$\alpha(\omega) = \frac{1}{L_z} \frac{e^2}{n\varepsilon_0 c m_0 \omega} \frac{P^2}{2m_0} \frac{m_c^* m_{HH}^{\perp}}{m_c^* + m_{HH}^{\perp}} \frac{1}{\hbar^2} Y(E_g + E_1 + HH_1 - \hbar\omega) \tag{18}$$

where Y is the step function $Y(x) = 1$ if $x > 0$, $Y(x) = 0$ if $x < 0$. For any III-V compound, $2P^2/m_0 \sim 23$ eV. If $m^{\perp}{}_{HH}$ is large compared to m_c^*, the reduced mass is not too different from m_c^*, which scales with the bandgap. Thus, we get a "universal" number $\alpha(\omega)\ L_z \simeq 6.10^{-3}$ per transition and per quantum well. The absorption manifests itself as a step at the energy $E_g + E_1 + HH_1$, which reflects the energy level quantization and the 2D density of states. The observation of step-like absorption spectrum with the absorption edge shifted toward higher energies compared to the bulk material (Dingle, 1974) [6] was the clearest experimental evidence of the quantization of the energy levels in real QW structure. If the structure consists in 10 QW's, a g = 10 degeneracy factor appears in the summation in Eq. 17, and we find $\alpha(\omega)L \sim 5\ 10^{-2}$, which indeed agrees very well with various experimental data. Transitions involving light holes are three times weaker (Table I) and should also exhibit a different reduced mass. The absorption in a type II system is generally so much smaller that the consideration of the order of magnitude of the absorption is usually a definitive argument to decide of which type a given system is [7].

Table I Polarization selection rule for interband transitions

Polarization	ε_x	ε_y	ε_z	
Propagation // z	$\frac{P}{\sqrt{2}}$	$\frac{iP}{\sqrt{2}}$		$HH_n \rightarrow E_m$
Propagation // x		$\frac{iP}{\sqrt{2}}$	forbidden	
Propagation // y	$\frac{P}{\sqrt{2}}$		forbidden	$\left\|\frac{3}{2},\frac{3}{2}\right\rangle = \frac{\|(X+iY)\uparrow\rangle}{\sqrt{2}}$
Propagation // z	$\frac{iP}{\sqrt{6}}$	$-\frac{P}{\sqrt{6}}$		$LH_n \rightarrow E_m$
Propagation // x		$-\frac{P}{\sqrt{6}}$	$\frac{-2iP}{\sqrt{6}}$	$\left\|\frac{3}{2},\frac{1}{2}\right\rangle =$
Propagation // y	$\frac{iP}{\sqrt{6}}$		$\frac{-2iP}{\sqrt{6}}$	$\frac{1}{\sqrt{6}}\|(X+iY)\downarrow\rangle - \sqrt{\frac{2}{3}}\|Z\uparrow\rangle$
Propagation // z	$-\frac{iP}{\sqrt{3}}$	$-\frac{P}{\sqrt{3}}$		$(\Gamma_7)_n \rightarrow E_m$
Propagation // x		$-\frac{P}{\sqrt{3}}$	$\frac{iP}{\sqrt{3}}$	$\left\|\frac{1}{2},\frac{1}{2}\right\rangle =$
Propagation // y	$-\frac{iP}{\sqrt{3}}$		$\frac{iP}{\sqrt{3}}$	$\frac{1}{\sqrt{3}}\|(X+iY)\downarrow\rangle + \frac{1}{\sqrt{3}}\|Z\uparrow\rangle$

Remarks :

It is true that heavy and light particles are completely decoupled at $\vec{k}_\perp = 0$. However, non parabolicity often affects significantly the calculation of the energy levels E_n, LH_m. This means that even at $\vec{k}_\perp = 0$, the admixture of $P_{1/2}(S)$ states in the conduction (light hole) wavefunctions is not negligible. For a light hole to conduction transitions, for instance, the optical matrix element becomes :

$$\left(< f^e_{S,n} \mid f^h_{P_{1/2},m} > + < f^e_{P_{1/2},n} \quad f^h_{S,m} > \right) \frac{P}{\sqrt{6}} \tag{19}$$

In fact, at $\vec{k}_\perp = 0$, $f^e_{p1/2}$ is proportional to the derivative $(\partial/\partial z)\, f^e_S$, and f^h_S proportional to $(\partial/\partial z) f^h_{p1/2}$. The two terms in Eq. (19) thus obey the same parity selection rule. The general result is that, for $\vec{\varepsilon} \perp z$ and at $\vec{k} = 0$, non-parabolicity does not affect the parity selection rule. On the other hand, at finite $\vec{k}_\perp$, light and heavy subbands are coupled and mix even in "parabolic" materials [3]. This means that the selection rules (both parity and polarization) are probably relaxed away from $\vec{k}_\perp = 0$. However , a forbidden transition will certainly not become abruptly allowed, nor an allowed one abruptly forbidden. This is why the selection rules which we have discussed remain certainly a sensible guide line.

3. Optical absorption in superlattices

A system of coupled QW's presents several additional features. The envelope functions $f_{n\nu k_\perp}(z)$ are now Bloch waves and depend on an additional quantum number, the superlattice wave vector $\vec{q}$ in the z direction. To this new translational invariance corresponds a selection rule $\Delta\vec{q} = 0$ equivalent to the selection rule $\Delta\vec{k}_\perp = 0$ found in section 2. $f_{\nu c}$ is solution of a 1x1 effective hamiltonian H_ν, obtained by projecting the 8x8 hamiltonian on the state $|\nu\rangle$. Consider two consecutive A, B layers and let P_A P_B be the planes bisecting the A, B layers. The product R_BR_A of two reflections R_A and R_B with respect to P_A and P_B, respectively, is equal to a translation τ_d of the superlattice period $d = L_A + L_B$ (Fig.1). R_A, R_B and τ_d commute with H_ν but not with each other, except for $q = 0$ and $q = \pi/d$. For these values, which correspond to standing Bloch waves, the $f_{\nu q}$'s are eigenfunctions of R_A and R_B, with eigenvalues $\pm$ 1 : they are even or odd with respect to P_A, P_B. The relation $R_BR_A\ f_{\nu q} = e^{iqd}\ f_{\nu q}$ shows that the parity with respect to the centers of one type of layers must be the same at $q = 0$ and $q = \pi/d$ while the parity with respect to the centers of the other type of layers must be opposite at $q = 0$ and $q = \pi/d$. This is easily seen in Table II. For an em wave propagating along the SL axis, the interband optical matrix element becomes, in the parabolic limit :

$$
\begin{aligned}
\langle f|\vec{p}|i\rangle &= \langle u_e|\vec{p}|u_h\rangle \int_{-d/2}^{+d/2} f^{*e}_{nqk_\perp} f^{h}_{mqk_\perp}\, dz \\
&= \langle u_e|\vec{p}|u_h\rangle M_{nm}(q,k_\perp)
\end{aligned}
\qquad (20)
$$

Table II Truth table of the parity statement

	R_A	R_B		R_A	R_B
$q = 0\ \ (R_BR_A = +1)$	+ 1	+ 1	or	- 1	- 1
$q = \pi/d(R_BR_A = 1)$	$\pm$ 1	$\mp$ 1		$\mp$ 1	$\pm$ 1

For type I systems, we assume that f^e_{nq} and f^h_{nq} both retain the same symmetry with respect to P_A at both $q = 0$ and $q = \pi/d$. Then if the transition is parity-allowed at $q = 0$ (n-m even), it will remain parity allowed at $q = \pi/d$, as illustrated in Fig. 2(a) for the ground wave function case n = m = 1. On the other hand, in a type II SL, $f^e_{n,q}$ (resp. $f^h_{m,q}$) is expected to retain the same parity with respect to P_A (resp. P_B) at both $q = 0$ and $q = \pi/d$. If the transition is parity allowed at $q = 0$ (say that $f^e_{n,q}$ and $f^h_{m,q}$ are even with respect to P_A, like in Fig. 2(b)), then this transition becomes parity forbidden at $q = \pi/d$ because in the integral in (20), $f^e_{n,q}$ remains even with respect to P_A whereas $f^h_{m,q}$ becomes odd with respect to the same plane. More generally, we find that in type I SL's, $M_{nm}(q,k_\perp)$ almost does not depend on $(k_\perp,q)$, is parity-forbidden if n-m is odd, and parity-allowed if n-m is even, the n=m transitions being by far the most intense. In type II SL's, the interband matrix element depends strongly on q. The transitions which are parity-allowed at $q = 0$ (n-m even) become parity-forbidden at $q = \pi/d$, and vice-versa. In a wide range of practical situations, we find :

$$
|M_{nm}(q,k)|^2 = \frac{1}{2}\,|M_{nm}(0,k)|^2(1 + (-1)^{n-m}\cos qd) \qquad (21)
$$

Note that transitions with n-m = 0 or n-m $\neq$ 0 will a priori have comparable strengths, always small compared to that of a n = m transition in a type I SL. Finally , in the case of type II SL's, the selection rule relies basically on the phase coherence of at least one of the Bloch envelope functions f^e, f^h, and thus could

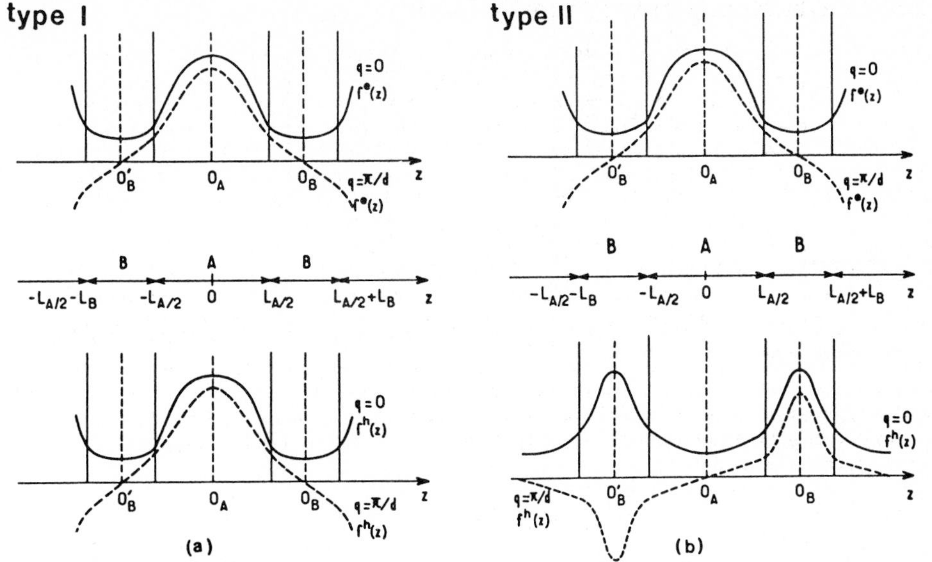

Fig. 2 Schematic envelope functions for conduction and valence ground subbands at q=0 (full lines) and q=π/d (dashed lines) in type I (Fig. 2a) and type II (Fig. 2b) SL's.

be relaxed if both the conduction and valence subband widths become smaller than their scattering induced broadening. Here, we recover the "no selection rule" statement of Section 2. On the other hand,the selection rules for Type I SL's are essentially those of the isolated quantum well.

The shape of the absorption edge will reflect the three-dimensional character of superlattices. For most realistic cases, the subband dispersion relations along the z directions are very well approximated by [4] :

$$
\begin{aligned}
E_m(k,q) &= E_m(k_\perp,0) - (-1)^m \Delta E_m \; (1-\cos qd)/2 \\
H_n(k,q) &= H_n(k_\perp,0) + (-1)^n \Delta H_n \; (1-\cos qd)2
\end{aligned}
\tag{22}
$$

where ΔE_m is the width of the conduction subband E_m and ΔH_n that of the heavy or light hole subband H_n. Using again the coarse "parabolic" approximation, we perform the sum over k and q and get, for the $HH_1 \rightarrow E_1$ transition :

$$
\begin{aligned}
\alpha^{I}(\omega) &= C\,\pi^{-1} \begin{cases} \text{Arc cos } (1-2\xi) & 0 < \xi < 1 \\ 1 & \xi > 1 \end{cases} \\
\alpha^{II}(\omega) &= 2C\mu\,\pi^{-1} \begin{cases} \text{Arc cos } (1-2\xi) + 2\sqrt{\xi(1-\xi)} & 0 < \xi < 1 \\ 1 & \xi > 1 \end{cases}
\end{aligned}
\tag{23}
$$

for type I and type II superlattices respectively. ξ is the reduced photon energy $(\hbar\omega - E_g^{SL})/(\Delta E_1 + \Delta HH_1)$, where E_g^{SL} is the SL bandgap ; μ is the overlap of the E_1 and HH_1 envelope functions at $q = 0$ near one interface, in a type II SL, and C is the prefactor in Eq. 18. The corresponding absorption profiles are shown in Fig. 3. Near the threshold ($\xi = 0$) we get an absorption profile corresponding to an anisotropic 3D material,while far above the threshold ($\xi > 1$) we recover the 2D

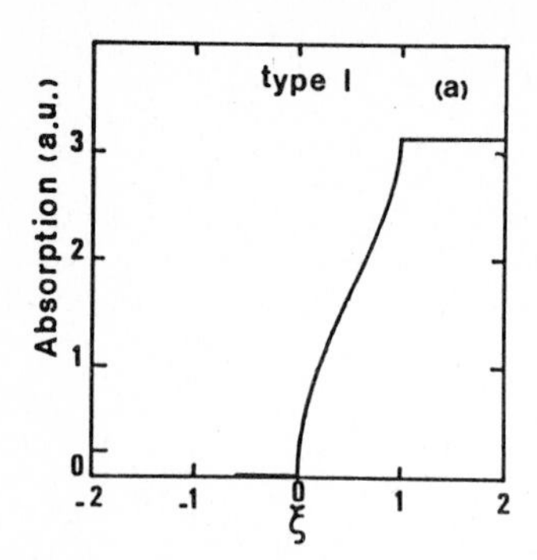

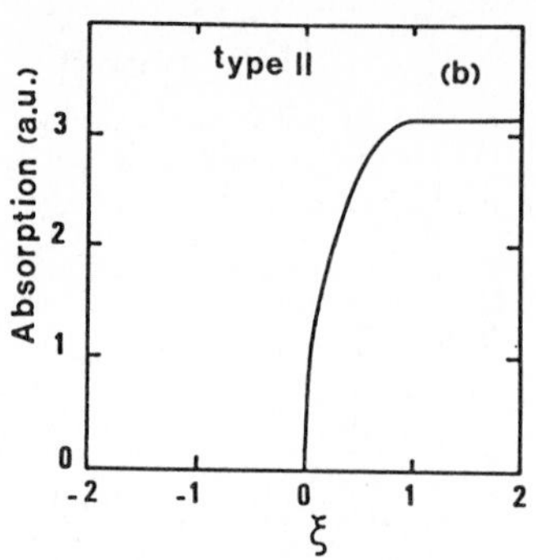

Fig.3 Absorption lineshapes in type I (Fig.3a) and type II (Fig.3b) SL's. ξ is the reduced photon energy.

plateau. The transition at $\xi = 1$ is eye-marked by a Van Hove singularity in type I SL's ; this singularity disappears in a type II SL, due to the q-dependence of the optical matrix element.

4. Excitonic effects

The theory of optical absorption that we have presented in Section I is oversimplified because it relies on a one-electron model. In fact, we should consider the initial state as N electrons in the valence band and the final state as N-1 electrons in the valence band plus one electron in the conduction band. As long as Coulomb interaction is neglected, this correct treatment gives exactly the same result as the one-electron model, though much more algebra is involved. Coulomb interaction in the final state leads to exciton states [8]. In the electron-hole formalism, and with many approximations, an exciton may be described by a two-particle envelope wave function in the form :

$$\psi_K^n(\vec{r}_e,\vec{r}_h) = \sum_k A_k^n \, e^{i\vec{k}(\vec{r}_e-\vec{r}_h)} \, e^{i\vec{K}(\vec{r}_e+\vec{r}_h)/2} \qquad (24)$$

The A_k^n's are the Fourier component of a function $\phi^n(\vec{r}_e-\vec{r}_h)$ which is solution of the Schrödinger equation of the hydrogen atom. The optical matrix element associated with the creation of an exciton is $\langle U_e|\vec{\varepsilon}.\vec{p}|U_h\rangle \, \phi^n(0)$, and the equivalent in terms of excitons of the selection rule $\vec{k}_e = \vec{k}_h$ is $\vec{K} = 0$. Thus, discrete bound states with S-type wave-functions and zero wave vector contribute to the optical absorption, and they manifest themselves as sharp lines below the one-particle band gap E_g. Their oscillator strength is proportional to $|\phi^n(0)|^2$ and it decreases rapidly with n ; usually, only the 1S exciton is clearly resolved. Beyond E_g, the absorption in the continuum corresponding to ionized pair states is still affected by the Coulomb interaction. The band to band absorption coefficient is multiplied by the "Sommerfeld factor" which considerably increases the absorption near the onset of the continuum. All these characteristics are common to 2D and 3D systems. The excitonic effects are enhanced by the two-dimensional character, as the series of binding energies in 3D and in purely 2D systems are respectively $\varepsilon_{3D}^n = -R^*/n^2$ and $\varepsilon_{2D}^n = -R^*/(n-1/2)^2$, where R^* is the 3D effective Rydberg energy :

$$R^* = \frac{e^4}{2\hbar^2\varepsilon_r^2\varepsilon_0^2} \, \frac{m_c^* m_H^*}{m_c^* + m_H^*} \qquad (25)$$

The oscillator strength $|\phi^n(0)|^2$ is also correlatively larger in the 2D than in the 3D regime.

Real type I QW structures represent an intermediate situation in which the density of states is purely 2D while the Coulomb interaction remains three dimensional. A variety of calculations of the exciton binding energies in quantum wells have been published [9-11], but up to now, all of them explicitly use parabolic in-layer dispersion relations, and their validity is probably questionable. In type II heterostructures, the spatial separation of the electrons and holes "kills" the Coulomb interaction, and the corresponding "interface exciton" would have a small binding energy [9].

5. Magneto-optical absorption

Whereas direct optical absorption brings rich information such as nature of the system, confinement energies of several pairs of subbands, etc ..., it is clear that this technique cannot adress the fundamental problem of the in-layer kinematics of the Γ_8 subbands in 2D systems : the only parameter directly related to this problem, the electron-hole pair reduced mass μ, is not measured with a sufficient accuracy. Magneto-optical absorption brings richer experimental data which may help for this purpose. When a magnetic field B is applied, parallel to the growth axis, the in-plane motion is quantized into Landau levels. As the electronic motion is now fully quantized, the density of states becomes a series of δ functions, as sketched in Fig. 4a.

We shall discuss briefly an extremely simplified situation, assuming parabolic, non-degenerate subbands. The energies of the Landau levels associated with the ground conduction and valence subbands are :

$$\begin{aligned} E^{\pm}_{1M} &= E_1 + (M + \tfrac{1}{2})\, \hbar\omega_c \pm \tfrac{1}{2}\, g_E \mu_B\, B \\ H^{\pm}_{1N} &= H_1 - (N + \tfrac{1}{2})\, \hbar\omega_v \pm \tfrac{1}{2}\, g_H \mu_B\, B \end{aligned} \tag{26}$$

where $\omega_i = eB/m_i$ is the cyclotron frequency, $g_{E(H)}$ the Lande factor of the corresponding band and μ_B the Bohr magneton. The associated wave functions are

$$\begin{aligned} \psi^{E\pm}_{1M} &= e^{ik_y y}\, \phi_M\left(\frac{x - x_0}{L}\right) |u_c \pm\rangle \\ \psi^{H\pm}_{1N} &= e^{ik_y y}\, \phi_N\left(\frac{x - x_0}{L}\right) |u_v \pm\rangle \end{aligned} \tag{27}$$

where $k_y = -x_0/L^2$. $L = (h/eB)^{1/2}$ is the magnetic length. The ϕ's are the eigenfunctions of the harmonic oscillator. In this simplified picture, optical transitions between the E_1 and H_1 Landau levels are allowed, with the selection rule N - M = 0 imposed by the overlap of the oscillator functions.

We shall observe transmission minima at phonon energies $h\nu_N = E_g + E_{1N} + H_{1N}$. This may be done by recording the transmission at fixed photon energy when sweeping the magnetic field, or at fixed B by sweeping the photon energy.These transition energies $h\nu_N$ extrapolate at zero magnetic field toward E_g and exhibit a slope :

$$\frac{\partial h\nu_N}{\partial B} = (N + \tfrac{1}{2})\, \hbar e \left(\frac{1}{m_c^*} \quad \frac{1}{m_h^*} \right) \tag{28}$$

Again, the electronic contribution to the reduced mass usually dominates, but the accuracy of magneto-optical data may allow the determination of m_H^*.

Non-parabolicity and Γ_8 subband mixing do not change this qualitative picture, but the quantitative interpretation of actual data will generally require a sophisticated calculation of the Landau level energies. In type I quantum wells, with large exciton binding energy E_b, the results are qualitatively changed in the

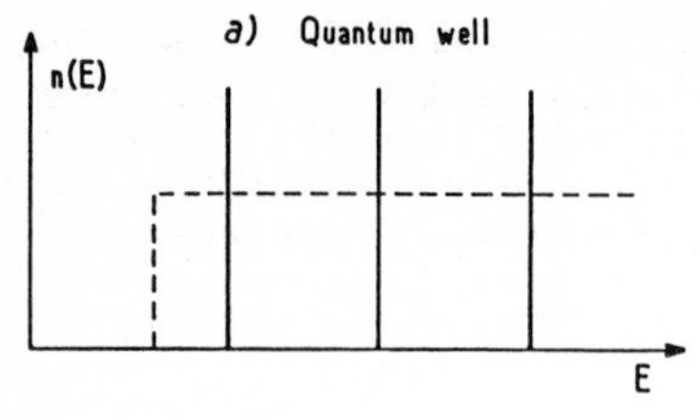

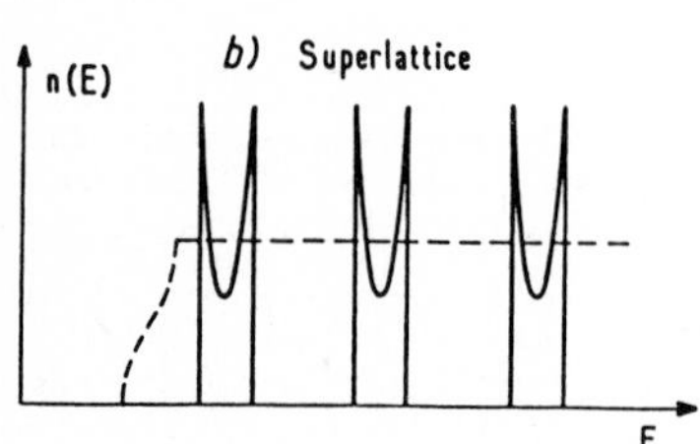

Fig.4 Schematic density of states for a QW (Fig. 4a) and a SL (Fig. 4b) in a strong magnetic field B parallel to the growth axis.

sense that the $h\nu_0$ transition rather extrapolates toward E_g-E_B at B = 0, while the other transitions $h\nu_N$, N > 1 still extrapolate toward E_g. This allows an accurate measurement of E_b, if a realistic theoretical model is used to interpret the data. A study of the excitation spectrum [12] under high magnetic fields was done recently in GaAs QW's [13], and the results indicated heavy hole masses much heavier - and exciton binding energies significantly larger - than expected from over simplified models.

If we are dealing with true superlattices having noticeable bandwidths, the Landau levels become Landau subbands. When the spacing $\hbar\omega_c$ is larger than the bandwidth E_1, the density of states of the conduction band presents two sharp maxima, at q = 0 and q = π/d, as shown in Fig. 4b. Magneto-optical transitions occurring at q = 0 and q = π/d may be observed simultaneously, allowing a determination of the bandwidth ΔE_1 [14].

Conclusion

Quasi two-dimensional heterostructures present a rich variety of optical properties [15] arising from their peculiar density of states and symmetry properties. Optical and magneto-optical absorption are very powerful experimental methods for the investigation of the band structure properties of these 2D systems. One of the most serious limitations of these methods lies in the fact that structures having at least 10 QW's are required for a type I system, and much more for a type II system.

Acknowledgements

These lecture notes owe very much to G. Bastard, who first described many of the results presented here.

References

1. G. Bastard: Phys. Rev. B 24, 5693 (1981).
2. G. Bastard: Phys. Rev. B 25, 7584 (1982).
3. M. Altarelli: this volume.
4. P. Voisin, G. Bastard and M. Voos: Phys. Rev. B 29, 935 (1984).
5. J.Y. Marzin: this volume.
6. R. Dingle, W. Wiegmann and C.H. Renry: Phys. Rev. Lett. 33, 827 (1974).
7. P. Voisin, G. Bastard, M. Voos, E.E. Mendez, C.A. Chang, LL. Chang and L. Esaki: J. Vac. Sci. Technol. B1, 409 (1983).
8. R. Knox: Theory of Excitons, Solid State Physics, Supp.5, Academic Press, 1963.
9. G. Bastard, E.E. Mendez, L.L. Chang and L. Esaki: Phys. Rev. B26, 1974 (1982).

10. R. Green, K. Bajaj and D. Phelps: Phys. Rev. B 29, 1807 (1984).
11. R.C. Miller, D.A. Kleinmann, W.T. Tsang and A.C. Gossard: Phys. Rev. B24, 1134 (1981).
12. C. Delalande: This volume.
13. J.K. Maan, G. Belle, A. Fasolino, M. Altarelli and K. Ploog: Phys. Rev. B 30, 2253 (1985).
14. J.K. Maan, Y. Guldner, J.P. Vieren, P. Voisin, M. Voos, L.L. Chang and L. Esaki: Solid State Comm. 39, 683 (1981).
15. P. Voisin: Surface Science, 142, 460 (1984).

Optical Properties of 2D Systems in the Far Infrared

E. Gornik and R.A. Höpfel

Institute of Experimental Physics, University of Innsbruck, Technikerstr. 15, A-6020 Innsbruck, Austria

1. Introduction

Spectroscopic investigations in the far infrared (FIR) reveal a great number of fundamental excitations in two-dimensional electron (2D) systems. In this lecture I will discuss the fundamental optical properties of a 2D electron layer embedded between a semiconductor and isolator. Fundamental excitations such as cyclotron resonance [1-7], 2D plasmons [8-10] excited via grating couplers and the blackbody emission [11,12] from hot 2D electrons will be discussed in a classical approach. Experimental work on 2D plasmon spectroscopy in absorption and emission and very recent results on hot electron emission will be presented.

2. General Optical Properties

In the following we discuss the optical properties of a free 2D electron system, in the absence of a magnetic field, considering only the lowest parabolic subband filled with n_s electrons of effective mass m^*. The optical properties of this system can be described by a complex dynamical conductivity $\sigma(\omega)$, where the specific form of $\sigma(\omega)$ depends on the approximation made. Starting with the reaction of a free electron system to a periodic electromagnetic wave $E(t) = \mathrm{Re}\, E(\omega)e^{-i\omega t}$ which is applied to the momentum equation

$$\frac{d\vec{p}}{dt} = -\frac{\vec{p}}{\tau} - e\vec{E}\ , \tag{1}$$

we obtain with Ohm's law $\vec{j}(\omega) = \sigma(\omega)\vec{E}(\omega)$ the Drude expression of the dynamic conductivity

$$\hat{\sigma}(\omega) = \frac{\sigma_o}{1 - i\omega\tau} \quad \text{and} \quad \sigma_o = \frac{n_s e^2 \tau}{m^*}\ , \tag{2}$$

where σ_o is the 2D d.c. conductivity and τ the momentum scattering time. σ_o is defined as the 2D areal conductivity according to $\sigma_o = g_D \cdot L/W$, where g_D is the source-drain conductance, L the length and W the width of the channel.

The optical properties of the 2D electron slab are calculated using the boundary conditions for the electric and magnetic field according to the geometrical situation shown in Fig. 1, where the thin electron slab is surrounded on the one side by silicon and on the other side by an oxide and vacuum (the oxide thickness d is neglected since the wavelength λ is much larger than d) [13].

$$\vec{E}_I + \vec{E}_R + \vec{E}_T \quad \text{and} \quad \vec{n} \times (\vec{H}_I + \vec{H}_R - \vec{H}_T) = \vec{j} \tag{3}$$

($\vec{j}$... surface current density, $\vec{n}$... surface vector). Using Ohm's law $\vec{j} = \sigma(\omega)\vec{E}_T$ and the following Ansatz for the incident (I) transmitted (T) and reflected (R) electromagnetic wave

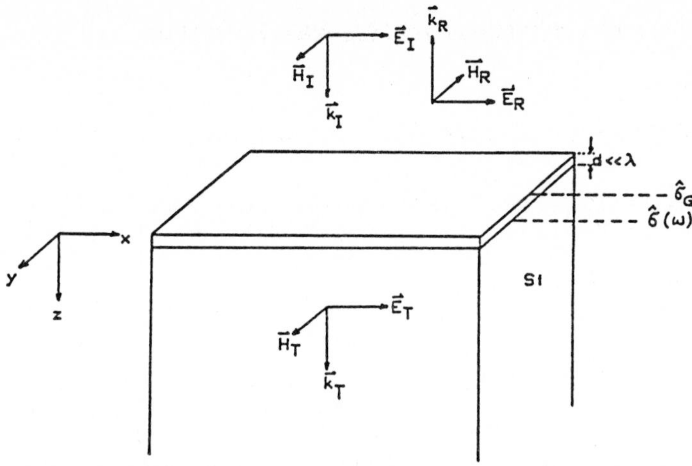

Fig.1 Geometry for the calculation of the dynamical conductivity defining the boundary conditions.

$$\vec{E}_I = \vec{E}_I^o . e^{i(kz-\omega t)}, \quad E_T = E_T^o . e^{i(\sqrt{\varepsilon_s}kz-\omega t+\phi_T)},$$

$$\vec{E}_R = \vec{E}_R^o . e^{i(-kz-\omega t+\phi_R)}$$

(ϕ_R, ϕ_T... are phase terms) we obtain from (3)

$$\frac{E_T^o}{E_I^o} = \frac{2.e^{i(\sqrt{\varepsilon_s}-1)kz}}{e^{i\phi_T}.(1+\sqrt{\varepsilon_s}+\frac{\hat{\sigma}(\omega)}{c\varepsilon_o})} \quad \text{and} \tag{4}$$

$$\frac{E_R^o}{E_I^o} = \frac{(1-\sqrt{\varepsilon_s}-\frac{\sigma(\omega)}{c\varepsilon_o}}{(1+\sqrt{\varepsilon_s}+\frac{\hat{\sigma}(\omega)}{c\varepsilon_o}\, e^{i(\phi_R-2kz)}} . \tag{4a}$$

The transmission, reflection and absorption coefficients are calculated according to

$$T = \frac{\sqrt{\varepsilon_s}|E_T^o|^2}{|E_I^o|^2} \; ; \; R = \frac{|E_R^o|^2}{|E_I^o|^2} \; ; \; A = 1 - R - T . \tag{5}$$

As a result we derive the following expressions :

$$T = \frac{4\sqrt{\varepsilon_s}}{|(1+\sqrt{\varepsilon_s}+\frac{\sigma(\omega)}{c\varepsilon_o})|^2}, \qquad R = \frac{|1-\sqrt{\varepsilon_s}-\frac{\hat{\sigma}(\omega)}{c\varepsilon_o}|^2}{|1+\sqrt{\varepsilon_s}+\frac{\hat{\sigma}(\omega)}{c\varepsilon_o}|^2},$$

$$A = \frac{4.\frac{\mathrm{Re}\hat{\sigma}(\omega)}{c\varepsilon_o}}{|1+\sqrt{\varepsilon_s}+\frac{\hat{\sigma}(\omega)}{c\varepsilon_o}|^2} . \tag{6}$$

However, we have to take into account the conductivity of the gate [denoted $\sigma_G(\omega)$, which is simply added to $\sigma(\omega)$]. Thus we obtain with $F = [\sigma(\omega) + \sigma_G(\omega)]/c\varepsilon_o$ for the absorptivity

$$A = \frac{4\mathrm{Re}\hat{F}}{|1 + \sqrt{\varepsilon_s} + \hat{F}|^2} \quad . \tag{7}$$

In real absorption experiments the relative change of transmission with and without inversion channel is measured. Thus the relative change in transmission is given by

$$-\frac{1}{2}\,\frac{\Delta T}{T}\cdot\left(1 + \frac{\hat{\sigma}_G}{c\varepsilon_o} + \sqrt{\varepsilon_s}\right) = \frac{\mathrm{Re}\sigma(\omega)}{c\varepsilon_o} \quad , \tag{8}$$

in agreement with [14] (with the assumption ReF, ImF << 1).

3. Cyclotron Resonance

We have seen in Chap. 2 that the absorption of light can be completely described if $\sigma(\omega)$ is known. The application of a magnetic field induces a new symmetry in the system, which results in a change of the isotropic conductivity $\sigma(\omega)$. The calculation of $\sigma(\omega)$ in the presence of a magnetic field is calculated for a classical situation.

We assume an isotropic effective mass m^* and a Faraday configuration $\vec{B}(0,0,B)$ giving an equation of motion for an incident electromagnetic wave

$$m^*\frac{\partial\vec{v}_D}{\partial t} + \frac{m\vec{v}_D}{\tau} = e(\vec{E} + \vec{v}_D \times \vec{B}), \tag{9}$$

where $\vec{E} = (E_{xy}, E_y)$ is the induced a.c. electric field and $\vec{V}_D = (v_D^x, v_D^y)$ is the drift velocity. For $|\vec{E}| \sim e^{i\omega t}$ and a current $|\vec{v}_D| \sim e^{i\omega t}$ we obtain

$$(im^*\omega + \frac{m^*}{\tau})\,\vec{v}_D = e(\vec{E} + \vec{v}_D \times \vec{B}) \tag{10}$$

or $\vec{V}_D = \mu(\vec{E} + \vec{V}_D \times \vec{B})$ with $\mu = \dfrac{\frac{e}{m}\tau}{1 + i\omega\tau}$.

Since $\vec{v}_D . \vec{B} = \mu(\vec{E}.\vec{B}) = 0$ we can rewrite (10) as

$$\vec{V}_D = \mu(\vec{E} + (\mu(\vec{E} + \vec{V}_D \times \vec{B})\times\vec{B})) = \mu\vec{E} + \mu^2\vec{E}\times\vec{B} + \mu^2\underbrace{\vec{v}_D \times \vec{B} \times \vec{B}}_{\vec{B}(\vec{V}_D.\vec{B}) - \vec{v}_D B^2} \ ,$$

$$\vec{v}_D = \frac{\mu\vec{E} + \mu^2\vec{E}\times\vec{B}}{1 + \mu^2 B^2} \quad . \tag{11}$$

With this result we obtain for the 2D current density

$$j_x = n_s e v_D^x = n_s e\,\frac{\mu}{1 + \mu^2B^2}\,E_x + n_s e\,\frac{\mu^2}{1 + \mu^2B^2}\,B\,E_y \ ,$$

$$j_y = n_s e v_D^y = n_s e\,\frac{\mu}{1 + \mu^2B^2}\,B\,E_y - n_s e\,\frac{u^2}{1 + \mu^2B^2}\,E_x \ , \tag{12}$$

and with Ohm's law

$$j_x = \sigma_{xx}E_x + \sigma_{xy}E_y ,$$

$$j_y = \sigma_{yx}E_x + \sigma_{yy}E_y ,$$

for the conductivity components

$$\sigma_{xx} = \sigma_{yy} = \frac{\sigma_o(1 + i\omega\tau)}{(1 + i\omega\tau)^2 + \omega_c^2\tau^2} ; \sigma_o = \frac{n_s e^2 \tau}{m^*} ,$$

$$\sigma_{xy} = - \sigma_{yx} = \frac{\sigma_o \omega_c \tau}{(1 + i\omega\tau)^2 + \omega_c^2\tau^2} ; \omega_c = \frac{e}{m^*} B . \tag{13}$$

In Faraday configuration the incoming light is oriented parallel to the magnetic field. We consider linear polarized light consisting of left and right circular polarized components

$$\vec{E} = \vec{E}^L + \vec{E}^R \quad \text{with} \quad \vec{E}^R = \begin{pmatrix} E_x^R \\ -iE_y^R \end{pmatrix}$$

$$\vec{E}^L = \begin{pmatrix} E_x^L \\ iE_y^L \end{pmatrix}$$

and obtain for

$$\sigma_+ (\omega,B) = \frac{j_x}{E_x^R} = \frac{\sigma_{xx}E_x^R + \sigma_{xy}E_y^R}{E_x^R} = \sigma_{xx} - i\sigma_{xy} = \frac{\sigma_o}{1 + i(\omega+\omega_c)} ,$$

$$\sigma_-(\omega,B) = \frac{j_x}{E_x^L} = \frac{\sigma_{xx}E_x^L + \sigma_{xy}E_y^L}{E_x^L} = \sigma_{xx} + i\sigma_{xy} = \frac{\sigma_o}{1 + i(\omega - \omega_c)} . \tag{14}$$

With the expressions for σ_+ and σ_- we can calculate the absorbed power

$$\overline{P}_{\pm}(\omega) = \frac{1}{2} |\vec{E}|^2 \mathrm{Re}\sigma_{\pm}(\omega,B) = \frac{1}{2} |E|^2 \frac{\sigma_o}{1 + (\omega \pm \omega_c)^2\tau^2} .$$

The total absorbed power is thus

$$\overline{P}(\omega) = \overline{P}_+(\omega) + \overline{P}_-(\omega) = P_o \frac{1 + (\omega^2+\omega_c^2)\tau^2}{|1 + (\omega^2-\omega_c^2)\tau^2|^2 + 4\omega_c^2\tau^2} , \tag{15}$$

with $P_o = \sigma_o |\overline{E}|^2$.

In the case of resonance $\omega = \omega_c$ and $\omega_c\tau >> 1$ the maximum absorbed power $\overline{P}(\omega) = P_o/2$. For $\omega << \omega_c$ we obtain a constant value

$$\overline{P}(\omega) = \frac{P_o}{1 + \omega_c^2\tau^2} , \text{depending on } \omega_c\tau.$$

If we include boundary conditions for the experimental situation with a semiconductor-air interface (neglecting the thin oxide) we obtain for the transmission, reflection and absorption the identical expressions as in (6) with $\sqrt{\varepsilon} = 1$.

$$T = \frac{4}{|2 + \frac{\sigma}{\varepsilon_o c}|^2}, \qquad R = \frac{|\sigma/\varepsilon_o c|^2}{|2 + \frac{\sigma}{\varepsilon_o c}|^2} \qquad A = \frac{4\mathrm{Re}\,\frac{\sigma}{\varepsilon_o c}}{|2 + \frac{\sigma}{\varepsilon_o c}|^2}, \tag{16}$$

where we have to include for σ the proper component :
for linear polarization $\sigma_{xx}(\omega,B)$, for circular polarization $\sigma_{xx} \pm i\sigma_{xy}$.

In the case where $\sigma(\omega,B)$ is calculated quantum-mechanically, the same formalism applies. A calculation of $\sigma_{yy}(\omega,B)$ for short-range scatterers in a 2D system was performed for the first time by Ando [1,4]. He obtains for a Gaussian distribution of scatterers

$$\frac{\mathrm{Re}\sigma_{xx}(\omega,B,N)}{\varepsilon_o c} = \frac{1}{137}(N+1)(f_N - f_{N+1})\,\hbar\omega_c\,\frac{\sqrt{\pi}}{\Gamma}\,\exp\left(\frac{-\Delta_N^2}{\Gamma^2}\right), \tag{17}$$

where N is the Landau level index, f_N the distribution function in the level N and

$$\Delta_N = \varepsilon_N + \hbar\omega - \varepsilon_{N+1} .$$

4. Two-dimensional Plasmons

For the classical electrodynamical calculation of the coupling photon-plasmon via a periodic grating structure, we have to calculate the response of a 2D system to a periodic electric field

$$\vec{E}_{ex}(\omega,\vec{k}) = \vec{E}_{ex}(\omega,\vec{k})\,e^{i(\vec{k}x-\omega t)} . \tag{18}$$

In the electrostatic limit ($c \gg \omega/k$) Poisson's equation

$$\nabla^2 \phi = -\frac{\rho_{3D}}{\varepsilon\varepsilon_o} \tag{19}$$

has to be solved in conjunction with the equation of continuity

$$\vec{\nabla}.\vec{j} = -\frac{\partial\rho}{\partial t} \qquad \text{in two dimensions.} \tag{20}$$

The local electric field E and the inversion layer and gate current are given by

$$\vec{E} = \vec{E}_{ex} + \vec{E}_{ind}, \quad \vec{j}_i = \hat{\sigma}(\omega).\vec{E}, \quad \vec{j}_g = \sigma_G.\vec{E}_{ind} . \tag{21}$$

The following boundary conditions apply

$$\begin{aligned} &E^x_{ind}(z = 0_+) = E^x_{ind}(z = 0_-), \\ &\varepsilon_{ox} E^z_{ind}(z = 0_+) - E^z_{ind}(z = 0_-) = \rho_g/\varepsilon_o, \\ &E^x_{ind}(z = d_+) = E^x_{ind}(z = d_-), \\ &\varepsilon_S E^z_{ind}(z = d_+) - \varepsilon_{ox} E^z_{ind}(z = d_-) = \rho_i/\varepsilon_o, \end{aligned} \tag{22}$$

where ρ_g, ρ_i are the surface change densities for the gate (g) and the inversion channel (i), which have the same periodicity as the incident wave :

$$\vec{j}_i = \vec{j}_i^{\,o}.e^{i(kx - \omega t)} \qquad \vec{j}_g = \vec{j}_g^{\,o}.e^{i(kx - \omega t)}$$
$$\rho_i = \rho_i^o.e^{i(kx - \omega t)} \qquad \rho_g = \rho_g^o.e^{i(kx - \omega t)} . \tag{23}$$

From the above equation we can derive a wave equation for the electrostatic potential ϕ (except at z = 0 and z = d)

$$\frac{\partial^2\phi(\omega, k, z)}{\partial z^2} - k^2 \phi(k, \omega, z) = 0 \tag{24}$$

and the following continuity equations :

$$k.j_i^x(\omega,k) = \omega\rho_i(\omega,k) ,$$
$$k.j_g^x(\omega,k) = \omega\rho_g(\omega,k) , \tag{25}$$

The solution for (15) are of the form

$$-\infty < z < 0 \qquad \phi_o(k, \omega, z) = A.e^{kz} ,$$
$$0 < z < \qquad \phi_{ox}(k, \omega, z) = B.e^{kz} + Ce^{-kz} , \tag{26}$$
$$d < z < \infty \qquad \phi_s(k, \omega, z) = D.e^{-kz} .$$

With (21), (22), (24) and (25) we have 12 equations for 13 variables A, B, C, D, ϕ_o, ϕ_{ox}, ϕ_s, j_i^x, σ_i, j_g^x, ρ_g, E_{ind}^x and E_{ind}^z, the 13th equation is the relation between ϕ and the electric field $\vec{E} = -$ grad ϕ.

The final result for $\vec{E}_{ind}$ can be written in the form

$$E_{ind}^x\ (z = d)$$
$$= \frac{-E_{ex}^x[k\hat{\sigma}(\omega)/i\omega][\varepsilon_o - k\hat{\sigma}_G/i\omega + \varepsilon_o\varepsilon_{ox}\coth(kd)]}{\varepsilon^2\varepsilon_{ox}^2 + [\varepsilon_o\varepsilon_s - k\sigma(\omega)/i\omega](\varepsilon_o - k\hat{\sigma}_G/i\omega) + [\varepsilon_o\varepsilon_s + \varepsilon_o - k\hat{\sigma}_G/i\omega - k\sigma(\omega)/i\omega]\varepsilon_o\varepsilon_{ox}\coth(kd)} . \tag{27}$$

Experimentally we have access to the absorbed energy, which is given by $\bar{p}_i = \mathrm{Re}E_{ex}^x.\mathrm{Re}j_i^x$, and has to be averaged over time and space

$$\bar{p}_i = 1/2\ |E_{ex}^x(\omega,\vec{k})|^2.\mathrm{Re}\hat{\sigma}(\omega,\vec{k}) \tag{28}$$

The dynamical conductivity $\hat{\sigma}(\omega,\vec{k})$ is derived from (21) and (22) together with (24) and (25) as

$$\hat{\sigma}(\omega,\vec{k}) = \hat{\sigma}\ (\omega)$$
$$. \frac{\varepsilon_o^2\varepsilon_{ox}^2 + \varepsilon_o\varepsilon_{Si}(\varepsilon_o - k\hat{\sigma}_G/i\omega) + (\varepsilon_o\varepsilon_{Si} + \varepsilon_o - k\hat{\sigma}_G/i\omega)\varepsilon_o\varepsilon_{ox}\coth(kd)}{\varepsilon_o^2\varepsilon_{ox}^2 + [\varepsilon_o\varepsilon_{Si} - k\hat{\sigma}(\omega)/i\omega](\varepsilon_o - k\hat{\sigma}_G/i\omega) + [\varepsilon_{Si}\varepsilon_o + \varepsilon_o - k\hat{\sigma}_G/i\omega - k\hat{\sigma}(\omega)/i\omega].\varepsilon_o\varepsilon_{ox}\coth(kd)}$$

(29)

$\hat{\sigma}(\omega,\vec{k})$ describes the reaction of the 2D-system to a time and space varying electromagnetic wave.

For a given $\vec{k}$-value resonances appear as a function of frequency. From the position of the resonances the plasmon dispersion can be derived.

Equation (29) can be simplified assuming

$$\left| \frac{k.\hat{\sigma}_G}{i\omega} \right| \gg \varepsilon_s \varepsilon_o , \quad \left| \frac{k.\hat{\sigma}(\omega)}{i\omega} \right| \quad \text{to}$$

$$\hat{\sigma}(\omega,k) = \frac{\hat{\sigma}(\omega)}{1 - \frac{\hat{\sigma}(\omega)}{i\omega} \frac{k}{\varepsilon_o(\varepsilon_s+\varepsilon_{ox}\coth(k.d))}} . \tag{30}$$

Figure 2 shows $\hat{\sigma}(\omega,\vec{k})$ as a function of frequency for a given density $n_s = 3.10^{12}cm^{-2}$ and a wave vector $\vec{k} = 2.10^4 cm^{-1}$ corresponding to a plasma wavelength of 3 µm. The resonance represents the 2D-plasmon. With (2) for $\sigma(\omega)$ and $\sigma_o = n_s e^2\tau/m^*$, (30) can be further simplified in the limit $\tau \to \infty$. The denominator of $\sigma(\omega,\vec{k})$ has zeros for

$$\omega^2 = \frac{n_s e^2}{m^* \varepsilon_o} \frac{k}{\varepsilon_s + \varepsilon_{ox}\coth(k.d)} , \tag{31}$$

which is the dispersion relation for longitudinal modes in a classical 2D electron gas.

Figure 3 shows the dispersion relation for 3 different densities and 2 different oxide thicknesses. The main features are that the

$\omega^2 \propto k$ for constant n_s

and $\omega^2 \propto n_s$ for constant k.

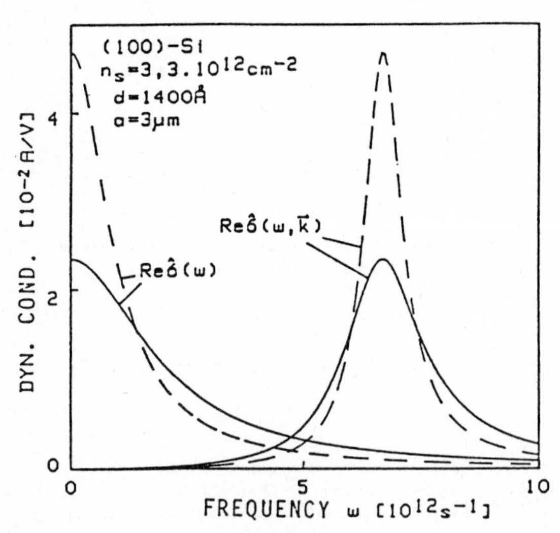

Fig.2 Dynamical conductivity $\hat{\sigma}(\omega)$ from (2) and $\hat{\sigma}(\omega,\vec{k})$ from (30) as a function of frequency for two values of τ : full curves $\tau = 0.5 \times 10^{-12}$s, dashed curves $\tau = 1 \times 10^{-12}$s.

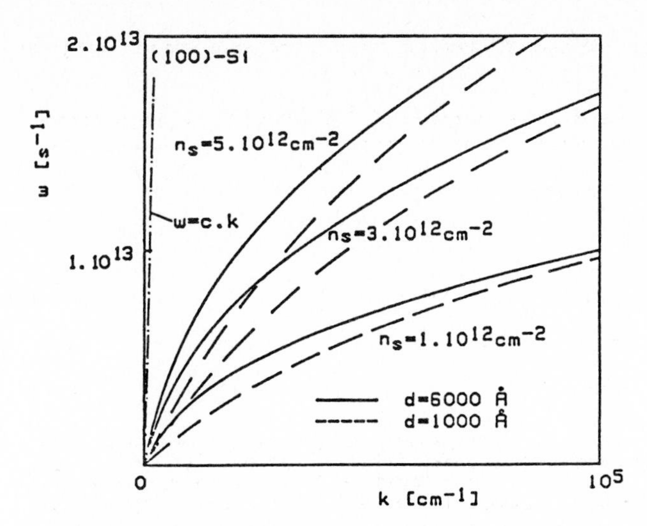

Fig.3 Dispersion relation for 2D-plasmons in a (100)-Si- MOS-structure for the indicated parameters.

The 2D-plasmon is a nonradiative mode since its wave vector is considerably larger than the light wave vector.

4.1 Grating Coupling

The coupling between the plasmon and photon can be achieved by using a grating on top of a semitransparent conducting gate. The structure is shown in Fig.4, where a is the period and σ_M the conductivity of the metallic grids. As a result of the grating structure the photon momentum $\hbar k_x = \hbar\ \omega/c \sin\gamma$ obtains an additional momentum $k_n = n.2\pi/a$. That means the electric field of the electromagnetic wave has the following periodicity in the near field:

$$\vec{E}(x,t) = \vec{E}_o . e^{i(k_x x - \omega t)} e^{ik_n x} . \tag{32}$$

This corresponds to a wave with a photon momentum of

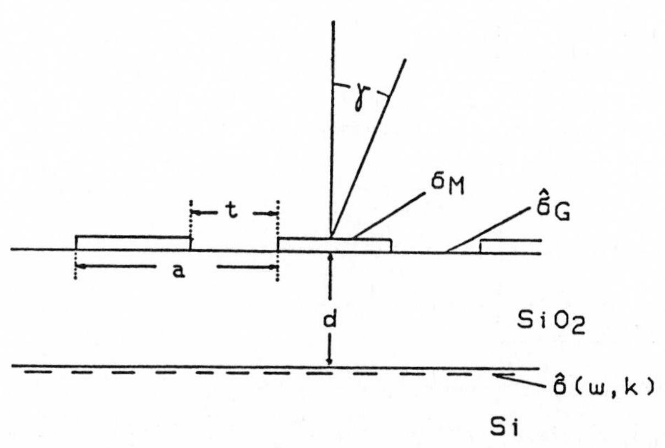

Fig.4 MOS-structure with a grid on top of a semitransparent gate. γ is the angle of incidence for the light.

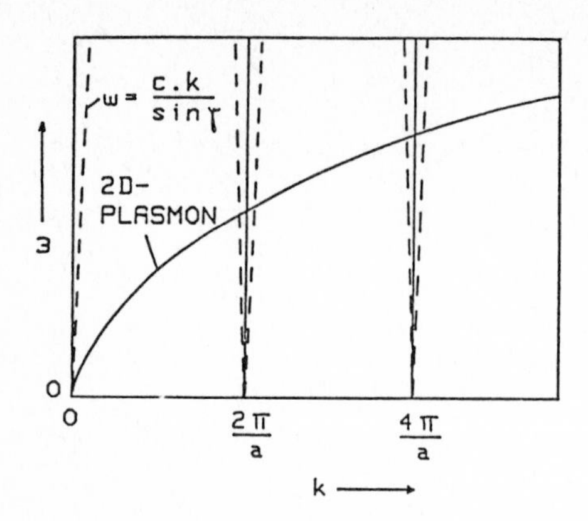

Fig.5 Plasmon dispersion relation demonstrating the actual change of light wave vector (dashed curve) in the presence of a grating with period a.

$$\hbar k_x' = \hbar k_x \pm n.\frac{2\pi\hbar}{a}, \tag{33}$$

which is demonstrated in Fig. 5.

Since the phase velocity of the plasmon is considerably smaller than c, the coupling relation can simply be written as $k = n.2\pi/a$.

To calculate the coupling in a quantitative manner, we have to calculate the Fourier components α_n of the electric field in the near field. For a structure shown in Fig. 4 we write the reciprocal conductivity as

$$\frac{1}{\sigma} = R_o + \sum_{n=1}^{\infty} R_n \cos(k_n x)$$

$$R_n = \frac{2}{n\pi} \frac{(\sigma_M - \sigma_G)}{\sigma_G \sigma_M} \sin(k_n x/2) \quad R_o = |(t/a)\sigma_M + (1-t/a)\sigma_G|/\sigma_G\sigma_M . \tag{34}$$

With the following assumptions

electrostatic limit $\omega/k \ll c$,

$$\left|\frac{a}{t}\sigma_G\right| \gg |\omega\varepsilon_{ox} a.c\varepsilon_o| \text{ and } \left|\frac{a}{t}\sigma_G\right| \gg |\hat{\sigma}(\omega)| ,$$

we obtain the following expression for the Fourier components α_n (in agreement with [8,9])

$$\alpha_n = \underbrace{\frac{2\sin^2(k_n t/2)}{(k_n t/2)^2} \left| \frac{\sigma_M - \sigma_G}{\sigma_M + \sigma_G(a/t-1)} \right|^2}_{\text{"1"}} \cdot \underbrace{\frac{\coth^2(k_n d)-1}{|\varepsilon_s/\varepsilon_{ox} + \coth(k_n d)|^2}}_{\text{"2"}} . \tag{35}$$

The term "1" represents the strength of the Fourier component as a function of geometry parameters. The term "2" describes the exponential decay of the field. For the case $\varepsilon_s = \varepsilon_{ox}$ the term "2" becomes e^{-2kd}.
The inversion layer experiences the electric field $E^x_{ext}(\omega,k_n) = \alpha_n^{1/2}.E^x_{rad}(\omega$, where $E^x_{rad}(\omega)$ is the incoming radiation amplitude. The absorbed intensity is given by

$$\overline{P}_i = \frac{1}{2} |E^x_{rad} (\omega)|^2 \cdot \sum_{n=1}^{\infty} \alpha_n \cdot \hat{\sigma}(\omega, k_n) . \tag{36}$$

The absorptivity is calculated according to (7). However, in the present situation F is the sum over all dynamical conductivities :

$$\hat{F} = [\hat{\sigma}(\omega) + \sum_{n=1}^{\infty} \alpha_n \cdot \hat{\sigma}(\omega, k_n)]/\varepsilon_o c \tag{37}$$

For small values of $F(|F| \ll 1)$, $A(\omega)$ can be simplified to

$$A(\omega) = [\frac{Re\hat{\sigma}(\omega)}{\varepsilon_o c} + \sum_{n=1}^{\infty} \alpha_n \frac{Re\hat{\sigma}(\omega, k_n)}{\varepsilon_o c}] \frac{4}{(\sqrt{\varepsilon_s} + 1)^2} . \tag{38}$$

This situation is present in most transmission experiments.

The FIR emission from the 2D system can be calculated if thermodynamic equilibrium is assumed : If electrons and plasmons are excited to a temperature T according to their quantum statistics, the emission from the system per solid angle in the direction perpendicular to the surface is given by

$$I(T,\omega) = I_{BB} (T,\omega) A (\omega) , \tag{39}$$

where $I_{BB}(\omega)$ is the emission of a blackbody of temperature T and A (ω) the absorption according to (38).

5. 2D-Plasmon Experiments

The experimental set-up for transmission and emission experiments is shown in Fig. 6. The left scheme is for transmission experiments showing a lightpipe connected with a laser (or Fourier spectrometer) as a light source; the sample is placed in the helium bath (and magnetic field if necessary), the transmitted light is measured with a broadband detector.

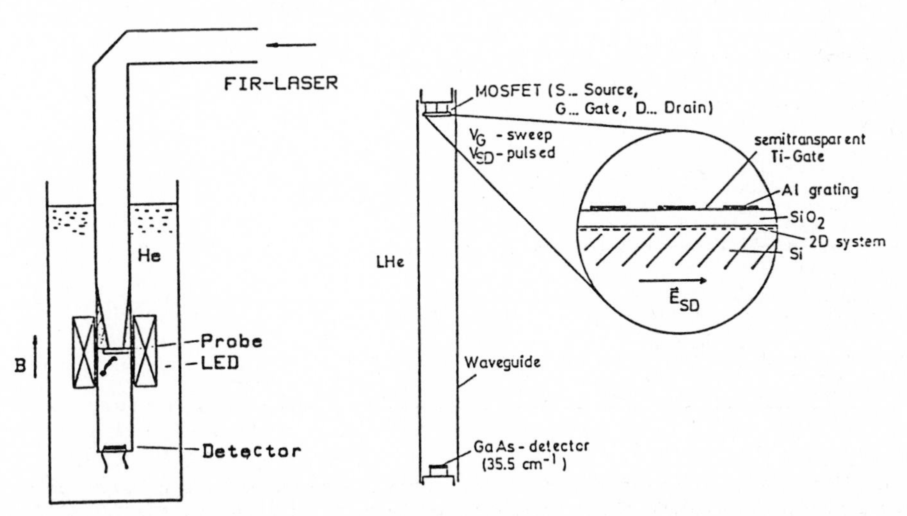

Fig.6 Schematic set-up for transmission (left side) and emission (right side) experiments. The insert shows an example of a Si-MOSFET structure.

The right part of the figure shows the emission technique set-up, with the sample and a FIR narrowband detector, both mounted in a brass lightpipe and inserted in the helium bath. A schematic cross-section of the MOSFET is shown in the insert : Upon the semitransparent Ti gate ($\sim$ 50 Å, area 2.5 x 2.5 mm^2), Al gratings (thickness 1000 Å) of periods 3, 2 and 1.5 µm are evaporated using optical contact lithography and lift-off technique. The grating wave vector defines the momentum of the plasmon which can be coupled out.

Examples of transmission spectra are shown in Fig. 7 according to Allen et al. [8], as a function of frequency obtained with a Fourier spectrometer. The relative change in transmission $\Delta T/T$ is plotted (open circles) for 2 different densities and a grating period of a = 3.5 µm. The full curves show fits according to (38), the dashed curves according to (8) . The agreement between the classical theory and the experiment is very good if the effective mass m^* is used as a fitting parameter and τ is taken from mobility measurements. In the emission experiments the gate voltage and thus the electron concentration n_s is varied, resulting in a tuning of the plasma frequency. This frequency is tuned through the narrowband resonance of a GaAs-extrinsic detector. The emitted radiation is generated by heating the 2D electrons via electric source-drain pulses. The detector used shows a narrowband photo response at 4.4 meV ($\sim$ 35.5 cm^{-1}) and a detectivity of 2 x 10^6 V/W [10].

Figure 8 shows the main results of the emission experiments : The detector signal is plotted as a function of the electron concentration for two values of grating wave vector. The heating pulses are 12 V/cm in both cases. Upon a background of broadband FIR emission from hot electrons -discussed in Sect. 6 - a peak from the 2D plasmon can be seen, that changes its position, linewidth and amplitude for different grating periods. From the $\omega(k)$ relation (31) it can be identified as the plasmon peak, which is described by (38) both in amplitude and in linewidth : The smaller plasmon signal at a = 1.5 ωm (lower curve) is due to a lower coupling efficiency and a lower electron concentration n_s, giving a smaller value of the resonance in Re $\sigma(\omega,k)$.

We performed experiments in the electron concentration range of 1 to 4 x 10^{12} cm^{-2} and in the electric field range of 2 to 40 V/cm. This corresponds to electron

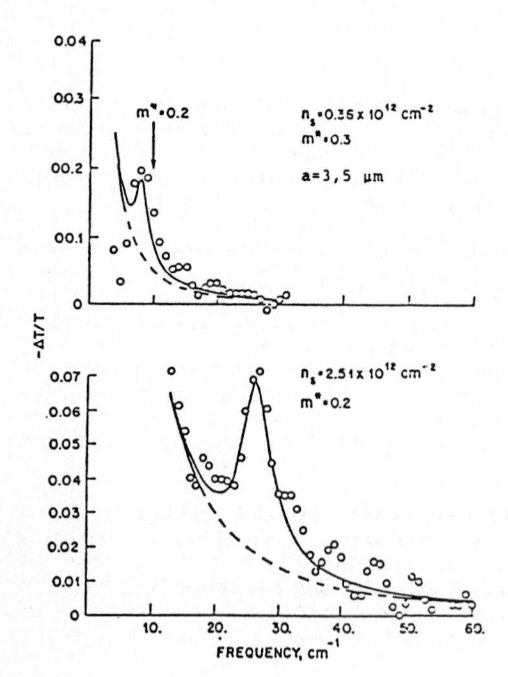

Fig.7 Relative change in transmission through a Si-MOSFET as shown in the insert of Fig.6, when the gate is turned on and off. The open circles are experimental results, the curves theoretical fits as explained in the text (after [8]).

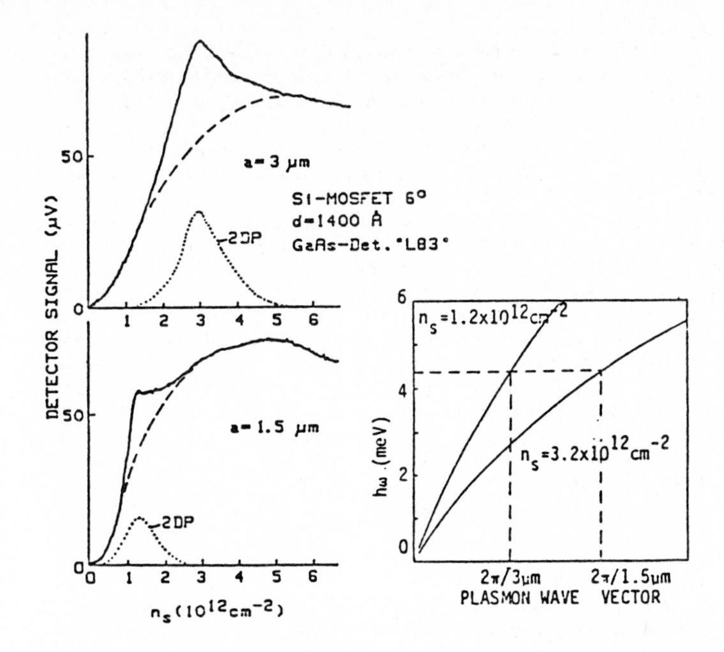

Fig.8 Emission spectra as a function of n_s for samples with a 3 μm and 1.5 μm grating period as observed with a GaAs detector at B = 0. The dotted curves indicate the pure plasmon emission. The insert shows the dispersion relation for two densities and explains the emission peaks at 4.4 meV for the two grating periods (after [15]).

temperatures of 2 up to 30 K above helium bath temperature. In this whole range the model of thermal excitation was found to describe well the experiment [16].

Similar results were found in inversion channels of GaAs/GaAlAs heterostructures : The dispersion relation is obeyed within experimental error, the plasmon emission intensity is found to depend on the amount of Coulomb scatterers present. With increasing mobility the emission intensity decays rapidly due to a decrease in Coulomb scattering [17].

6. Broadband Emission from Hot Carriers

6.1 Electron Temperatures in Si-MOSFETs

FIR emission from the inversion layer of the Si-MOSFETs brings a new method of determining electron temperatures. Since the broadband FIR absorption by the 2D electron system is known experimentally [14] and theoretically, the absolute emission intensity of FIR radiation from the 2D electron system according to (29) represents a direct measure for the electron temperature T [16].

In these experiments we used MOSFETs with transparent gate electrodes and a GaAs detector, whose responsivity at 35.5 cm^{-1} has been determined accurately by using a carbon-glass bolometer as a reference source : At 35.5 cm^{-1} the absorptivity of the whole bolometer arrangement is measured to be about 0.7; in addition, the absolute intensity is determined by comparison with InSb cyclotron emission at saturation electron temperature [18]. The 2D electron system is heated by applying electric source-drain pulses of 0.5 V up to 10 V to the MOSFETs. Simultaneously the current-voltage characteristics of the MOSFETs is measured to determine the n_s- and field-dependent mobilities of the samples for the calculation of the input power $e\mu E^2$ and the dynamical conductivity Re $\sigma(\omega)$.

We investigated MOS samples with different mobilities from 2000 up to 10 000 cm^2/V.s. In Fig. 9 the electron temperatures, evaluated according to (29), are plotted versus the input power $e\mu E_{SD}^2$ which, in equilibrium situation, is equal to the power loss of the hot electron gas. Also in the figure, the slope for $\Delta T \propto (e\mu E^2)^{1/2}$ is indicated. As a result, the electron heating ΔT is in the whole range of temperature exactly proportional to the square root of the input power

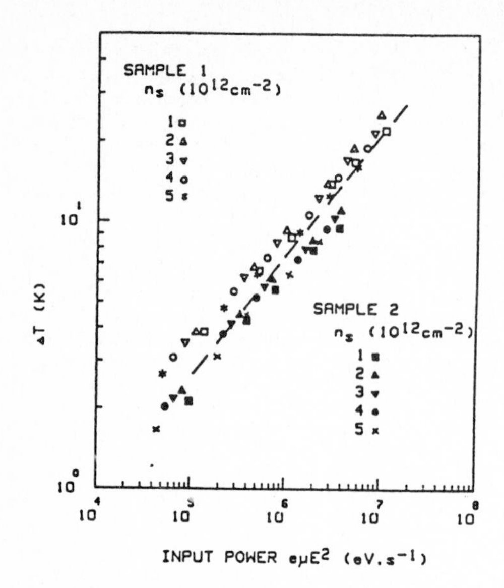

Fig.9 Electron temperatures determined from the absolute intensity of the FIR emission. The two samples and the related electron concentrations are indicated in the figure. The slope of ΔT (electron temperature minus 4.2 K) proportional to $(e\mu E^2)^{1/2}$ is shown by the dashed line (after [11]).

respectively the power loss $e\mu E^2$. We can therefore express the electron heating as a function of the input power $e\mu E^2$ quantitatively as

$$\Delta T/(e\mu E^2)^{1/2} = (0.9 \pm 0.2) \times 10^{-2} \mathrm{K}\ \mathrm{s}^{1/2}\ (\mathrm{eV})^{-1/2} \qquad (40)$$

This result is somewhat lower than the electron temperature measurements made at low electron temperatures by FANG and FOWLER [19], KAWAJI and KAWAGUCHI [20] and HÖNLEIN and LANDWEHR [21], and close to those obtained from subband emission [24] FIR emission is the first method that can give results in the whole temperature range of 2 K up to more than 30 K. The linear dependence of ΔT on $(e\mu E^2)^{1/2}$ is predicted by recent theories of acoustic phonon scattering in 2D systems [23,24] up to electron temperatures of a least 50 K.

6.2 Electron Temperatures in GaAs/AlGaAs Heterostructures

FIR broadband emission was also investigated in MBE grown GaAs/AlGaAs single heterostructures [12]. The radiation is analyzed at two different frequencies (35 cm^{-1} and 100 cm^{-1}). From the relative dependence of the intensities on the applied electric field, hot electron temperatures are determined. The results are shown in Fig. 10, where the electron heating is plotted versus input power per electron : The electron heating increases with the input power and reaches 100 K when the input power exceeds 10^7eV.s^{-1} per electron. The values of heating for the two different values of carrier concentration and mobility are surprisingly close together, although the electron concentration varies by a factor of three and the mobility by a factor of 100. Also the slopes of the curves are similar except sample 2 (light) which shows a somewhat stronger increase of heating with input power. The change of slope, however, is too weak and almost within the error bars, so there can be no conclusions drawn from this behaviour. These results show that the heating in GaAs/AlGaAs heterostructures in this range of lattice temperature and carrier heating is a function of the input power per carrier rather than an explicit function of carrier density and mobility.

This observation is similar to the results obtained in Si-MOSFETs [11] and is also consistent with magneto-conductivity data of SAKAKI et al. [25] showing also only a small dependence on carrier concentration and mobility. This behaviour is expected when acoustic phonon scattering determines the energy loss of the carriers to the lattice [25,24]. The absolute values obtained by this optical emission method are in good quantitative agreement with the results of photoluminescence experiments by

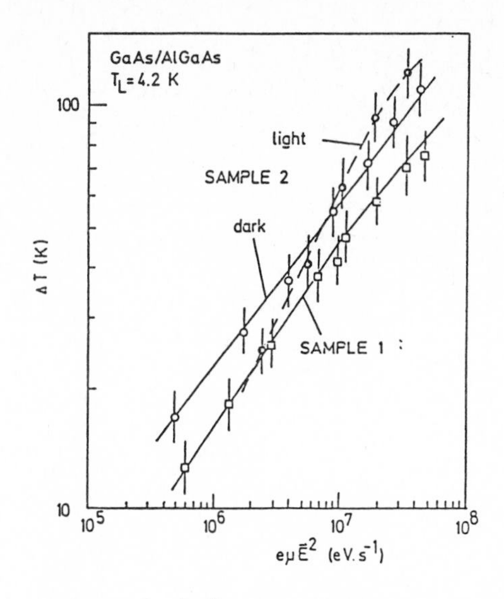

Fig.10 Electron heating for 2D electrons in GaAs obtained from evaluating the emission intensity at two frequencies (35 cm^{-1}, 100 cm^{-1}).
Sample 1 : n_s = $8.7 \times 10^{11} cm^{-2}$ and μ = $1.6 \times 10^4 cm^2/Vs$
Sample 2 : n_s = $2.4 \times 10^{11} cm^{-2}$ and μ = $1.0 \times 10^6 cm^2/Vs$ (after [12]).

SHAH et al. [26] at high electron temperatures, and the values obtained by SAKAKI et al. [25] from magnetoconductance measurements at low electric fields.

In conclusion, we emphasize that FIR emission from heated carriers in 2D systems is a powerful tool to study electronic excitations in two dimensions, such as plasmons, subband excitations etc., since only the electron system is heated, and other excitations of the device are not disturbing in the experiments. Up to now only spontaneous emission from thermally excited plasmons could be experimentally achieved. This means that in all experiments thermodynamic equilibrium of plasmons and electrons was given. A non-equilibrium specific excitation of plasmons would be of interest, since the radiative decay efficiency is quite high. There are theoretical calculations of amplification of plasmons in 2D systems. Promising possibilities of non-equilibrium excitation would be optical excitation and, especially, drifting carriers - exceeding the phase velocity of the plasma wave or interacting strongly with the grating, in analogy to travelling wave tubes. High drift velocities in GaAs/AlGaAs heterostructures up to $3 \times 10^7 cm/s$ have been reported, which are already equal to the phase velocities of 2D plasmons at low electron concentrations. Striking experiments on plasmon generation by drifting carriers can therefore be expected in the near future.

References

1. T. Ando : J. Phys. Soc. Jpn. 38, 989 (1975)
2. R.J. Wagner, T.A. Kennedy, B.D. Mc Combe, D.C. Tsui : Phys. Rev. B22, 945 (1980)
3. J. Kotthaus, G. Abstreiter, J.F. Koch, R. Ranvand : Phys. Rev. Lett. 34, 151 (1974)
   J. Allen, D.C. Tsui, J.V. Dalton : Phys. Rev. Lett. 32, 107 (1974)
4. T. Ando, A.B. Fowler, F. Stern : Rev. Mod. Phys. 54, 437 (1982)
5. W. Seidenbusch, G. Lindemann, R. Lassnig, J. Edlinger, E. Gornik : Surf. Sci. 142, 375 (1984)
6. K. Muro et al. : Surf. Sci. 142, 394 (1984)
7. Z. Schlesinger, J.C.M. Hwang, S.J. Allen : Surf. Sci. 142, 423 (1984)
8. J. Allen, D.C. Tsui, F. De Rosa : Phys. Rev. Lett. 35, 1359 (1975)
9. T.N. Theis : Surf. Sci. 98, 515 (1980)
10. R.A.. Höpfel, E. Vass, E. Gornik : Phys. Rev. Lett. 49, 1667 (1982)
11. R.A. Höpfel, E. Vass, E. Gornik : Solid-State Comm. 49, 501 (1984)
12. R.A. Höpfel, G. Weinmann : Appl. Phys. Lett. (1985)
13. R.A. Höpfel : "FIR emission from 2 D-plasmons and hot electrons in Si-MOSFETs", Thesis, Univ. of Innsbruck, Austria (1983)

14. J. Allen, D.C. Tsui, R.A. Logan : Phys. Rev. Lett. 38, 980 (1977)
15. E. Gornik : Physica 127B, 95 (1984)
16. R.A. Höpfel, E. Gornik : Surf. Sci. 142, 412 (1984)
17. R.A. Höpfel, E. Gornik, A.C. Gossard, W. Wiegmann : Physica 117/118B, 646 (1983)
18. E. Gornik : J. Magn. Mat. 11, 39 (1979)
19. F.F. Fang, A.B. Fowler : J. Appl. Phys. 41, 1825 (1970)
20. S. Kawaji, Y. Kawaguchi : Lecture Notes in Physics, Vol. 177, ed. by G. Landwehr (Springer Berlin, Heidelberg 1983), p. 53
21. W. Hönlein, G. Landwehr : Surf. Sci. 113, 260 (1983)
22. E. Gornik, D.C. Tsui : Solid-State Electron. 21, 139 (1978)
23. Y. Shinba, J. Nakamura : J. Phys. Soc. Jpn. 50, 114 (1981)
24. E. Vass : to be published
25. H. Sakaki et al.: Surf. Sci. 142, 306 (1984)
26. J. Shah et al.: Appl. Phys. Lett. 44, 322 (1984)

Raman Scattering by Free Carriers in Semiconductor Heterostructures

G. Abstreiter

Physik-Department, Technische Universität München,
D-8046 Garching, Fed. Rep. of Germany

The application of inelastic light scattering to the investigation of electronic properties of semiconductor heterostructures is discussed. Some aspects of the theory of Raman scattering by free carrier excitations are presented. Special emphasis is laid on the difference between two- and three-dimensional systems. Experimental results of both single-particle and collective excitations are discussed for various semiconductor systems involving single heterostructures, multiquantum well structures, and doping superlattices. Most of the structures are grown by molecular beam epitaxy based on GaAs as the prominent example.

Introduction

Elementary excitations of solid-state electron plasmas have been discussed extensively in the literature [1]. Contrary to most of the highly diluted gaseous plasmas, in semiconductors it is possible to achieve cold degenerate plasmas in which electron-electron interactions play a substantial role. The plasma frequencies of free carriers in semiconductors are in the range of conventional Raman spectroscopy. The carrier density can be varied over a wide range by doping and by thermal or optical excitations. The first observation of light scattering by a solid-state plasma was reported by MOORADIAN and WRIGHT [2] in doped n-type GaAs. The observed plasmons were coupled to LO-phonons. Mooradian also reported light scattering by single-particle excitations of a degenerate electron gas in GaAs [3]. This pioneering work has stimulated a lot of theoretical and experimental research in this field. A review on light scattering by free carrier excitations in semiconductors has been published recently by ABSTREITER, CARDONA, and PINCZUK [4]. This article contains a chapter on light scattering by two-dimensional systems which deals with one of the most exciting new developments in solid-state physics. The present lecture is based on Ref. 4. In Sec. 2 we discuss in a comprehensive form some basic concepts of light scattering by free carriers. Collective and single-particle excitations, coupling to LO-phonons, the resonance behavior, and the characteristic differences of two- and three-dimensional electron systems are treated theoretically. In sec. 3 we discuss selected experimental results which have been obtained for various types of semiconductor structures. Scattering of plasmon-like excitations leads to information on carrier concentration or effective mass and electron damping. Recently, however, exciting results have been obtained for two-dimensional carrier systems which can exist at semiconductor-semiconductor interfaces, in metal-insulator-semiconductor systems, and in multi layer structures. Those are discussed in the main part of this lecture.

2. Theoretical Aspects

The theory of light scattering by electron plasmas in solids was developed in the early sixties [1]. The Hamiltonian that represents the interaction of the incident and scattered light fields with electrons in the semiconductor can be written as [4]

$$H_{ep} = H'_{ep} + H''_{ep} , \quad \text{where} \tag{1}$$

$$H''_{ep} = \frac{e^2}{2m} \sum_j [\vec{A}(\vec{r}_j)]^2 \tag{2}$$

and

$$H'_{ep} = \frac{e}{2m} \sum_j [\vec{p}_j \, A(\vec{r}_j) + \vec{A}(\vec{r}_j)\vec{p}_j]. \tag{3}$$

$\vec{A}(\vec{r}_j)$ represents the sum of the vector potentials of the incident and scattered fields. The summation includes all the electrons. H''_{ep} is of second order in the fields, and therefore leads to light scattering in first-order perturbation theory. H'_{ep} is of first order in the fields. It gives rise to light scattering in second order perturbation theory. Both intraband and interband matrix elements of $\vec{P}_j$ enter in eq. 3. The second order perturbation of H'_{ep} has a resonant denominator which is approximately equal to $(E_g - \hbar\omega_L)$ where E_g is the optical energy gap of the semiconductor. Under resonance condition interband contributions to H'_{ep} are dominant. Resonance conditions are important for light scattering by two-dimensional carrier systems which exhibit a very small scattering volume. The scattering cross-section obtained from these interaction Hamiltonians is found to be related to the spectrum of density fluctuations. At higher densities the one-electron excitations are modified by dynamical screening effects with the longitudinal polarization of the plasma. Under resonance condition, however, the band structure influences light scattering in various ways. The resonance behavior of the scattering cross-section and the spin orbit interaction of the valence bands have opened up the possibility of observing excitations with single-particle character also at high electron densities.

The frequencies of elementary excitations of a degenerate electron gas as a function of wavevector are shown in Fig. 1. The hatched region represents the continuum of single-particle excitations. The plasma frequency ω_p is shown with a weak quadratic dispersion. The finite q-vector, which is necessary for the observation of single-particle excitations, is provided by the wavevector of the incident and scattered light $k_s \approx k_i = 2\pi n/\lambda_L$, where n is the refractive index and λ the wavelength of the laser light. The dynamical structure factor relevant for these excitations is proportional to the imaginary part of the susceptibility $\mathrm{Im}[X(q,\omega)]$ which is related to the density of states for single-particle excitations of the free electron gas. In the limit that the Fermi energy $E_F \gg k_B T$, the energies of possible single-particle excitations are within the hatched region of Fig. 1. For a three-dimensional system, $\mathrm{Im}[X_e(q,\omega)]$ has a triangular shape, as indicated in the figure. In a high density plasma, electron-electron interactions play a dominant role. Single-particle excitations are then only observed via spin density fluctuations which can be excited under resonance conditions due to spin orbit interaction. Scattering due to charge density fluctuations on the other hand

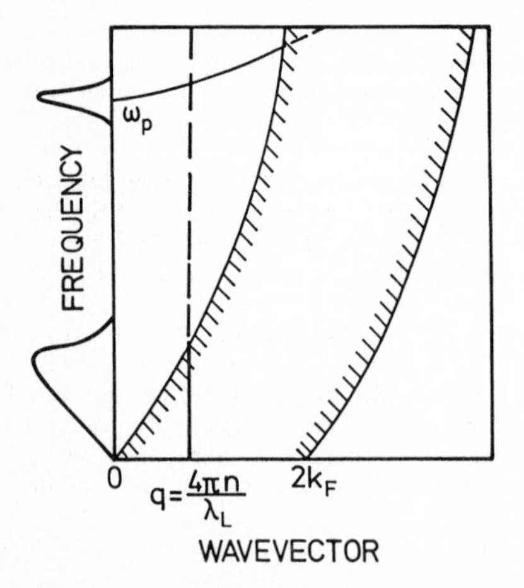

Fig.1 Frequencies of elementary excitations of a degenerate electron gas as a function of wavevector.

leads to a cross-section which is determined by the full longitudinal response function $\mathrm{Im}[-1/\varepsilon(q,\omega)]$ where

$$\varepsilon(q,\omega) = \varepsilon_\infty + \chi(q,\omega) + \chi_L(\omega) \tag{4}$$

is the total longitudinal dielectric function with

$$\chi_L(q,\omega) = \varepsilon_\infty \frac{\omega_{LO}^2 - \omega_{TO}^2}{\omega_{TO}^2 - \omega^2} \tag{5}$$

being the contribution of the polar lattice to the electric susceptibility. ω_{TO} and ω_{LO} are the transverse and longitudinal optical phonons. Structure in the light scattering spectra is expected at energies related to those of the maximum in $\mathrm{Im}\{-1/\varepsilon(q,\omega)\}$. Outside the region of single-particle excitations $\mathrm{Im}\,\chi(q,\omega)$ is small and the peaks in $\mathrm{Im}\{-1/\varepsilon(q,\omega)\}$ occur at frequencies and wavevectors that are determined by

$$\mathrm{Re}\{\varepsilon\,(q,\omega)\} = 0\ . \tag{6}$$

Equation (6) is the condition which gives the plasmon frequency of the electron gas. Its dispersion can be approximated by

$$\omega_p^2(q) \simeq \omega_p^2\,(0) + \frac{3}{5}\,q^2 v_F^2\ , \quad \text{where} \tag{7a}$$

$$\omega_p(0) = \left(\frac{Ne^2}{\varepsilon_o \varepsilon_\infty m^*}\right)^{1/2}\ , \tag{7b}$$

N is the carrier density, and v_F is the Fermi velocity. In polar semiconductors the contribution of the lattice to the longitudinal dielectric function leads to coupled LO-phonon plasmon modes. With the familiar Drude form of the electronic dielectric function

$$\chi\,(q \to 0,\omega) = -\,\varepsilon_\infty \frac{\omega_p^2}{\omega^2} \tag{8}$$

one obtains from (4) and (6),

$$\omega_\pm^2 = \frac{1}{2}\,\{(\omega_p^2 + \omega_{LO}^2) \pm [(\omega_p^2 - \omega_{LO}^2)^2 + 4\omega_p^2(\omega_{LO}^2 - \omega_{TO}^2)]^{1/2}\} \tag{9}$$

Light scattering by coupled LO-phonon plasmons has been studied in several semiconductors. For further references see [4].

The strong resonance enhancement of scattering by single-particle and collective excitations as observed by PINCZUK et al. [5] in n-GaAs led to the proposal [5,6] that resonant light scattering is a sensitive method for studies of the elementary excitations of two-dimensional electron systems at semiconductor surfaces and interfaces. BURSTEIN et al. [7] pointed out that within the effective mass approximation the mechanisms and selection rules for light scattering by two-dimensional semiconductor plasmas are similar to those of three-dimensional systems. Two-dimensional systems are characterized by the separation of motion parallel and perpendicular to the surface or interface. The usual dispersion of the bands is maintained in the parallel direction. Perpendicular to the potential well the carriers are bound in subbands with minimum energy E_o, E_1, E_2,... The total energy is given by

$$E = E_i + \frac{\hbar k_{/\!/}^2}{2m^*}\ . \qquad i = 0,\ 1,\ 2,\ \ldots \tag{10}$$

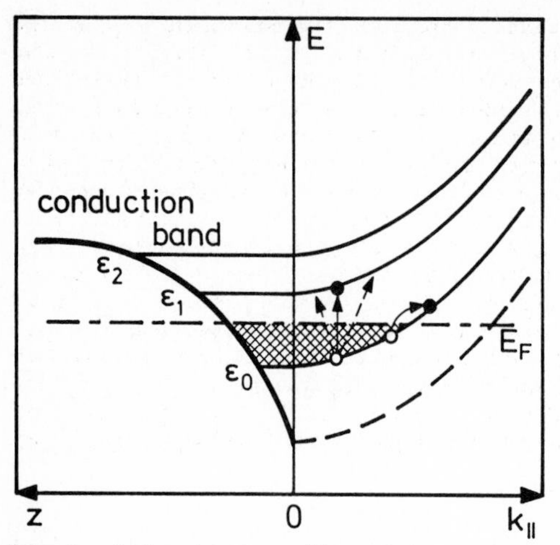

Fig.2 Potential well and $k_{//}$ -dispersion of subbands at semiconductor heterojunctions. Also shown are intra- and intersubband excitations relevant for light scattering experiments.

In Fig. 2 a potential well characteristic for a semiconductor heterostructure, the subbands E_0, E_1, E_2, and their dispersion in $k_{//}$ are shown schematically. The Fermi energy is chosen such that only the lowest subband is occupied. Depending on the scattering wavevector one can create both intra- and intersubband excitations. This separation of electronic excitations is caused by the separation of motion due to quantization in one direction. Excitations within one subband (intra) are only possible for configurations with a component of the scattering wavevector $\vec{q}$ in the direction $\vec{k}_{//}$. Similarly to the three-dimensional case, one can create both single-particle and collective excitations. Because of the two-dimensional nature of the Fermi surface, however, the lineshape of the single-particle excitations is different. This is shown schematically in Fig. 3. The collective intrasubband excitations are two-dimensional plasma oscillations parallel to the surface or interface. The frequency of this plasmon mode tends to zero with decreasing $q_{//}$.

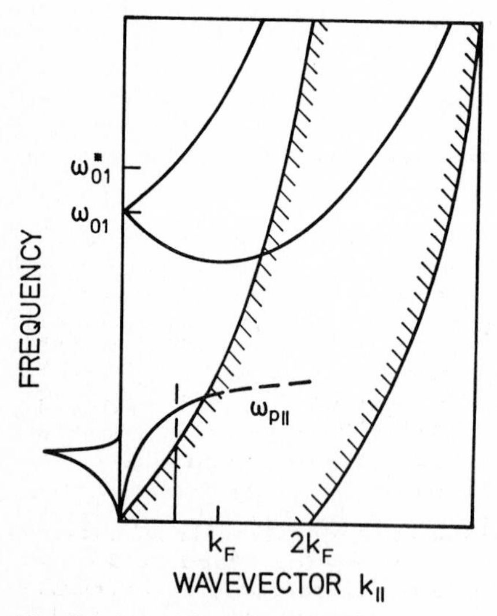

Fig.3 $k_{//}$ -dispersion of electronic excitations in two-dimensional carrier systems.

In backscattering geometry, which is usually applied, the component of $\vec{q}$ in the $\vec{k}_{//}$ direction is negligibly small. Therefore the majority of investigations have so far been concentrated on intersubband excitations. Single-particle intersubband transitions are vertical transitions of electrons below the Fermi energy in a lower subband to an empty state in a higher subband. As in the three-dimensional case, these unscreened excitations are observed under resonance conditions when scattering occurs via spin-density fluctuations. The measured energies correspond directly to the subband splitting, when small shifts due to final state interactions (exciton-like shift) are neglected. An additional component of the wavevector parallel leads to a strong broadening (Figs. 2 and 3). Collective intersubband excitations involve charge density fluctuations, and are consequently screened dynamically. This Coulomb screening causes a shift to higher energies. It describes the dielectric response of the thin layer of carriers to the subband excitation and can be written in terms of an "effective" plasma frequency ω_p^* normal to the interface. In a way this reflects the finite extension of the carrier system in the direction of quantization. If only two subbands are involved, one finds [8,9] :

$$\omega_p^{*2} = \frac{2N_s e^2}{\hbar \varepsilon_o \varepsilon_\infty} \omega_{01} f_{11} , \tag{11}$$

where f_{11} is the matrix element of the Coulomb interaction of the two subbands, N_s is the two-dimensional carrier concentration, and $\hbar\omega_{01}$ is the bare energy splitting $(E_1 - E_0)$. The frequency of the collective subband excitation is given by

$$\omega_{01}^{*2} = \omega_{01}^2 + \omega_p^{*2} . \tag{12}$$

As discussed for the three-dimensional case, in polar semiconductors the collective excitations interact with the LO-phonon. The frequencies of the coupled modes are determined by the zeros of the real part of the dielectric function. The Drude-like form of the electronic dielectric function is maintained by replacing (8) with

$$\chi(\omega) = - \varepsilon_\infty \frac{\omega_p^{*2}}{\omega^2 - \omega_{01}^2} . \tag{13}$$

For the coupled modes one obtains

$$\omega_\pm^2 = \frac{1}{2} (\omega_{01}^2 + \omega_{LO}^2 + \omega_p^{*2}) \pm \frac{1}{2} [(\omega_{01}^2 + \omega_{LO}^2 + \omega_p^{*2})^2 - 4 (\omega_{01}^2 \omega_{LO}^2 + \omega_{TO}^2 \omega_p^{*2})]^{1/2} \tag{14}$$

For $\omega_{01} << \omega_{LO}$ one finds $\omega_- \simeq \omega_{01}^*$; for $\omega_{01} >> \omega_{LO}$ the high-frequency mode ω_+ approaches ω_{01}^*. Typical for the two-dimensional behavior is also the crossing of the ω_--mode with ω_{TO} when $\omega_{01} = \omega_{TO}$. The frequencies $\omega_\pm$ for the three- and two-dimensional case are shown in Figs. 4 and 5 for comparison.

3. Experimental Results

There exist extensive studies of light scattering by three-dimensional carrier systems in semiconductors which can be created by doping, temperature, or photoexcitation [4]. In this section we concentrate on selected experimental results where Raman scattering has been used to investigate free carrier excitations in single heterostructures, multiquantum wells, and doping

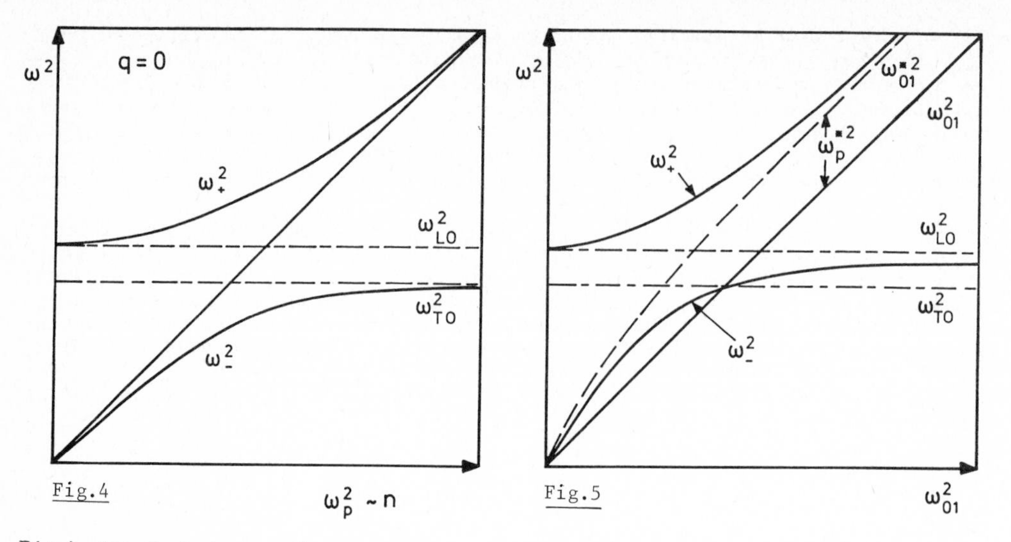

Fig.4 Coupled phonon-plasmon modes of a three-dimensional carrier system in a polar semiconductor (schematically).

Fig.5 Single-particle and coupled phonon-intersubband modes of a two-dimensiona carrier system in a polar semiconductor (schematically).

superlattices. The first observations of light scattering spectra of intersubband excitations between discrete energy levels has been reported for electrons in GaAs/(Al_xGa_{1-x})As heterostructures [10]. Light scattering turned out to be an excellent method for the investigation of single-particle and collective excitations in two-dimensional carrier systems. In subsequent studies the technique has been applied to study subband splittings and coupling to LO-phonons in various semiconductor structures involving GaAs, (Al_xGa_{1-x})As, InP, InAs, Ge, and Si. At semiconductor-semiconductor interfaces charge carriers may be transferred from the wide-gap material to the semiconductor with the smaller forbidden energy gap. Raman spectra obtained from such structures are shown in Fig. 6. The collective and single-particle excitations are separated experimentally by analyzing the polarization properties of the scattered light. While for the coupled modes incoming and scattered light are polarized parallel to each other, spin-flip single-particle excitations are observed with perpendicular polarization. The selectively doped GaAs/$Al_xGa_{1-x})$As heterostructure used for the experiments shown in Fig. 6 contains a two-dimensional carrier concentration $N_S \simeq 7x10^{11}$ cm^{-2}. Two subbands are occupied in this system. Therefore the displayed spectra exhibit quite complicated structures. Various transitions are observed. The arrows in Fig. 7 mark the subband separations, as indicated in the insert.

In Fig. 7 spectra are shown as obtained with a GaAs/-(Al_xGa_{1-x})As$_o$ multiquantum well structure. The thickness of the GaAs layers is d_{GaAs} = 200 Å. Each layer contains $N_s \simeq 3x10^{11}$ cm^{-2} carriers. Only the lowest subband is occupied. Therefore the two-level model described in the theoretical part is a good approximation. The spectrum for crossed polarizations $(z(xy)\bar{z})$ exhibits one single peak labeled ω_{01}. The energy of this peak is found to be in excellent agreement with self-consistent calculations [11]. The parallel polarized spectrum $[z(YY)\bar{z}]$ shows three peaks, a sharp LO-phonon line, and the coupled modes. From the energies of these modes ω_p^* = 16.7 meV and f_{11} = 10 Å is deduced. This compares well with model calculations for the system. Such spectra have been measured for various different kinds of samples in order to determine quantitatively subband splittings, ω_p^*, and the Coulomb matrix elements. Comparison of those quantities with self-consistent calculations leads also to a direct determination of the band-offset ΔE_c in the system.

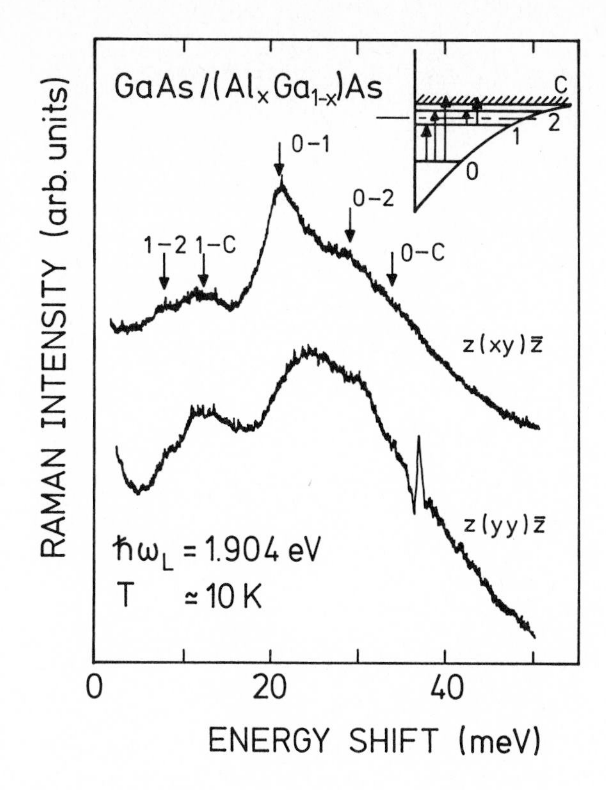

Fig.6 Electronic Raman spectra of a GaAs/(Al$_x$Ga$_{1-x}$)As single heterostructure.

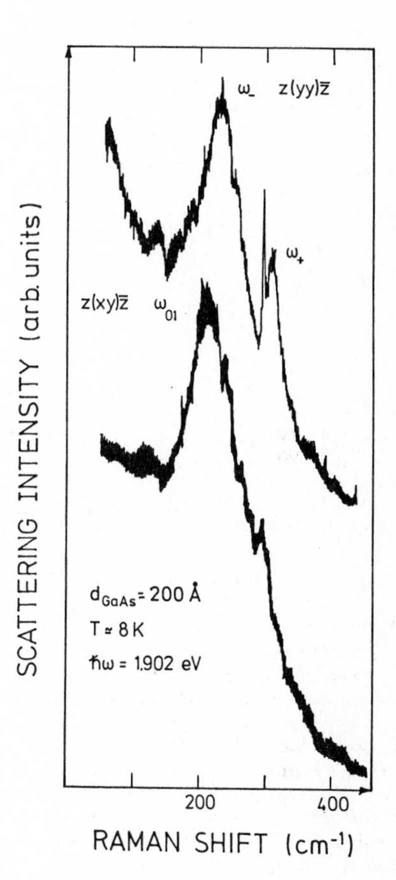

Fig.7 Electronic Raman spectra of a modulation doped GaAs/(Al$_x$Ga$_{1-x}$)As multilayer structure.

As a last example we now discuss the recent light scattering work in periodic multilayer structures, so-called "nipi"-crystals. This type of superlattice has been proposed and studied theoretically by DÖHLER [12]. The fascinating electrical and optical properties of those novel superlattices will be discussed by Maan in this School [13] (see also Ref. 14). A doping superlattice is composed of aperiodic sequence of ultrathin n- and p-doped layers of a semiconductor (GaAs in our case). The electrons from the donors are attracted by the acceptors in the p-type layers, resulting in a periodic rise and fall of the conduction and valence band. Due to the induced space charge potential, excited electrons and holes are separated in space and may have recombination lifetimes of orders of magnitude longer than in homogeneous bulk crystals. The reduced effective band gap, which depends on the nonequilibrium electron and hole concentrations, results in a strong tunability of the optical absorption and luminescence [14]. To obtain direct information on the quantization of photo-excited carriers in doping superlattices, resonant inelastic light scattering experiments have been performed [15-18]. Spin-flip single-particle intersubband excitations were studied, using different power densities of the incident laser. At low excitation intensity, several distinct peaks have been observed. These are identified as $\Delta = 1$, $\Delta = 2$, $\Delta = 3$ intersubband transitions of

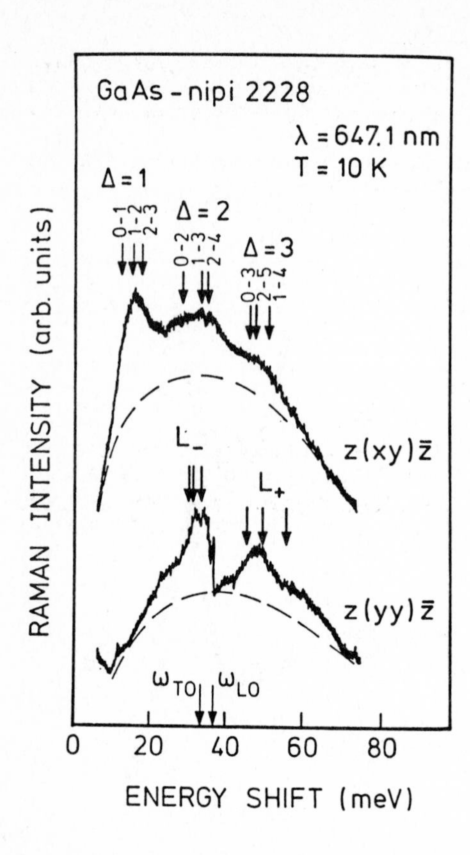

Fig.8 Electronic Raman spectra of a "nipi"-crystal.

photo-excited electrons in the nipicrystal (Fig. 8). $\Delta = 1$ means all possible transitions between first nearest subbands, $\Delta = 2$ between second, and $\Delta = 3$ between third nearest. The peaks shift with excitation intensity. They are in excellent agreement with self-consistent calculations of the subband splittings. An example is shown in Fig. 8, where also collective excitations are included. The experimentally determined carrier concentration is $2.8\mathrm{x}10^{12}\ \mathrm{cm}^{-2}$. The theoretical values for various transition energies are marked by arrows. For the collective excitations only the strongest lines are indicated. With increasing excitation intensity a change to three-dimensional behavior is observed [17].

4. Concluding Remarks

The usefulness of resonant inelastic light scattering for the investigation of various electronic properties of semiconductor heterostructures and doping layers has been demonstrated with GaAs as an example. Among the most spectacular recent developments in this field are excitations with finite $k_{//}$, with and without applied magnetic field, δ-doped layers, graded junctions, and heterostructures based on different semiconductor materials including strained layer superlattices. The light scattering processes are quite well understood by now. So resonant electronic light scattering acts as a powerful tool to analyze new structures and materials.

References

1. See for example D. Pines : Elementary Excitations in Solids, (Benjamin, New York 1963)
2. A. Mooradian, G.B. Wright : Phys. Rev. Lett. 16, 999 (1966)
3. A. Mooradian : Phys. Rev. Lett. 20, 1102 (1968)
4. G. Abstreiter, M. Cardona, A. Pinczuk : In Light Scattering in Solids IV

(Topics Appl. Phys. Vol. 52), Ed. M. Cardona, G. Güntherodt (Springer, Berlin Heidelberg (1984), p.5
5. A. Pinczuk, G. Abstreiter, R. Trommer, M. Cardona : Solid-State Comm. 30, 429 (1979)
6. E. Burstein, A. Pinczuk, S. Buchner : In Physics of Semiconductors 1978, Ed. by B.L. Wilson (The Institute of Physics, London 1979) p. 1231
7. E. Burstein, A. Pinczuk, D.L. Mills : Surf. Sci. 98, 451 (1980)
8. S.J. Allen, D.C. Tsui, B. Vinter : Solid-State Commun. 20, 425 (1976)
9. T. Ando : Solid State Comm. 21, 133 (1977)
10. G. Abstreiter, K. Ploog : Phys. Rev. Lett. 42, 1308 (1979)
11. T. Ando, S. Mori : J. Phys. Soc. Jpn. 47, 1518 (1979)
12. G.H. Döler : Physica Status Solidi (b) 52, 79, 533 (1972)
13. J. Maan : these proceedings
14. K. Ploog, G.H. Döhler : Adv. Phys. 32, 285 (1983)
15. G.H. Döhler, H. Künzel, D. Olego, K. Ploog, P. Ruden, H.J. Stolz, G. Abstreiter : Phys. Rev. Lett. 47, 864 (1982)
16. Ch. Zeller, B. Vinter, G. Abstreiter, K. Ploog : Phys. Rev. B26, 2124 (1982)
17. Ch. Zeller, B. Vinter, G. Abstreiter, K. Ploog : Physica B117/118, 729 (1983)
18. G.Abstreiter : In Two-Dimensional Systems, Heterostructures, and Superlattices Springer Ser. Solid-State Sci. Vol. 53, Ed. by G. Bauer, F. Kuchar, H. Heinrich (Springer, Berlin, Heidelberg 1984) p. 232

Confined and Propagative Vibrations in Superlattices

B. Jusserand and D. Paquet

Laboratoire de Bagneux, Division PMM, Laboratoire associé au CNRS (L.A. 250), Centre National d'Etudes des Télécommunications, 196 rue de Paris, F-92220 Bagneux, France

Lattice dynamics and electronic properties of superlattices present striking similarities. We show, on the basis of one-dimensional models, that optical vibrations can be confined in "phonon quantum wells" and that the superperiodicity induces a folding of the acoustical modes. Raman scattering experiments are presented and discussed in the light of these simple models.

Electronic properties of superlattices made of a periodic stacking of layers of two semiconductors have been extensively studied in the past ten years and are, at least qualitatively, very well understood. On the contrary, vibrational properties of these structures have been sparsely investigated and progress in the understanding appear only recently. In these lecture notes we shall try to review the present knowledge on lattice dynamics of superlattices, emphasizing as much as possible the similarities between electronic and vibrational properties.

The paper is organized as follows : in the first part we give a qualitative description of the influence of superperiodicity on the lattice dynamics. Part 2 is devoted to one-dimensional models describing the folding of acoustical modes and the confinement of optical ones. In part 3 we recall simple notions on phonon Raman scattering. The experimental results are described and analysed in part 4. General trends and further developments are commented upon in the conclusion.

1. Qualitative analysis of vibrations in superlattices

In what follows, we shall deal only with the almost unique case experimentally studied, i.e. superlattices with grown direction along [001] (as referred to the bulk zinc blende cristalline axis). Along this direction the bulk III-V compounds can be viewed as a stacking of planes made of atoms belonging alternatively only to the III or V column. Let us denote a_1, respectively a_2, the smallest distance between planes of same chemical nature in the two materials constituting the superlattice, and n_1a_1, respectively $n_2 a_2$, the respective thicknesses of the layers. The actual periodicity of the structure along the growth direction becomes : $d = n_1a_1 + n_2a_2$. In first approximation, due to lattice matching growth conditions, we can identify a_1 and a_2. As a consequence, the extent of the Brillouin zone along [001], as compared to the bulk one, is reduced by a factor $n_1 + n_2$. On the other hand, the superlattice unit cell contains $2(n_1 + n_2)$ atoms and one expects thus the existence of $2 (n_1 + n_2)-1$ zone center longitudinal optical modes. Due to the one-dimensional character of the structure along the growth direction, these longitudinal modes will be correctly described within linear chain models. Moreover, most Raman scattering experiments, the unique optical technique used up to now, have been performed along the [001] direction. In this configuration, only longitudinal modes can be detected. Thus one-dimensional models will be sufficient to analyse most of the experimental results.

To get a first insight on the lattice dynamics, we shall start from analogies with the electronic properties of superlattices. On Fig. 1a we have sketched the near zone center electronic dispersion curves of both constituting bulk materials 1 and 2, as a function of $k_\perp$, the wavevector along [001]. We have also plotted in dashed lines the imaginary wavevector for energies within the gaps. On Fig. 1b is drawn the traditional picture representing the extrema of both valence and conduction bands, in the real space, along the growth direction. Thick hatched

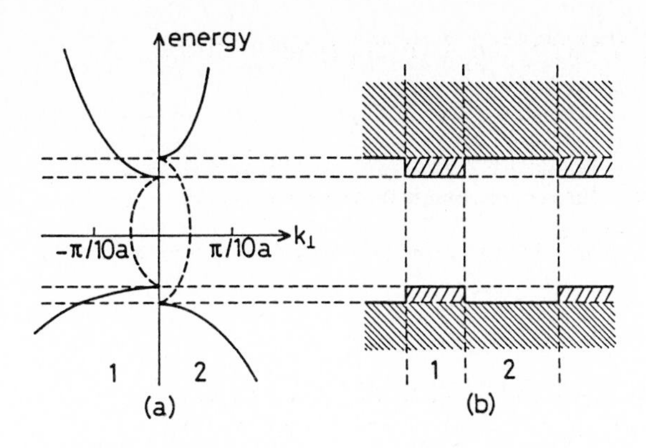

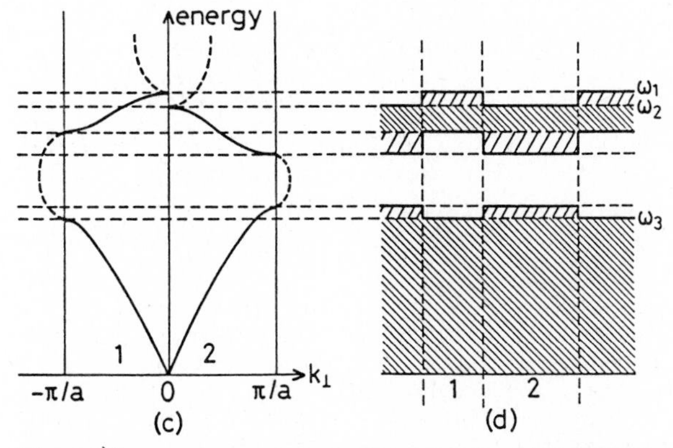

Fig.1 a)Near zone center electronic dispersion curves for two semiconductors : compound 1 on the left, compound 2 on the right. Full line : real dispersion curves. Hatched line : complex one within the gaps. b) Variation of the electronic band extrema along the growth direction of a superlattice made with compounds 1 and 2. Thick hatched regions : electron or hole confinement in the wells (compound 1). Thin hatched regions : scattering states. c) and d) same sketches as in a) and b) for phonons. Note the confinements coming here from near zone center and near zone edge states of bulk materials.

regions correspond to quantum wells where either electrons or holes are confined as long as the barrier thickness is larger than the penetration length given by the imaginary part of the wavevector $k_{\perp}$. Typical values for this length lay between 20 and 50 Å. Otherwise, due to tunneling through the barriers, phase coherence between adjacent wells sets up, thus leading to dispersive states. At energies allowed for both compounds (thin hatched regions), propagative states exist in the superlattice.

The lattice dynamics of superlattices can be understood on the same basis. On Fig.1c are sketched the longitudinal phonon dispersion curves of both compounds as a function of $k_{\perp}$ on the whole Brillouin zone. On contradistinction with the electronic case, the band widths are rather small as compared with the band offsets. It is therefore relevant to draw, as done in Fig.1d, the two extrema of each band in the real space, displaying energy regions associated with confined states or propagative states. Concerning superlattice modes at frequencies near the bulk zone center ones, two extreme cases always happen :

i) within the optical phonon frequency-range ω_1-ω_2 (see figure 1) the situation is similar to hole quantum wells. The curvature of the bulk dispersion curves are negative and the vanishing length at frequencies a few wavenumbers above ω_2 amounts only to a few Angströms. As a consequence, between ω_1 and ω_2, only discrete frequencies are allowed, leading to non-dispersive modes.

ii) within the acoustical mode frequency-range ($0\ \omega\ \omega_3$), there is evidently no band offset. As a consequence all states are propagative. The influence of the superperiodicity can be understood in terms of a folding of an average dispersion curve,which induces new zone center modes and minigaps between minibands.

To sum up, in any superlattice, both confined and propagative models always coexist.

2. One-dimensional Kronig-Penney models

2.1 Long wavelength acoustical modes : an effective medium approximation

To keep on the analogy with the standard treatment of an electron in a superlattice, we shall briefly review Rytov's treatment (1) of elastic waves in a periodic layered medium. For wavelength quite larger than the lattice parameter, the local unit cell displacement can be considered as a continuous field u. Consequently the two compounds can be viewed as effective media characterized by their respective densities ρ_1, ρ_2 and their elastic constants C_1 and C_2. For the electronic case, the effective media would have been defined by the effective masses and the potentials and the energy spectrum would be deduced from the Kronig Penney model.

Considering only propagation along the z direction, which is perpendicular to the layer planes, the Lagrangian density writes :

$$\mathcal{L} = \frac{1}{2}\,\rho(z)\left(\frac{\partial u}{\partial t}\right)^2 - \frac{1}{2}\,C(z)\left(\frac{\partial u}{\partial z}\right)^2$$

where the functions ρ and C take the values ρ_1, ρ_2 (respectively C_1, C_2) according to whether z stands in medium 1 or in medium 2.

The equation of motion takes the form :

$$\frac{\partial}{\partial t}\left[\rho(z)\,\frac{\partial u}{\partial t}\right] - \frac{\partial}{\partial z}\left[C(z)\,\frac{\partial u}{\partial z}\right] = 0 \qquad (1)$$

which reduces, within each layer, to :

$$\rho_{1,2}\,\frac{\partial^2 u_{1,2}}{\partial t^2} = C_{1,2}\,\frac{\partial^2 u_{1,2}}{\partial z^2} \qquad (2)$$

One recovers the bulk equations, which lead to the linear dispersion curves :

$$\omega = \sqrt{\frac{C_{1,2}}{\rho_{1,2}}}\,.\,k$$

Equation (1) which can be understood as a conservation relation, leads to the following interface continuity equation :

$$C_1\,\frac{\partial u_1}{\partial z} = C_2\,\frac{\partial u_2}{\partial z} \quad \text{(stress continuity)}$$

which holds together with the condition of displacement continuity $u_1 = u_2$.

Furthermore, due to the periodicity, the displacement field fulfills the Bloch relation :

$$u(z+d) = u(z)e^{iQd}$$

where Q is a superlattice wavevector.

Equation (2) implies that any displacement field oscillating with pulsation ω may be written in each layer as a linear combination of a forward and a backward propagating planewave with the same vector k_1 or k_2 :

$$u_{1,2}(z) = (\lambda_{1,2}e^{ik_{1,2}z} + \mu_{1,2}e^{-ik_{1,2}z})e^{-i\omega t}$$

The continuity equations take the form :

$$\left.\begin{aligned} &\lambda_1 + \mu_1 = \lambda_2 + \mu_2 \\ &C_1(\lambda_1-\mu_1)k_1 = C_2(\lambda_2-\mu_2)k_2 \end{aligned}\right\} \text{ at } z = 0$$

$$\left.\begin{aligned} &\lambda_2 e^{ik_2d_2} + \mu_2 e^{-ik_2d_2} = (\lambda_1 e^{-ik_1d_1} + \mu_1 e^{ik_1d_1})e^{iQd} \\ &C_2k_2(\lambda_2 e^{ik_2d_2} - \mu_2 e^{-ik_2d_2}) = C_1k_1(\lambda_1 e^{-k_1d_1} - \mu_1 e^{ik_1d_1})e^{iQd} \end{aligned}\right\} \text{ at } z = d_2$$

where d_1 and d_2 are the respective thicknesses of each layer. Evidently one gets $d = d_1 + d_2$. After some tedious algebra, the secular equation leads to the following superlattice dispersion relation :

$$\cos Qd = \cos k_1d_1 \cos k_2d_2 - \frac{C_1\rho_1 + C_2\rho_2}{2[C_1\rho_1C_2\rho_2]^{1/2}} \sin k_1d_1 \sin k_2d_2 \qquad (3)$$

For completeness we give also the eigendisplacement fields :

$$u_1 = C_1k_1\sin k_2d_2\cos k_1z - C_2k_2\cos k_2d_2 \sin k_1z + C_2k_2e^{iQd} \sin k_1(z + d_1) \qquad -d_1 \leqslant z \leqslant 0$$

$$u_2 = C_1k_1e^{iQd} \cos k_1d_1 \sin k_2z + C_2k_2e^{iQd} \sin k_1d_1\cos k_2z - C_1k_1\sin k_2 (z-d_2) \qquad 0 \leqslant z \leqslant d_2$$

The dispersion relation (3) can be rewritten :

$$\cos Qd = \cos (k_1d_1 + k_2d_2) - \frac{\varepsilon^2}{2} \sin k_1d_1 \sin k_2d_2 \quad \text{where}$$

$$\varepsilon = \frac{C_1\rho_1 - C_2\rho_2}{C_1\rho_1 + C_2\rho_2}$$

qualifies the normalized difference between the two materials. For III-V compounds, ε^2 is always small and the superlattice dispersion curve shown on Fig. 2 looks like

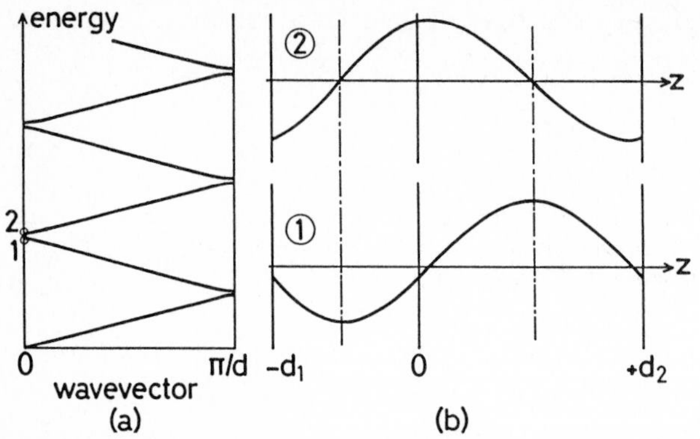

Fig.2 a) Schematic folded acoustical dispersion curves. b) Eigendisplacement fields for modes labelled 1 and 2 in (a). d is the superlattice period, d_1 and d_2 the thicknesses of the two constitutive layers.

the folding of an effective bulk dispersion curve :

$\omega = v_s k$ with

$$\frac{d}{v_s} = \frac{d_1}{v_1} + \frac{d_2}{v_2} \tag{4}$$

$v_{1,2}$ are the sound velocities in bulk compounds 1,2. The difference between the two constituents induces only minigaps at zone center and zone edge. One therefore finds new zone center optical modes of acoustical origin which, as will be shown later, can be observed by Raman scattering. The two new zone center modes associated to the first folding have been pointed out on the typical dispersion curve drawn on Fig. 2a. Their eigendisplacements (Fig. 2b) look like a sinusoidal wave with wavelength d, the modulation of the sound velocity inducing some distortion and a pinning of the phase relative to the interfaces, such that the waves are either symmetric (1) or antisymmetric (2) relative to the midlayer planes.

The keypoint of this whole treatment is the reduction of the eigenvalue problem to interface continuity conditions. This method has been used in many contexts, for instance to study the dielectric constant of a superlattice (2). We shall now use the same technics to treat on the same footing both acoustical and optical modes.

2.2 Linear chain model

A very simple and useful model of the longitudinal lattice dynamics along the growth direction is the linear chain model with only nearest neighbour interactions (3). For III-V superlattices, one must consider two situations, which are sketched on Fig. 3. In the first case Fig.3a the two constituents AB and AC, having a common chemical species, share a common atom A at the interface. In the second one (Fig.3b) the two constituents AB and CD having no common atom, the interfaces are of BC or DA type and abnormal bounds appear. Due to the nearest neighbour approximation, the equation of motion of each atom, except the interface ones, is the same as in the bulk. As a consequence, at a given pulsation ω, the superlattice eigendisplacement reduces again in a given layer to a linear combination of the +k and -k bulk eigenmodes at the same pulsation. Note that k may be complex. The dispersion relations of the superlattice are now obtained.

- first by using the Bloch relation, which introduces the superlattice wavevector Q
- second by writing compatibility relations for the interface atoms.

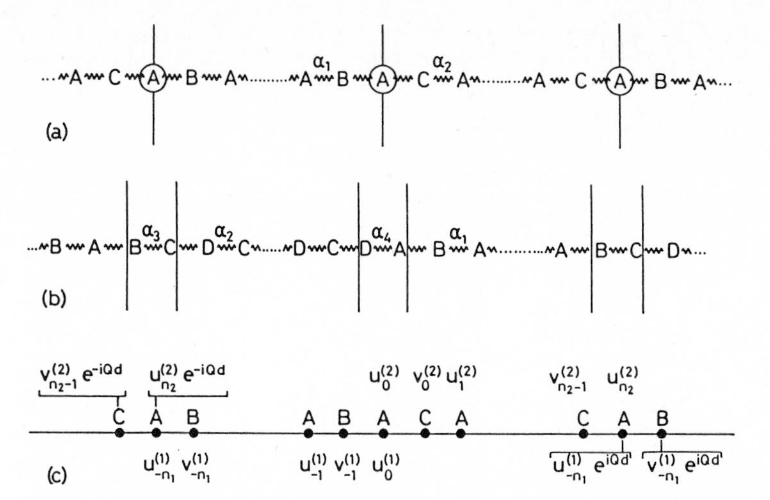

Fig.3 Linear chain models for superlattices : a) the two constituting compounds share a common atom, b) they have no common atom. c) definition of the atomic displacements for case (a) as used in the text. α_1, α_2, α_3 and α_4 are different spring constants.

In the first case (Fig.3a.), one identifies the two displacements of the interface atom A obtained from the two adjacent layers and imposes its equation of motion to be fulfilled. In the second case (Fig.3b, interface BC) the motion of B atom (respectively C) is derived from the bulk AB (respectively CD) layer, and the two equations of motion of the interface atoms must be simultaneously fulfilled. In what follows, we will treat explicitly the first case, assuming for simplicity a common force constant α for both compounds. This approximation is extremely good for GaAs/AlAs superlattices which are, up to now, the most studied ones.

Let us first consider some textbook results on the lattice dynamics of a linear chain with two atoms by unit cell (masses m_A and m_B) and only nearest neighbour interaction (force constant α, lattice constant a). The complex dispersion curve is given by :

$$\cos ka = 1 - \frac{m_A + m_B}{\alpha}\,\omega^2 + \frac{m_A m_B}{2\alpha^2}\,\omega^4$$

and if one denotes u and v the respective longitudinal displacements of atoms A and B, the related eigendisplacement within a given unit cell writes up to a common factor :

$$u = \alpha\,(1 + e^{-ika}) \qquad\qquad v = 2\alpha - m_A\omega^2$$

Using the notations explained in Fig. 3c, the atomic displacements write in medium 1 :

$$\begin{aligned} u_j^{(1)} &= \lambda_1\,\alpha\,(1 + e^{-ik_1a_1})\,e^{ijk_1a_1} \\ &\quad + \mu_1\alpha\,(1 + e^{ik_1a_1})\,e^{-ijk_1a_1} \\ v_j^{(1)} &= \lambda_1\,(2\alpha - m_A\omega^2)\,e^{ijk_1a_1} \\ &\quad + \mu_1(2\alpha - m_A\omega^2)\,e^{-ijk_1a_1} \end{aligned} \tag{5}$$

and are obtained for medium 2 by changing index 1 into 2. The superlattice eigenmodes are defined as long as one knows the four coefficients λ_1, μ_1, λ_2 and μ_2.

The interface compatibility equations take the form :

$$u_o^{(1)} = u_o^{(2)} \tag{6a}$$

$$-m_A\omega^2 u_o^{(1)} = \alpha\,[v_o^{(2)} + v_{-1}^{(1)} - 2u_o^{(1)}] \tag{6b}$$

$$u_{n_2}^{(2)} = u_{-n_1}^{(1)}\, e^{iQd} \tag{6c}$$

$$-m_A\omega^2 u_{n_2}^{(2)} = \alpha\,[v_{n_2-1}^{(2)} + v_{-n_1}^{(1)}\, e^{iQd} - 2u_{n_2}^{(2)}] \tag{6d}$$

Equations (6b) and (6d) may be simplified if one notices that $u_o^{(1)}$ and $u_{n2}^{(2)}$ fulfil:

$$-m_A\omega^2\, u_o^{(1)} = \alpha\,[v_o^{(1)} + v_{-1}^{(1)} - 2u_o^{(1)}]$$

$$-m_A\omega^2\, u_{n_2}^{(2)} = \alpha\,[v_{n_2-1}^{(2)} + v_{n_2}^{(2)} - 2u_{n_2}^{(2)}]$$

because they correspond to eigenmodes of the bulk materials. $v_o^{(1)}$ and $v_n^{(2)}$ are fictive displacements obtained by continuation of one medium within the other. As a consequence, (6b) and (6d) are replaced by :

$$v_o^{(1)} = v_o^{(2)} \tag{6b'}$$

$$v_{-n_1}^{(1)}\, e^{iQd} = v_{n_2}^{(2)} \tag{6d'}$$

Substituting (5) into (6a), (6b'), (6c), (6d'), one gets four linear equations connecting $_1$, μ_1, λ_2 and μ_2. The secular equation provides after a rather tedious calculation the following dispersion relation :

$$\cos Qd = \cos n_1k_1a \cos n_2k_2a - \frac{1 - \cos k_1a \cos k_2a}{\sin k_1a \sin k_2a} \sin n_1k_1a \sin n_2k_2a$$

Note that :

i) the secular equation is a 4x4 determinant whatever the dimension of the superlattice unit cell.

ii) if one expands up to second order the quantities $\sin k_{1,2}a_{1,2}$ and $\cos k_{1,2}a_{1,2}$ around $k_{1,2} = 0$, one evidently recovers Rytov's equation (3).

Typical dispersion curves can be seen on Fig. 4. They were obtained with the assumption that the common atom A is the heaviest one, which implies that the two bulk zone edge acoustical frequencies are equal. For such a case, there is no "acoustical well" around ω_3 (Fig. 1). Case drawn in Fig. 4a corresponds to a superlattice designed with two bulk materials whose optical frequency ranges do not overlap. One recovers folded acoustical modes and non-dispersive optical modes

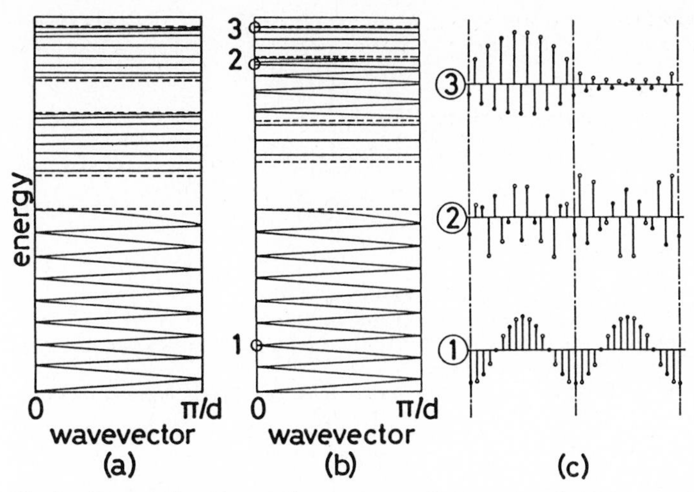

Fig.4 Schematic dispersion curves of two superlattices made of bulk materials with no overlapping (a) or overlapping (b) optical bands. c) typical eigendisplacements for modes labelled in (b). Empty and filled circles denote the two types of atom in the bulk unit cell.

confined either in AB or in AC. For case drawn in Fig. 4b, the two bulk optical frequency ranges overlap. Outside the overlap region, the modes are again non dispersive and confined, they are otherwise dispersive and propagative. Typical eigendisplacements are also shown.

Although this model may seem rather naive, it provides a quantitative description of the experimental results,as will be shown later. More sophisticated calculations have been performed, using more realistic bulk lattice dynamics : short-range interactions in three dimensions leading also to the transverse modes (4) and furthermore, treatment of the long-range Coulomb forces with the bond charge model providing estimates of longitudinal - transverse splittings (5).

3. Simple notions on Raman scattering

A typical Raman scattering experiment consists in radiating a sample with polarized monochromatic light and analysing the scattered light as a function of its wavelength, polarization and direction. The macroscopic process involved is the following Fig. 5a : the incident photon creates a virtual electron hole pair which interacts with the lattice vibrations thus creating or annihilating a phonon. Then, the pair recombines emitting the scattered photon. A detailed balance of energy and momentum shows that the difference of energy (momentum) between the scattered and the incident photons equals the energy (momentum) of the created or annihilated phonon. In the following we shall deal only with the phonon creation (Stokes) process, which is the most efficient. The scattered photon energy is thus lower than the incident one. The most commonly used configuration for studying III-V semiconductors is the so-called backscattering one, where the incident and scattered lights propagate within the crystal in opposite directions perpendicular to the sample surface. The energy of the incident photon being much larger than the semiconductor gap, the light penetrates on less than one micron; this allows to study epitaxial layers without testing the substrate. Furthermore, taking advantage of the high value of the refractive index, one rather uses a Brewster incidence for the incoming light, thus spatially separating the scattered light from the reflected one. Within the sample, the incoming and detected light propagate approximately perpendicular to the sample surface (Fig. 5b). As the typical energy of an optical phonon (say 30 meV) is quite smaller than the incident photon one (say 2.5 eV), the modulus of the scattered and incident photons wavevector are essentially equal. Thus, in backscattering configuration, the wavevector Q of the involved phonon

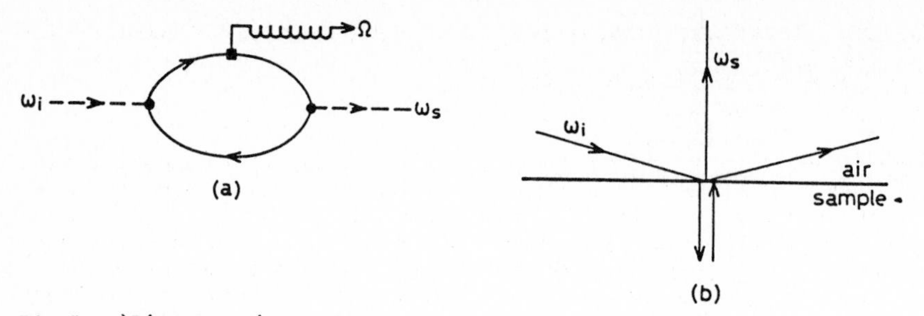

Fig.5 a) Diagrammatic representation of a Raman process.
Dashed line : photon. full line : electron or hole
Wavy line : phonon. full circle : electron-photon interaction. full square : electron-phonon interaction.
b) Schematic representation of the experimental Raman backscattering configuration.

takes the approximate value :

$$Q = \frac{4\pi n(\lambda)}{\lambda} \tag{7}$$

where $n(\lambda)$ is the refractive index at wavelength λ. Typical values for Q range around $10^6 cm^{-1}$ which is rather small compared with the bulk Brillouin zone extent ($\sim 10^8 cm^{-1}$). As a consequence, Raman scattering usually provides informations about zone center optical phonons.

From a macroscopic point of view, Raman scattering can be understood as a non linear process. The incident electric field E_j^{inc} and the normal mode m displacement Q_α^m create a polarization field P_l at frequency $\omega_{inc} \pm \omega_m$:

$$P_\ell = \sum_{j,\alpha} R^\alpha_{\ell j} \, Q^m_\alpha \, E^{inc}_j \tag{8}$$

where R^α_{lj} is the Raman tensor.

The scattered light is polarized parallel to this field. Relation (8) must be invariant relative to the point symmetry of the crystal, which imposes some selection rules on the relative polarizations of the incident and scattered beams. The actual selection rules for various experimental configurations can be found for instance in Ref. 6. To be specific, for a backscattering experiment on a (001) surface of a bulk III-V compound, the transverse optical (TO) modes are forbidden for any polarization configuration. The longitudinal optical (LO) mode is allowed for an incoming light polarized along x and a scattered light polarized along y ($z(x,y)\bar{z}$ configuration) and forbidden in $z(x,x)\bar{z}$ configuration. In these notations x,y,z denote the crystalline axis. For the case of a superlattice grown along [001], the point symmetry of the crystal is no longer cubic but becomes quadratic (D_{2d}). In the same backscattering configuration, TO modes remain forbidden. Concerning the $2(n_1 + n_2)-1$ LO modes, two cases must be distinguished :

i) A_1 modes, which are invariant relative to the 2-fold rotation axis perpendicular to the growth direction and are only allowed ni $z(x,x)\bar{z}$ configuration.
ii) B_2 modes, which are antisymmetric relative to the same axis and only allowed in $z(x,y)\bar{z}$ configuration.

Let us finally return to the comparison with the electronic case. Whereas any optical experiment involves electronic transitions between different quantum states, Raman scattering provides absolute values of phonon energies. As a consequence, for vibrational properties, the awful problem of band offset determination disappears.

4. Experiments

The first experimental studies of lattice dynamics of superlattices have been performed on short period GaAs/GaAlAs samples (7,8,9,10,11). They provide the first experimental evidence of the folding of the acoustical dispersion curves, either by identifying new zone center modes (11) or new gaps within the acoustical frequency range (9). At the same time, new lines in the optical frequency range were observed in resonant conditions, whose assignment remained controversial (8,10). These problems have been reexamined the two past years, leading to advances in the experimental study of the acoustical properties (12,13,14,15) and to the unambiguous evidence of optical modes confinement (14,16,17,18). These studies have been extended recently to other systems, like GaSb/AlSb (19), GaAs/InGaAs (20), CdTe/CdMnTe (21).

4.1 Folded acoustical modes

Typical Raman spectra in the low-frequency range, obtained on a few GaAs/GaAlAs superlattices, are shown in Fig. 6. New lines clearly appear, which do not exist for a bulk material, and are assigned to the folding of LA modes. The onset of such new lines in the low-frequency range is systematically observed in any superlattice whatever its period, ranging from 20 to 500 Å. One gets typically four new lines. In some samples (Fig. 7) we got up to nine lines, in resonant conditions. For not too large periods, the new lines appear as doublets (Fig. 6a)

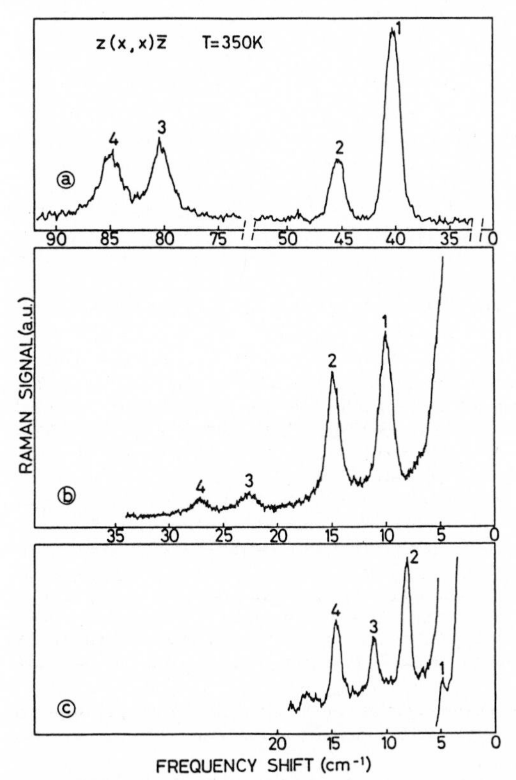

Fig.6 Raman spectra in the $z(x,x)\bar{z}$ configuration at low-frequency shift on (a) sample S1, (b) sample S2, (c) sample S3 (see Table I for parameters of the samples).

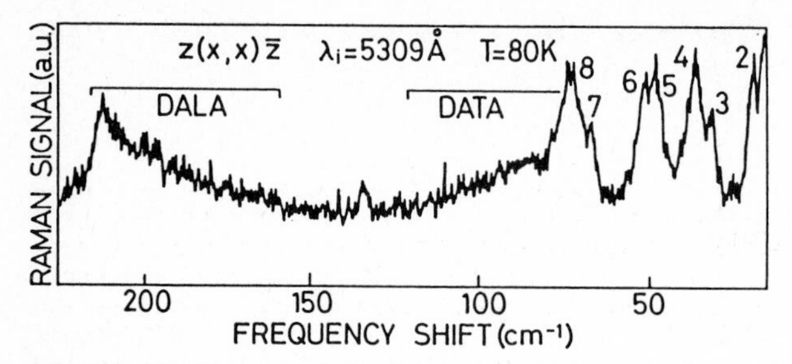

Fig.7 Raman spectrum on sample S4 (see Table I) in resonant conditions. Lines 2 to 8 are assigned to folded LA phonons and coexist with disorder activated bands (denoted DATA and DALA).

whose intensity, measured in non-resonant conditions, decreases as a function of their average frequency ω_{av}. Furthermore, the latter quantity fits the approximate relation :

$$\omega_{av} = v_s \frac{2\pi n}{d}$$

where n = 1,2,3,...and v_s is defined by (4).

Some independant information about this average sound velocity has been obtained by surface wave Brillouin scattering (12). Note that in GaAs/$Ga_{1-x}Al_xAs$ superlattices, the intensity of the folded lines increases with the Aluminum contents x. In other words, it is inasmuch larger as the two constituents are different.

Although the average frequency fits correctly the predictions, the calculated doublet splitting is rather too small. For large period samples, one cannot safely define any doublet (Fig.6c). This apparent discrepancy can, however, be overcome as soon as one realizes that the actual extent of the superlattice Brillouin zone becomes of the same order of magnitude as the wavevector of the created phonon. Consequently, the observed lines do not correspond to zone center modes. Taking in account this effect (see formula 7), we recover, in all cases, a good fit between experiment and Rytov's model predictions. From another point of view, the measurement of the folded acoustical frequencies provides a good determination of the period of the sample.

Moreover, for large period superlattices (and thus small Brillouin zones), one can take advantage of the dependence of the created phonon wavevector on the incident wavelength to determine experimental dispersion curves (13,14). Indeed, contrary to what happens in bulk materials, the Raman shifts vary for different incident laser lines (see Fig. 8). Although the sound velocity in AlAs is poorly known, one gets again a fairly good fit of the dispersion curves (see Fig. 9). Thus, due to the small extent of the Brillouin zone, photons can be used as a probe to get the superlattice dispersion curves, like neutrons are for bulk materials.

Note that for rather small period samples, one reaches the limits of Rytov's model, the average of the second doublet becoming slightly smaller than the linear elasticity predictions : one must take into account the curvature of the bulk acoustical dispersion curves. Note also that the Raman spectra for superlattices containing the alloy $Ga_{1-x}Al_xAs$ display, beside the narrow folded phonon lines, broad lines similar to those found (22) in the bulk alloy (Fig. 7). These broad lines, attributed to the influence of alloy disorder, are thus found to be unsensitive to the superperiodicity (13,14).

We have shown, up to now, spectra obtained in $z(x,x)\bar{z}$ configuration. In the same energy range, nothing is observed in the $z(x,y)\bar{z}$ one. This result is puzzling

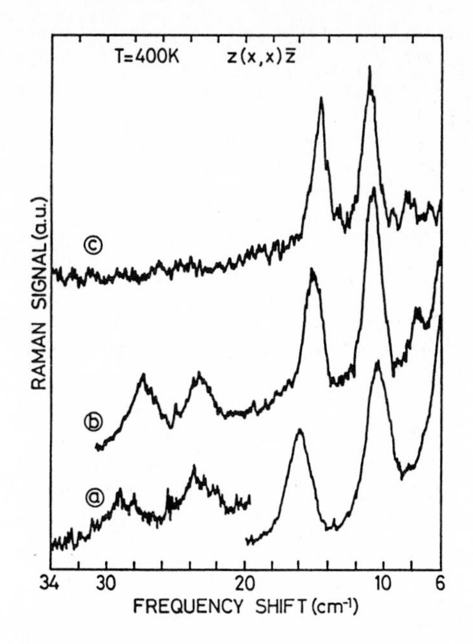

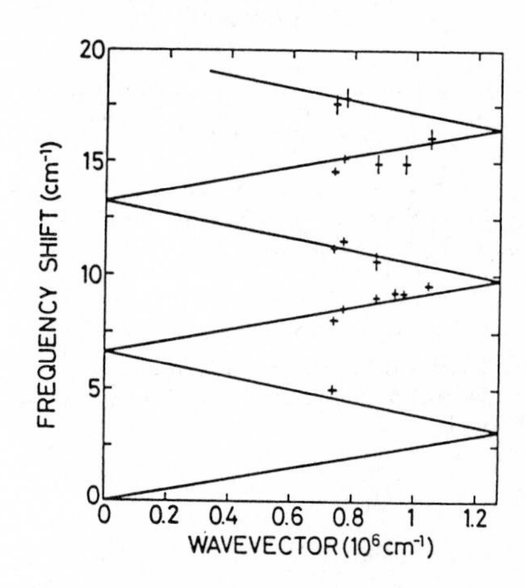

Fig.8 Raman spectra in $z(x,x)\bar{z}$ configuration on sample S2 (see Table I) at different incident wavelengths : (a) 4765 Å, (b) 5682 Å, (c) 6764 Å [From Ref.13].

Fig.9 Experimental frequencies of folded acoustical modes at different superlattice wavevector on sample S1 (see Table I) compared with the theoretical dispersion curves.

because a symmetry analysis of the eigenmodes of Rytov's model predicts that each doublet, at zone center, is made of a A_1 and a B_2 modes. As explained before, Raman selection rules would impose the observation of the A_1 mode only in $z(x,x)\bar{z}$ and of the B_2 mode only in $z(x,y)\bar{z}$. A tentative explanation of this discrepancy can be found in Ref. 12 : the scattering mechanism is understood as a Brillouin one, involving the photoelastic effect. The corresponding selection rules, different from the Raman ones, are then in agreement with the experiments.

4.2 Confined optical modes

4.2.1 GaAs/AlAs superlattices

Fig.10 displays Raman spectra obtained on a GaAs/AlAs superlattice in the frequency range of the GaAs bulk LO mode, for $z(x,y)\bar{z}$ (Fig.10b) and $z(x,x)\bar{z}$ (Fig.10c) configurations, compared to a bulk GaAs spectrum (Fig.10a). Once again new Raman lines appear for the superlattice, all of them being shifted towards smaller frequency as compared to the bulk one. The higher energy one, which is also the most intense one, suffers only a very slight shift. Let us outline that such a $z(x,x)\bar{z}$ spectrum can only be obtained in resonant conditions, non-resonant spectra never exhibiting any structure in this polarization configuration. The lines obtained for the two different configurations do not coincide. As the Raman shift decreases, one encounters alternatively (x,y) and (x,x) lines i.e., from selection rules, B_2 and A_1 modes. Furthermore,the spectrum in the AlAs LO mode frequency range displays two weak lines. These spectra can be qualitatively understood while comparing the LO dispersion curves of GaAs and AlAs. The two LO energy bands being separated by a gap of about 70 cm^{-1}, the vibrations remain confined in one compound or in the other one and the superlattice energy spectrum consists of two series of non-dispersive frequency levels. A quantitative analysis of the experiments can be made on the basis of the linear chain model, using effective linear chains for bulk materials. In the spirit of an effective medium approximation, the characteristics

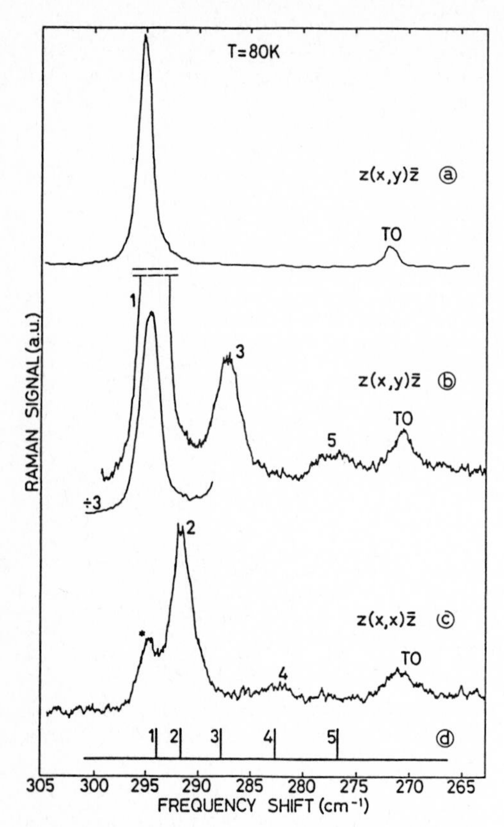

Fig.10 : Raman scattering spectra on pure GaAs (a) and on superlattice S5 ((b) and (c)) in the GaAs-like LO phonon frequency-range compared with the predicted frequencies (d). Modes 1, 3 and 5 appear in $z(x,y)\bar{z}$ configuration (a), modes 2 and 4 in $z(x,x)\bar{z}$ configuration (b). The line labelled by a star is a small forbidden contribution of mode 1.

(i.e. zone center frequency and curvature) of the bulk chains are deduced from the experimental LO dispersion curves. The results of such a computation are reproduced on Fig. 10d and favourably compare with experiment. A drawback of this calculation is to introduce effective atomic masses which are fairly different from the real ones. A best model would be to account for electronic long-range forces, as done in (5). This, however, implies three-dimensional lattice dynamics and heavy computations, but leads to equally good fits. Some attempts for a simple treatment of long-range interactions have been reported (10), using effective dielectric media and interface compatibility relations. Their predictions concerning the anisotropy of polar optical phonons in superlattices must be tested experimentally. The only experiment we know which studies phonons propagating sideways relative to the growth direction (23) have been performed in GaAs/GaAlAs structures, and give from our point of view rather puzzling results.

4.2.2 GaAs/GaAlAs superlattices

Lattice dynamics of this type of microstructure appears together more complex and more interesting. $Ga_{1-x}Al_xAs$ is an alloy exhibiting a two-mode bahaviour. The GaAs type optical frequency band in the mixed crystal partially overlapping the bulk GaAs one, confined and propagative optical modes coexist.

Let us first recall some general ideas on two-mode behaviour. For a bulk alloy like $Ga_{1-x}Al_xAs$, the Raman spectra always display, whatever the concentration x, two LO-TO pairs roughly located around the GaAs and AlAs optical frequencies. Actually the alloy frequencies depend on the Aluminium contents as shown in Fig.11. A careful study of the asymmetry of these lines (22) as well as Coherent Potential Approximation computation (24) show that, although the compound is disordered, one can still define thick dispersion curves whose curvature around zone center remains negative. The zone center GaAs type LO frequency within the alloy lies below the GaAs bulk one. Thus, as already explained in Part I, the electronic analogous would be a hole multiquantum well, GaAs (respectively $Ga_{1-x}Al_xAs$) acting here as a phonon well (respectively barrier).

Typical Raman spectra around the GaAs optical phonon Raman shift can be seen on Fig. 12. The scattering experiments were made on samples with x around 0.3 (see Table I). As in the GaAs/AlAs case, one gets a series of lines (1,3,5,7) at frequency shifts lower than the bulk GaAs LO one. The number of such lines and their frequency shifts decrease as the well thickness decreases. Furthermore, the lowest energy line, denoted P, corresponds to an almost constant frequency shift whose fluctuations seem uncorrelated with the behaviour of the other lines. This rather complex feature can be understood on the basis of the effective linear chain model : we fit the force constant and masses for GaAs as done previously; concerning $Ga_{1-x}Al_xAs$, we drop the confined AlAs like modes and consider the alloy as an effective ordered compound mimicking the GaAs-type mode behaviour. We drew on

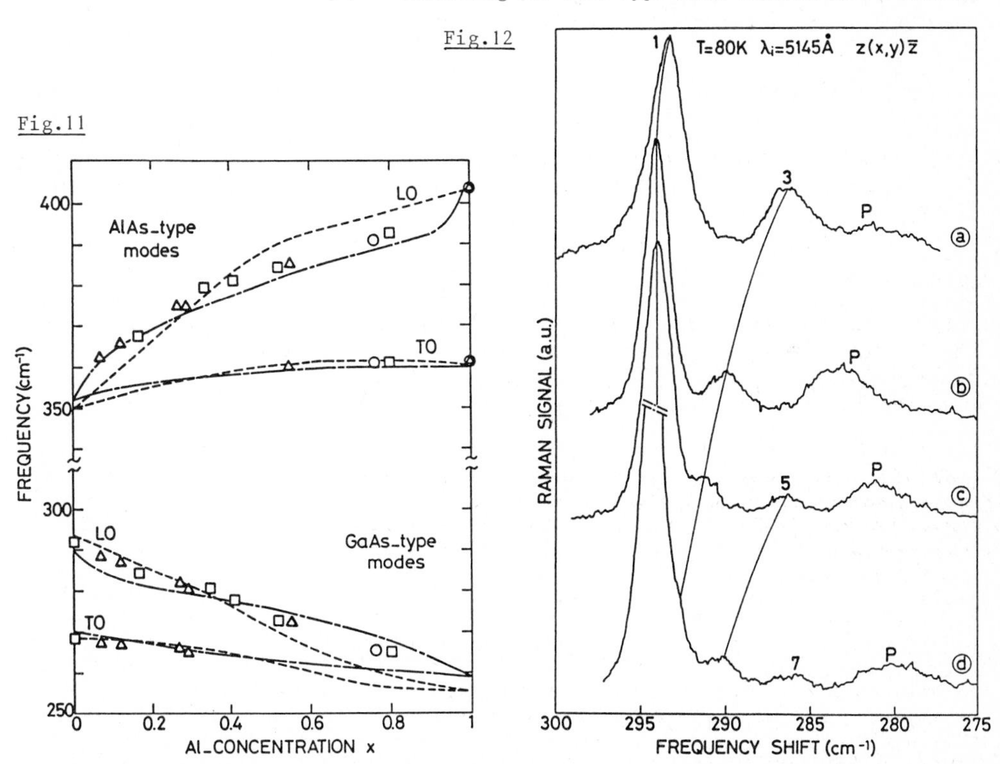

Fig.11 Composition dependence of the zone center optical frequencies in $Ga_{1-x}Al_xAs$. Squares, circles, triangles and the dashed lines denote experimental results. Mixed lines denote calculated ones (see Ref. 22).

Fig.12 Raman spectra in the GaAs optical frequency range on samples (a) S6, (b) S7, (c) S8, (d) S9. The line indexes are defined in the text and the samples described in Table I.

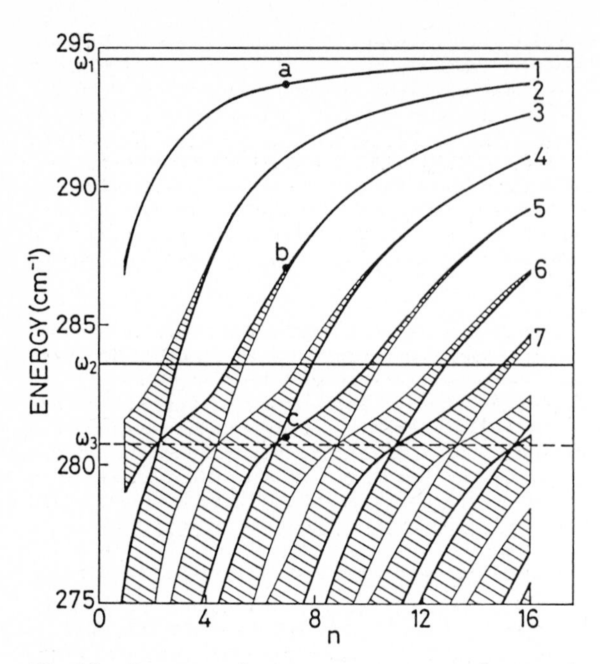

Fig.13 Energy of the zone center (thick lines) and zone edge (thin lines) superlattice eigenmodes as a function of the well width n. The hatched regions correspond to the allowed energy bands. The parameters of the calculation, the labels 1 to 7, ω_1, ω_2, ω_3 are defined in the text ; a,b,c refer to Fig. 14. The calculation making sense only for integer values of n, continuous lines are only a guide for the eyes.

Fig. 13 the allowed energy bands as a function of the number n of unit cells in the well, assuming this number to be constant and equal to 5 within the barrier. We chose a typical concentration for the alloy leading to a zone center LO frequency ω_2 = 283.6cm^{-1}. The corresponding GaAs frequency ω_1 amounts to 294.6cm^{-1} thus leading to a well depth of 11cm^{-1}. In the well (i.e. between ω_1 and ω_2) the phonon branches are almost undispersive and their energy decreases as the well width decreases. At zone center these modes display alternating B_2 like (modes 1,3,5,...) and A_1 like (modes 2,4,6,...) symmetry, if one refers to the three-dimensional point group D_{2d}. As the energy decreases towards ω_2, they smoothly gain some dispersion and continuously transform, around ω_2, into dispersive energy branches. More precisely, one can define a frequency ω_3, smaller than ω_2, where all the branches "anticross". Above ω_3 the dispersion curves come from a small coupling between the wells. Below ω_3 the states are scattering states separated by minigaps induced by the superperiodicity perturbations. In the same way the eigenmodes smoothly evolve from confined to propagative. Fig. 14 displays the eigendisplacements of the common atom for a confined (14a) an intermediate (14b) and a propagative (14c) case. To estimate the influence of the superlattice parameters on the zone center frequencies, we varied the depth of the well and the thickness of the barrier. As long as the considered modes frequency is far from ω_3, they hardly depend on the thickness and slightly on the well depth. Coming back to the experiments, we assign lines 1,3,5,7 (B_2 symmetry) to confined states and line P to the higher energy zone center propagative one. The whole experimental data (except P lines) are gathered on Fig. 15 as a function of the GaAs layer thickness and compared with two calculations using two extreme values of ω_2. A rather good fit is obtained.

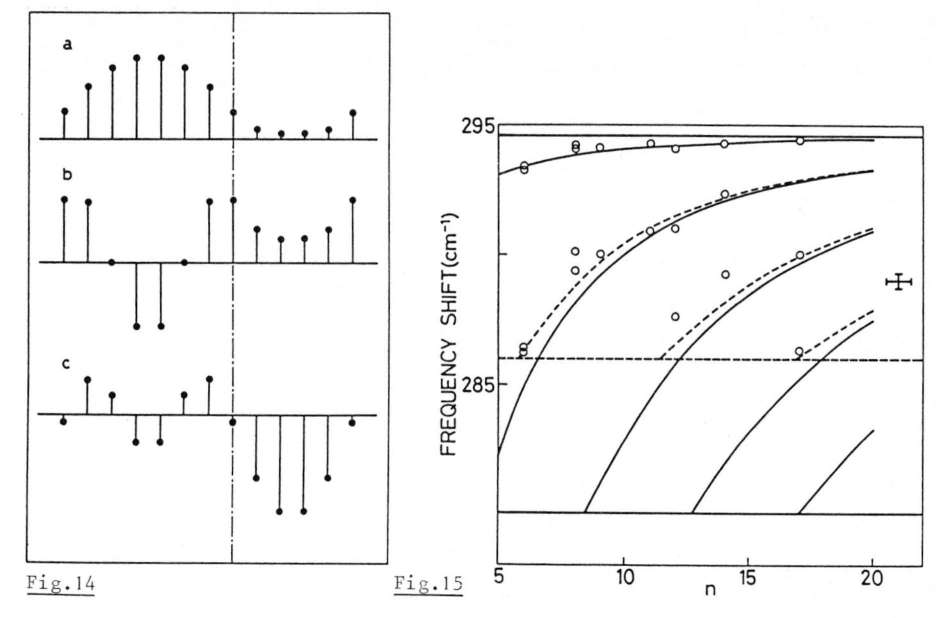

Fig.14 Eigendisplacement of the common atom in the superlattice unit cell for three zone center B_2 modes defined by the same labels a, b, c in Fig. 13.

Fig.15 Experimental frequency shifts (circles) within the GaAs optical mode frequency range, for a series of GaAs/GaAlAs superlattices compared as a function of the well width n with two calculations : $\omega_2 = 280\ cm^{-1}$ (solid line); $\omega_2 = 286\ cm^{-1}$ (dashed line). [From Ref. 16].

4.2.3. An estimate of the Raman tensor

An exact calculation of the Raman intensities would need a precise estimate of the microscopic process sketched in Fig. 5a, which implies a knowledge of the electronic wave functions and of the electron-phonon and electron-photon matrix elements. We shall here use the semi-classical approach (25) introduced for superlattices in Ref.7. Each covalent tetrahedral bond is assumed to be polarizable independently of the others and characterized by its polarizability along the bond $\alpha_{\shortparallel}$ and perpendicular to it $\alpha_{\perp}$. Furthermore, these two polarizabilities vary with the bond length l with two coefficients $d\alpha_{\shortparallel}/dl$ and $d\alpha_{\perp}/dl$. Let us consider only a longitudinal vibration and an electromagnetic wave $\vec{E}$ both propagating along z. The electric polarization $\vec{P}$ at Stokes frequency is obtained by adding the contribution of all the bonds. This calculation leads to :

$$\begin{pmatrix} P_x \\ P_y \end{pmatrix} = \begin{pmatrix} \sqrt{R_{xx}} & \sqrt{R_{xy}} \\ \sqrt{R_{xy}} & \sqrt{R_{xx}} \end{pmatrix} \cdot \begin{pmatrix} E_x \\ E_y \end{pmatrix}$$

R_{xx} (respectively R_{xy}) is the Raman intensity per bulk unit cell in $z(x,x)\bar{z}$ (respectively $z(x,y)\bar{z}$ configuration. **Using notation of Fig. 3c one gets :**

$$R_{xx} = \frac{1}{n_1 + n_2} \left[\sum_{i=-n_1}^{n_2-1} \left\{ a^{(j)} (u_i^{(j)} - v_{i-1}^{(j)}) - a^{(j')} (u_i^{(j')} - v_i^{(j')}) \right\} \right]^2 \tag{9}$$

$$R_{xy} = \frac{1}{n_1 + n_2} \left[\sum_{i=-n_1}^{n_2-1} \left\{ b^{(j)} (u_i^{(j)} - v_{i-1}^{(j)}) + b^{(j')} (u_i^{(j')} - v_{i-1}^{(j')}) \right\} \right]^2$$

In these formulas

i) the displacements are normalized as :

$$\sum_{i=-n_1}^{n_2-1} (u_i^{(j)2} + v_i^{(j)2} = 1$$

ii) $a^{(j)}$ and $b^{(j)}$ are linear combinations of bond characteristics in compound (j).
iii) the electric field has been assumed to be constant within a supercell.

In agreement with zone center symmetry requirements, R_{xx} appears to vanish for B_2 modes as does R_{xy} for A_1 modes. Furthermore for A_2 modes, R_{xx} writes :

$$R_{xx} = \frac{4 u_o^2 (a^{(1)} - a^{(2)})^2}{n_1 + n_2}$$

the only contribution coming from the interface atoms. Consequently the corresponding Raman intensity remains small, which explains the usual non observation of new lines in $z(x,x)\bar{z}$ configuration. On Fig. 16 is drawn R_{xy} for some B_2 modes as a function of the well thickness, assuming for convenience $b^{(1)} = b^{(2)}$.

Formula (9) shows that the Raman intensity comes from interferences between amplitudes emitted by each bulk unit cell. It reaches thus a large value only when the local displacements within a given layer are in phase, i.e. when one of the local wavevector modulus $k_{1,2}$ is small. Consequently the Raman intensity remains always large for mode 1 and comes uniquely from the GaAs layer ($k_1 \sim 0$). Concerning

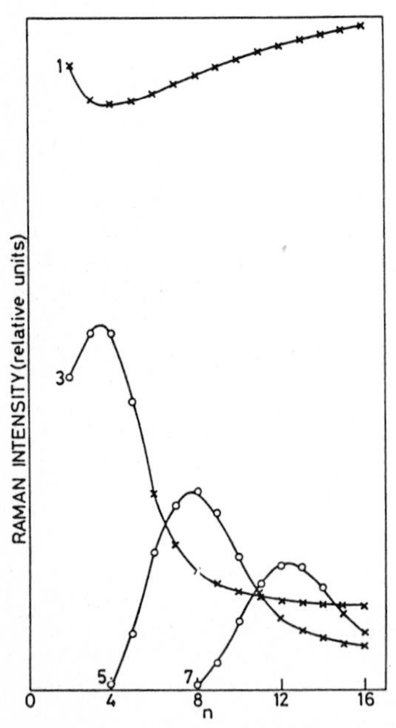

Fig.16 Calculated Raman intensity R_{xy} as a function of the well width n for four B_2 zone center eigenmodes already labelled in Fig. 14. The calculated values are represented by crosses or circles whether the related mode frequency lies above or below ω_2.

Table 1 : Parameters of the studied samples.

	d_1	x_1	d_2	x_2
S1	26.	0.	14.	1
S2	56.	0.	78.	0.77
S3	150.	0.	107.	0.36
S4	50.	0.5	50.	1.
S5	20.	0.	20.	1.
S6	17.	0.	12.	0.3
S7	25.5	0.	25.5	0.29
S8	34.	0.	20.5	0.37
S9	48.	0.	34.	0.35

the other modes, they reach their maximum intensity when k_2 is small, i.e. for frequencies near ω_2, the Raman intensity coming then from the GaAlAs layer. This describes correctly the experimental trends.

5. Conclusion

We hope to have shown the strong similarities between electrons and vibrations in superlattices. In these structures, two types of vibrations coexist : confined ones and propagative ones. In the most representative case, when the optical frequency bands of the two constituents overlap, one of the layers acts as an "optical phonon quantum well", leading to the onset of new non-dispersive modes. In the overlap frequency range, new propagative modes appear. On the other hand, the superperiodicity induces the folding of the acoustical branches, thus creating new zone center "optical modes" of acoustical origin. These new features have been evidenced by Raman scattering experiments and the variation of their frequency as a function of the superlattice parameters are well described by one-dimensional models.

Unfortunately, little experimental information is up to now available on transverse vibrations and on phonons propagating sideways. Such information would allow to tackle the problem of long-range Coulomb forces in superlattices. Another future experimental development would be the study of the resonant Raman cross section,which must reflect the cross-confinement of electrons and vibrations, a relevant point to understand transport properties (26).

Finally, let us outline that Raman scattering is a powerful technique to determine the structural parameters : the acoustical folding provides the period of the superlattice, the frequencies of the confined optical modes provide the thickness of the well. Furthermore,the confined modes being very sensitive to limiting conditions, they can be used as a probe of interdiffusion profiles at the interfaces (27).

Acknowledgments

We thank A. Regreny and F. Alexandre for providing us with high quality samples, P. Auvray and G. Le Roux for skilful X-ray characterization and J.Y. Marzin and J. Sapriel for fruitful discussions.

References

1. S.M. Rytov, Akust. Zh.: 2, 71 (1956) Sov. Phys. Acoust. 2, 68 (1956)
2. H. Shi and C.H. Tsai: Solid State Comm. 52, 953 (1984). R.E. Camley and D.L. Mills, Phys. Rev. B29, 1695 (1984)
3. R. Tsu and S.S. Jha : Appl. Phys. Lett. 20, 16 (1972)
4. L. Dobrzynski, B. Djafari-Rouhani and O. Hardouin-Duparc: Phys. Rev. B29, 3138 (1984)
5. S. Yip and Y.C. Chang: Phys. Rev. B30, 7037 (1984)
6. M. Cardona in Light Scattering in Solids II, edited by M. Cardona and G. Güntherodt (Springer Verlag, Berlin Heidelberg New York, 1982), p.49
7. A.S. Barker Jr., J.L. Merz and A.G. Gossard: Phys. Rev. B17, 3181 (1978)
8. G.A. Sai Halasz, A. Pinczuk, P.Y. Yu and L. Esaki : Solid State Comm. 25, 381 (1978)
9. V. Narayamurti, H.L. Störmer, M.A. Chin, A.C. Gossard and W. Wiegmann: Phys. Rev. Lett. 43, 2012 (1979)
10. R. Merlin, C. Colvard, M.V. Klein, H. Morkoç, A.Y. Cho and A.C. Gossard: Appl. Phys. Lett. 36, 43 (1980)
11. C. Colvard, R. Merlin, M.V. Klein and A.C. Gossard: Phys. Rev. Lett. 45, 298 (1980)
12. J. Sapriel, J.C. Michel, J.C. Tolédano, R. Vacher, J. Kervarec and A. Regrény: Phys. Rev. B28, 2007 (1983)
13. B. Jusserand, D. Paquet, A. Regrény and J. Kervarec: Solid State Comm. 48, 499 (1983)
14. B. Jusserand, D. Paquet, A. Regrény and J. Kervarec: J. Phys. (Paris) C5, 145 (1984)
15. J. Sapriel and J.C. Michel: Superlattices and Microstructures 1, (1985)
16. B. Jusserand, D. Paquet and A. Regrény: Phys. Rev. B30, 6245 (1984)
17. B. Jusserand, D. Paquet and A. Regrény: Superlattices and Microstructures 1, 61 (1985)
18. C. Colvard, R. Fischer, T.A. Gant, M.V. Klein, H. Morkoç and A.C. Gossard, Superlattices and Microstructures 1, 81 (1985)
19. B. Jusserand, P. Voisin, M. Voos, L.L. Chang, E.E. Mendez and L. Esaki : Appl. Phys. Lett
20. M. Nakayama, K. Kubota, H. Kato and N. Sano: Solid State Comm. 51, 343 (1984)
21. S. Venugopalan, L.A. Kolodziejski, R.L. Gunshor and A.K. Ramdas: Appl. Phys. Lett. 45, 974 (1984)
22. B. Jusserand and J. Sapriel: Phys. Rev. B24, 7194 (1981)
23. J.E. Zucker, A. Pinczuk, D.S. Chemla, A. Gossard and W. Wiegmann: Phys. Rev. Lett. 53, 1280 (1984)
24. B. Jusserand, D. Paquet and K. Kunc: in Proceedings of the 17th International Conference on Physics of Semiconductors, San Francisco 1984 (to be published)
25. G. Placzek: in Marx Handbuch der Radiologie (2d Ed.) vol. 6 (Akademische Verlagsgesellschaft, Leipzig, 1934), p. 209
26. J.F. Palmier: lecture in this school
27. B. Jusserand: (to be published)

Electrical Transport Perpendicular to Layers in Superlattices

J.F. Palmier

Centre National d'Etudes des Télécommunications, 196, rue de Paris, F-92220 Bagneux, France

Fifteen years ago, Esaki proposed to build artificial semiconductor superlattices to obtain new transport properties. In Esaki's pioneering work [1-2], electrons in the superlattice were supposed to be completely delocalized. The main purpose was to obtain oscillatory features with Bragg reflections at electron minizone boundary, i.e. at π/d in the kz direction, d being the super-period. Although the pure Bloch oscillation in this system has been discussed by Zak [3], the existence of extended states in the growth direction is presently a relevant question, since many future devices may use electrical transport perpendicular to these superlattices. The present paper starts from the most ideal situation on theoretical aspects, and ends with experimental results and discussion. The first part deals with the near equilibrium mobility as limited by polar optical phonons, acoustical phonons, and epitaxial irregularities. Two different approaches of that mobility are made : either extended state diffusive transport in ideal superlattices, or hopping, i.e. phonon-assisted tunneling from layer to layer considering one or a finite number of periods. The next part considers the problem of limited thickness for real epitaxial layers. Since the superlattice is imbedded with other contact layers, a careful analysis has to be made based upon an effective medium theory. These considerations will help to extract "debugged" (as far as possible) information giving microscopic parameters of the superlattice. The main purpose is to measure the mobility in the growth direction ; this parameter may be obtained from two different types of information : the current-voltage variation and the photocurrent response to short optical pulses. These experimental methods will be described in the last part and comparison of experiment and theories will be the final conclusion.

1. Microscopic description

Several simplified theories can predict the rough behaviour of electron or hole states in the superlattice system. Two of the best known are the Kronig-Penney model [4]-[7], and the envelope theory based upon the Kane band theory of III-V compounds as given by Bastard [8] [9]. In this course, for the sake of simplicity, we shall limit our treatment to the plane wave expansion of the envelope wavefunction. Our results may be generalized to include Bastard's theory. A weakness of the model is that it is limited to materials for the barrier and for the well of comparable effective masses. So we shall restrict our treatment to electron conduction, since for electrons this restriction is more easily satisfied. Moreover, we consider a single band, i.e. we suppose a relatively narrow well, and a distribution function corresponding to a low occupation of the second miniband when it exists. We shall mainly discuss the near-equilibrium transport along the growth direction z. Two different situations are possible : either the Bloch propagation of carriers allows for the classical diffusive transport theory via the Boltzmann equation, or localized states dominate, even if the localization is only restricted to the z direction (pure 2D systems with weak coupling). This latter theory must be the limit obtained for rather large periods, or more precisely, large barrier widths.

1.1. Diffusive transport of Bloch electrons [11]

The Boltzmann equation is the basic equation for most transport studies in semiconductors. It can be used for the derivation of either linear or nonlinear transport parameters, approximate expressions being valid in the first case. As we wish to obtain tractable expressions for transport parameters, a simplified model of superlattice band structure is used. The total one-electron Hamiltonian is :

$$H = H_o + H_1 + H_F = \frac{p^2}{2m^*} + V_{SL} + H_1 + H_F \tag{1}$$

in which :

- $p^2/2m^*$ is the kinetic energy term, m^* being an averaged effective mass. While oversimplified with respect to correct expressions, it allows quantitative further discussions.

- V_{SL} is a superlattice potential which takes into account the conduction band discontinuities (generally reduced to a crenel-like function).

- H_1 is the sum of all the perturbation Hamiltonians as electron-phonon terms, electron-impurities, potential irregularities etc ...

- H_F is the field term i.e. $-e\vec{E}.\vec{r}$. It will be useful to expand V_{SL} into Fourier series :

$$V_{SL} = \sum_G V_G \, e^{iGz} \tag{2}$$

That expansion and the very simple form of H allow for the wavefunction expression :

$$\psi(\vec{k}) = \phi_o(\vec{k}_\perp) \sum_G C_G \, e^{ik_2 z} \tag{3}$$

The eigenvalues and eigenstates of (3) with the unperturbed Hamiltonian then lead to minibands of width 2Δ and a set of C_G coefficients. It can be verified that 5 to 7 plane waves are sufficient to obtain a result close to the Kronig-Penney model to within 1%. To derive the near-equilibrium mobility tensor perturbation methods are efficient. Starting from :

$$e\frac{\vec{E}}{\hbar}.\vec{\nabla}_k f(\vec{k}) = \sum_{k'\alpha} W_\alpha(\vec{k}'\to\vec{k}') f(\vec{k}') - W_\alpha(\vec{k}\to\vec{k}') f(\vec{k}) \tag{4}$$

in which k is the electron momentum, E the applied electric field, e the electron charge, $\hbar$ the Planck constant, $W_\alpha(k\to k')$ the transition probability per unit time of the α-type collision. General solutions of (1) can be obtained considering the Legendre expansion of the electron distribution function $f(\vec{k})$:

$$f(\vec{k}) = \sum_n f_n(k) P_n(\cos\theta) = f_o(k) + f_1(k)\cos\theta + \ldots \tag{5}$$

in which θ is the angle $(\vec{k},\vec{E})$. Near-equilibrium transport requires only two terms of that expansion : the zero order term, which may be supposed close to the Fermi-

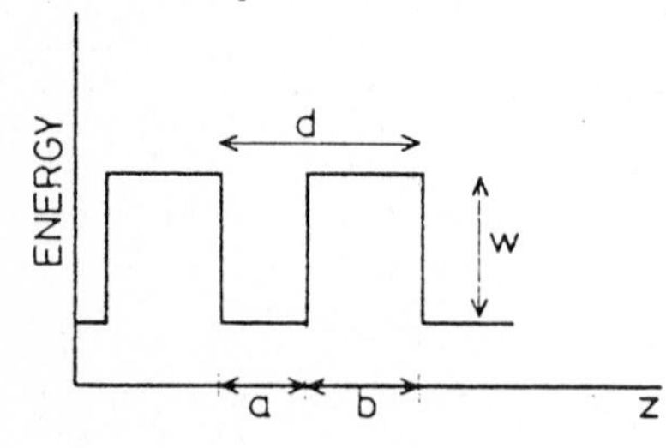

Fig.1 Basic cell and parameters of the superlattice.

Dirac ($\overline{f}$) solution at equilibrium, and a term proportional to E. To go further it is essential to know the expressions W in the collision terms. We make use of the Fermi golden rule :

$$W_\alpha(\vec{k}\rightarrow\vec{k}') = \frac{2\pi}{\hbar} \mid <\vec{k}\,|H_\alpha|\,\vec{k}'>|^2 \; f(\varepsilon'_{tot} - \varepsilon_{tot}) \tag{6}$$

The term $\langle\vec{k}|Hi|\vec{k}'\rangle$ lead to terms like $\langle\vec{k}|\exp(i\vec{q}.\vec{r})|\vec{k}'\rangle$ which reduce to selection rules as

$$|<\vec{k}\,|\,e^{\pm i\vec{q}\vec{r}}\,|\vec{k}'>|^2 = f(\vec{k}' - \vec{k} \pm \vec{q}) \tag{7}$$

in usual semiconductor cases. Here the situation is somewhat different, and we make use of (3). Meanwhile the selection rule (7) becomes :

$$|<\vec{k}\,|\,e^{iqr}\,|\,\vec{k}'>|^2 = I^{\pm}(\vec{k},\vec{k}',\vec{q}) = \delta(\vec{k}'_\perp - \vec{k}_\perp \pm \vec{q}_\perp) \sum_G \Big(\sum_{G'} C^*_{G'-G}\, C_{G'}\Big)^2 \delta(k'_2 - k_2 \pm q_2 - G) \tag{8}$$

in which Gn are simply given by 2.n.π/d, d being the superlattice period. We are now in position to look at solutions of (4) near equilibrium. The final density of states term in (6) is generally given by a simple expression as :

$$\delta(U^{\pm}) = \delta(\varepsilon(\vec{k}') - \varepsilon(\vec{k}) \pm \hbar\omega(\vec{q})) \tag{9}$$

Two classes of problems can be considered : elastic collisions and inelastic ones.

For elastic collisions, the near-equilibrium solution (4) limited to the two first terms of (5) is :

$$f(\vec{k}) \simeq \overline{f}(\varepsilon) - \frac{e\vec{E}}{\hbar} \cdot \vec{\nabla}_k \overline{f}\, \tau_R(\vec{k}) \tag{10}$$

$$\tau_R^{-1} = \sum_{k'} W(\vec{k}\rightarrow\vec{k}')\,(1 - \cos(\vec{k},\vec{k}'))$$

Then, we have all the linear transport parameters via the distribution function f(k). For example, the mobility tensor is :

$$<v>_i = \mu_{ij}\, E_j = \frac{1}{\hbar} \int \frac{\partial\varepsilon}{\partial k_i} f(\vec{k})\, d^3\vec{k} \tag{11}$$

Now, if we consider the tetragonal point group of the superlattice, only two terms are non zero : $\mu_{xx} = \mu_{yy}$ and μ_{zz}. We also expect that at rather large d values $\mu_{xx} \gg \mu_{zz}$ and close to the bulk value in the well.

For inelastic collisions, the near-equilibrium distribution function cannot be relaxed to some isotropic part. If we chose the polar optical phonon scattering, which is the main scattering process in AlGaAs and other III-V materials at room temperature, the formula (10) is still valid except that the derivation of τ_R is a little more complicated. However, for an order of magnitude estimate, the very simple τ can be used :

$$\tau^{-1} = \sum_{k'} W(\vec{k}\rightarrow\vec{k}') \tag{12}$$

For polar optical phonon scattering, the Fröhlich Hamiltonian will be used. This is a critical point in our approach, and a precise analysis would require to build a new Hamiltonian as done by Lassnig [10]. However, we deal with the unmodified

Hamiltonian in the present course. The two main transition probabilities we shall treat are :

- Polar optical phonons :

$$W^{\pm}_{PO} = \frac{\pi}{\hbar V} \sum_q \left(N_q + \frac{1}{2} \pm \frac{1}{2}\right) \frac{e^2 \hbar \omega_o}{q^2} \left(\frac{1}{K_\infty} - \frac{1}{K_o}\right) \delta(U^{\pm}) I^{\pm}(\vec{k}', \vec{k}, \vec{q}) \tag{13}$$

- Acoustical phonons (deformation potential) :

$$W^{\pm}_{a} = \frac{2\pi}{\hbar V} \sum_q \left(N_q + \frac{1}{2} \pm \frac{1}{2}\right) \frac{E_1^2 \, h\, \omega(q)}{2\rho\, s^2} \delta(U^{\pm}) I^{\pm}(\vec{k}, \vec{k}, \vec{q}) \tag{14}$$

The derivation of the electron mobility by (10 - 14) is then a question of integration techniques, the reader being able to find the details in appendix A. Meanwhile, we focus on the main differences between τ values in superlattices with respect to bulk values.

a) <u>Acoustical phonons</u>

From (11-14) with the technique given in A we find :

$$\frac{1}{\tau_a} = \frac{E_1^2 \, m^* \, k_B T}{2\pi \rho\, s^2 \hbar^3} \int_{-k_1}^{k_1} \frac{I^+ + I^-}{2} \, dk'_z \tag{15}$$

in which $I = \sum (An). \delta(k'-k+-q-Gn)$. It is really essential to consider the Umklapp term I in (15). Suppose that we neglect it ; let us recall the 3D value for τ_a :

$$\tau_a^{3D} = \frac{\pi \rho\, s^2 \hbar^4 \varepsilon^{1/2}}{\sqrt{2}\, E_1^2 \, k_B T \, (m^*)^{3/2}} \tag{16}$$

So, the comparison of (15) and (16) would give :

$$\frac{\tau_a^{2D}}{\tau_a^{3D}} \overset{?}{=} \frac{k\, d}{\pi} \tag{17}$$

This would imply much longer relaxation times for the superlattice with respect to the 3D case. For a single well of large dimension this is not possible. As shown in fig. 2, we can either consider the sum of U-processes as in (17) or modify the interaction Hamiltonian in (14) to work with the folded phonons [12] [13], and a multiple scattering possibility at a given k in the mini-zone. Another interesting limit is the small period superlattice ; it can be considered as a material of tetragonal symmetry with a conduction band minimum having two effective masses : m* for (kx,ky) directions, and M along kz. The following formula is found immediately from (15) :

$$\tau_a(d \to 0) \to \frac{\pi \rho\, s^2 \hbar^4 \varepsilon^{-1/2}}{\sqrt{2}\, E_1^2 \, k_B T \, m^* \sqrt{M}} \tag{18}$$

which reduces to (16) when m*=M. These results can be compared with calculations of Hess [14] or Vinter [24], for two-dimensional systems.

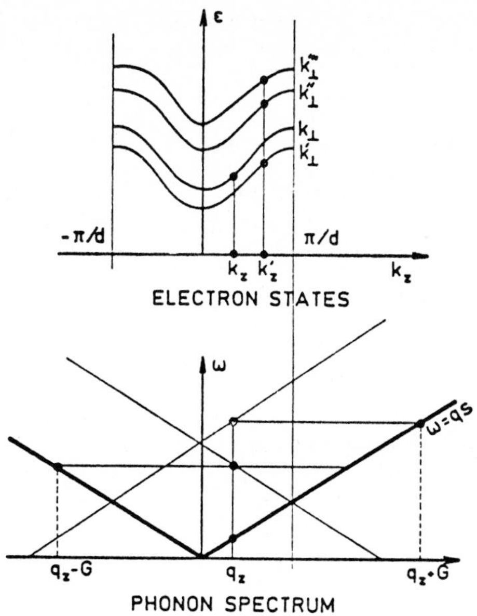

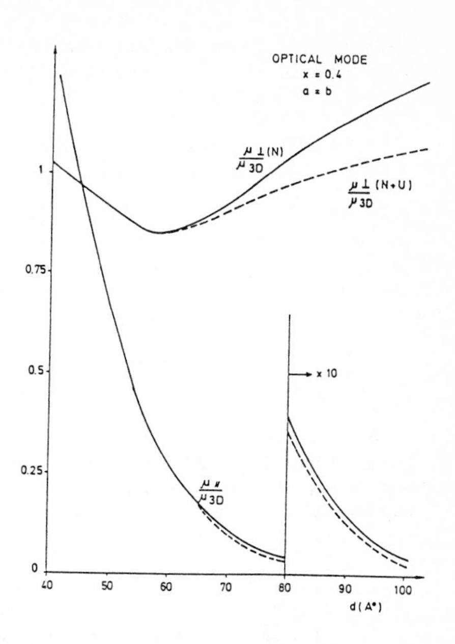

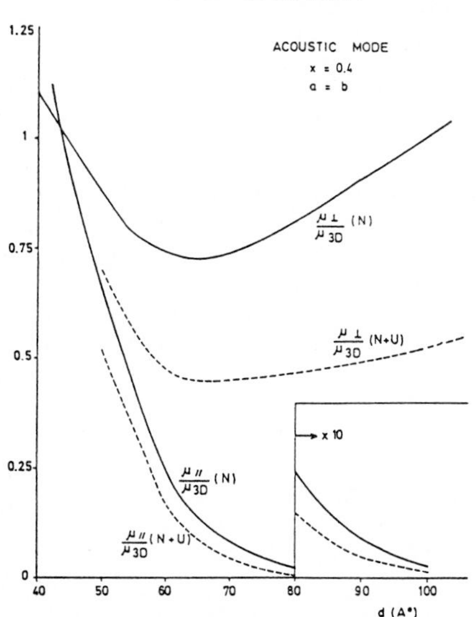

Fig.2 Equivalence between U-processes considered in the text and scattering with folded phonon branches. Note that k" corresponds to k'z = kz + qz -G and k"' to k'z = kz + qz + G.

Fig.3 Near-equilibrium mobility values (acoustic modes) versus d assuming a = b. U-processes are somewhat important.

Fig.4 Mobility limited by polar-optical modes at room temperature, for an AlGaAs superlattice versus d, with a = b.

b) Polar optical phonon modes

Following integration details given in Appendix A, we find :

$$\frac{1}{\tau_{PO}} = \frac{\gamma_1}{k_\perp} \sum_n [(N_o + 1) J_n^+ + N_o J_n^-] \tag{19}$$

in which J+ and J- are integrals over kz and γ_1 a numerical constant depending on material constants. Variations of the low field mobilities are drawn in fig. 3 and 4. Now, looking at the quantitative predictions made by that derivation, the mobility in the growth direction decreases exponentially with d, but stays in reasonable values when d is less than 60 A. This fact may be of great importance in future developments, when almost-perfect superlattices may be successfully grown.

We also expect that the barrier thickness has a major contribution in the variation with d. Finally let us discuss the relative contribution to the mobility limitation by growth irregularities.

Interface defects due to irregular growth can contribute to limit the mobility. A simple modelling of the near equilibrium mobility contribution of a particular type of defect has been done in ref. [15]. It consisted in cylindrical constant potential areas of width η and diameter D, oriented with their axis parallel to z. These defects introduce random variations of the perfect-superlattice potential. A Born approximation was made to calculate the relaxation time due to collisions with these defects. Using perturbation matrix elements such as (6), with H = +-W in the cylinder and 0 elsewhere, elastic collisions allow for the following expression :

$$\tau_R^{-1} = \frac{V_d}{V} \frac{W^2 m^* \eta}{\pi \hbar^3} \int_{-k_1}^{k_1} |B|^2 dk'_z \int_0^{2\pi} \frac{J_1^2(qR)}{q^2} (1 - \cos(k,k'))\, d\theta' \tag{20}$$

in which V_d/V is the ratio the total volume occupied by defects to the sample volume, $|B|$ a square matrix element such as :

$$|B|^2 \sim |< \Psi_z(k_z)\; e^{iqz}\; |\Psi_z(k'_z)>|^2 \tag{21}$$

$$(q = k_z - k'_z)$$

J1 is a Bessel function. Applied to the classical AlGaAs system with typical parameters, the numerical application of (20) shows that :

- At room temperature the mobility is as well limited by phonons as by the defects if their concentration exceeds 5%.

- At low temperature it may be important even at low defect concentrations.

In the above discussion it was assumed the carrier states were not affected by disorder (in which case the Born approximation is valid). Now, drastic changes may occur if the potential disorder is sufficient to give strong localisation in the wells. Completely different results for the mobility could be obtained, as will be shown hereafter.

1.2. Phonon-assisted tunneling (Hopping)

The main criticism one can make against the preceding approach is that the potential fluctuations could prevent Bloch propagation. We do not intend to discuss here the conditions which could create a system of localized electrons in layers, but only describe its consequences on the perpendicular transport.

TSU and DOHLER have proposed a hopping theory [16] to modelize the high field transport between Stark subbands in superlattices. Now, in the low field case, CALECKI, PALMIER and CHOMETTE have proposed a similar approach based upon a phonon assisted tunneling [17]. We always start from the same electron-phonon scattering perturbation Hamiltonian as in part 1, but strongly modify the electron state system. Considering that basic electron states are localized in the wells of the superlattice, the basic cell for transport is drawn in fig. 5. Electron states are still the product of two-dimensional Bloch states in xy directions and localized quantum well states Rn and Ln. A very simple and quite rigorous calculation of the current density can be done : let us put a surface S between L and R. We can count upstream and downstream electrons going from L to R and from R to L. The current density is thus :

$$J_Z = \frac{-e}{S} \left(\frac{\text{number of el } (L \to R)}{\text{unit time}} - \frac{(R \to L)}{\text{u.t.}} \right) \tag{22}$$

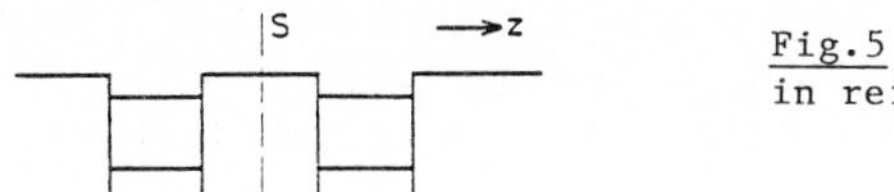

Fig.5 System of two wells described in ref. [7]

or, in other words :

$$J_Z = \frac{-e}{S} \sum_{i \neq j} W(iL \to jR)\, f(iL)(1 - f(jR)) - W(jR \to iL)\, f(jR)(1 - f(iL)) \tag{23}$$

in which $W(iL \to jR)$ are elementary transition probabilities which we can decompose over the phonon spectrum. It may seem curious to consider localized electrons and completely delocalized phonons, but great modifications to the perturbation Hamiltonians are not in the present possibilities. So the transitions can be obtained :

$$W(iL \to jR) = W((\vec{k}_\perp, n, L) \to (\vec{k}'_\perp, n', R)) \tag{24}$$

in which $\vec{k}_\perp$ and $\vec{k}'_\perp$ are the transverse electron wavevectors and n, n' quantum numbers of the well states. Equation (21) has been rigorously demonstrated in the present case [17]. At low applied electric fields, the development of the distribution function f(k,Rn) leads to:

$$J_Z = \frac{n_s e^2 d \langle W \rangle}{k_B T} E \tag{25}$$

with

$$\langle W \rangle = k_B T \frac{\sum_{k'_\perp k_\perp n' n} W(Lnk_\perp \to kn'k'_\perp) \left(- \frac{\partial f_o(\varepsilon_n k)}{\partial \varepsilon} \right) (1 - f_o(\varepsilon_n k_\perp) - f_o(\varepsilon_{n'} k'_\perp))}{\sum_{nk_\perp} f_o(\varepsilon_n k_\perp)} \tag{26}$$

$\langle W \rangle$ is clearly a mean transition probability per unit time, and is a measurement of the "connection" between both wells. The calculation of $\langle W \rangle$ is very tedious but straight forward. With the Fröhlich Hamiltonian already used in part 1, we find :

$$W(Lnk_\perp \to Rn'k'_\perp) = \sum_{\vec{q}} \frac{\gamma |\langle L_n| e^{iqz} |R_{n'} \rangle|^2}{q_\perp^2 + q_Z^2} [N_o\, \delta(\varepsilon_n(k) - \varepsilon_{n'}(k') + \hbar W_o) \ldots$$

$$\ldots + (1 + N_o)\, \delta(\varepsilon_n(k) - \varepsilon_{n'}(k') - \hbar W_o)] \tag{27}$$

The square matrix elements in (27) can be calculated analytically, the work being lengthy, it has been reported in appendix B. It is not surprising to find a $\langle W \rangle$ variation with b exponentially decreasing, as a result of the exponential decay of the ground state wavefunction in the barrier area. A main feature of (24) is that the conductivity, now, is proportional to $\langle W \rangle$ whereas it was the contrary in the Bloch case. This difference in functional forms reflects the opposite assumptions made in both unperturbed systems. In the former system, scattering events tended to destroy the conductivity, whereas the same events tend to enhance the conductivity in localized systems, making more and more transitions. Transport parameters are also modified by the lattice temperature. In the Bloch case, μz decreased with T as the phonon population was enhanced. Now, to increase the NO number of occupied

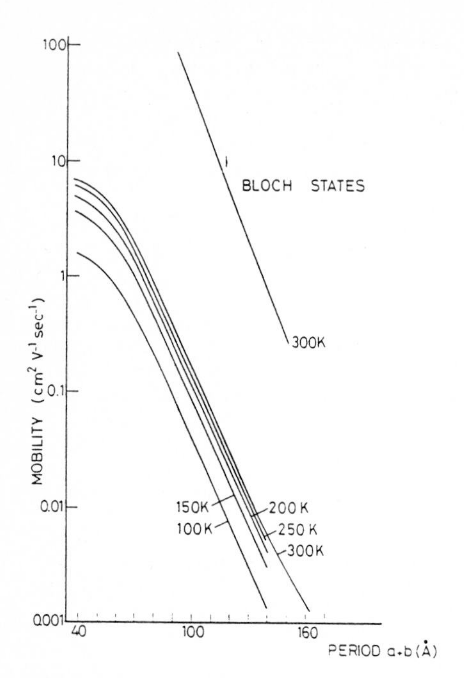

Fig.6 Variation of μ versus d(a=b) for the hopping case. Here W =.3 e.v. The Bloch electron calculation of section 1.1 is also drawn as reference

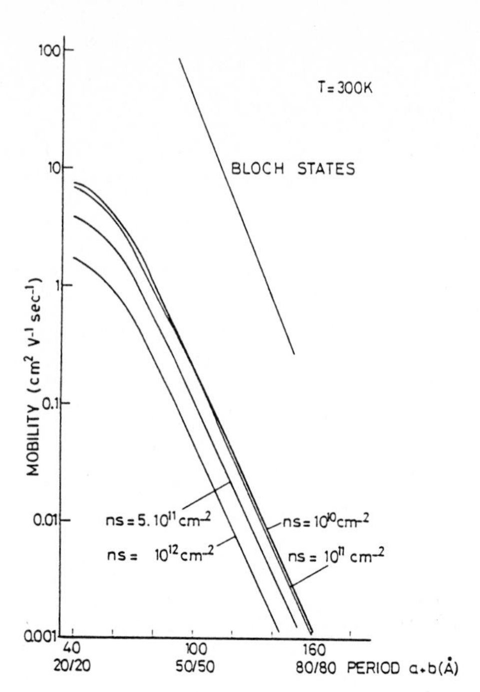

Fig.7 Variation of the hopping mobility versus d assuming a = b different ns.

phonon states can increase the conductivity. Another difference is exhibited : the mobility variation with ns. In usual semiconductor systems, the average number of carriers only marginally affects the mobility by renormalization effects or screening and at relatively high densities. Now, in filling the subbands, we leave less and less empty states available for hopping. Both T and ns effects are described in fig. 6,7. Compared to the results of fig. 3 these mobilities, while varying qualitatively with the same decrement with d, are at least two orders of magnitude lower. A qualitative discrepancy occurs for the lowest paths, which we shall see now in taking into account the multiple hopping.

To generalize the preceding approach, we consider that carriers can hop from one well not only to the next one but also to any well in the superlattice layer. We have to define the following transitions :

$$W(i,n,k_\perp \rightarrow i+1,n',k'_\perp) \quad W(i,n,k_\perp \rightarrow i+2,n'',k''_\perp) \quad W(i,n,k_\perp \rightarrow i+3,n''',k'''_\perp) \tag{28}$$

and so on.

Then, the current density is given by :

$$\begin{aligned} J_Z = -\frac{e}{V}\frac{d}{2} \sum_{\substack{n,n',n''\ldots \\ k_\perp,k'_\perp,k''_\perp\ldots}} \Big[& (f_i(1-f_{i+1})\, W_{i,i+1} - f_{i+1}(1-f_i)\, W_{i+1,i}) \ldots \Big] \\ & + 2(f_i(1-f_{i+2})\, W_{i,i+2} - f_{i+2}(1-f_i)\, W_{i+2,i}) \\ & - (f_i(1-f_{i-1})\, W_{i,i-1} - f_{i-1}(1-f_i)\, W_{i-1,i}) + \ldots \end{aligned} \tag{29}$$

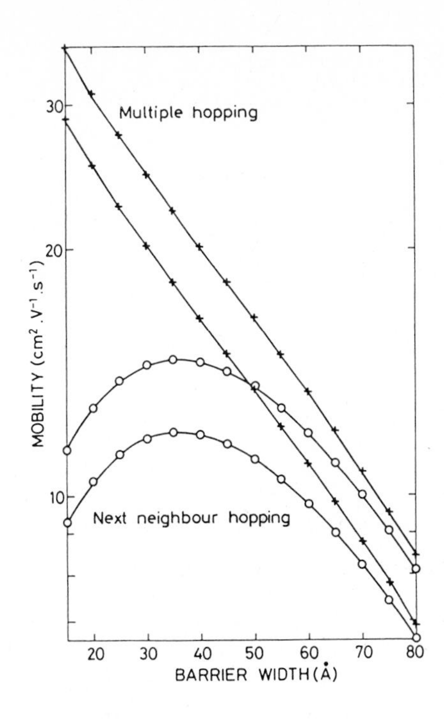

Fig.8 Improvement to the hopping theory in adding multiple hopping. Here we show a calculation for light holes in Al Ga As in assuming 40% of the gap difference as the valence-band offset.

The expansion of f_i, f_{i+1}, f_{i+2}, ... can be made in linear power of f. It results that the mean transition probabilities must be replaced by

$$W_{LR} \not\rightleftarrows \sum_{0}^{\infty} (j+1)^2 \; W_{L\,n\,k\,;\,R+j\,n'\,k'} \tag{30}$$

in which L and R+j replace L and R of the next-neighbour approach. The stronger modification has been found for the light hole hopping in the AlGaAs system (see fig. 8). The interesting result is that the multiple hopping is in qualitative agreement with an exponential decrease of μ_z with b, even at the lowest values of b.

2. Study of real structures

2.1. Basic macroscopic transport equations

In the preceding part, expressions of μ_z have been derived, assuming the number of periods was infinite. Now, in real situations, this cannot be done, as the superlattice total thickness cannot exceed 1-2 microns. So we have to take in account the effects of injection, space charge limited currents, compensation of the active layer, thermoionic emission at hetero-interfaces. Let us suppose that the experimental situation consists of one regular superlattice imbedded between two access layers. Two main structures are currently studied : the n+/SL/n+ structure and the p/SL/n one. To get started with any transport interpretation in such systems, it is essential to replace the SL by an effective medium having well chosen macroscopic parameters. These parameters are :

- The band parameters with given extrema of both conduction and valence bands.

- An effective mobility for electrons, heavy holes and light holes.

- A fixed thermodynamic non equilibrium situation, as given by quasi Fermi levels or electron temperature.

We shall also make use of the Einstein relation, which follows from the existence of a thermalised population, even though it has not been proven that in the hopping case, this relation is still valid. Thus, the classical drift-diffusion equations can be considered :

$$\frac{\partial n}{\partial t} = \frac{1}{e}\frac{\partial}{\partial}\left(D_n \frac{\partial n}{\partial z} - \mu_n n \frac{\partial \phi}{\partial z}\right) - R(z) + G(z) \tag{31}$$

$$\frac{\partial p}{\partial t} = \frac{1}{e}\frac{\partial}{\partial z}\left(D_p \frac{\partial p}{\partial z} + \mu_p p \frac{\partial \phi}{\phi z}\right) - R(z) + G(z) \tag{32}$$

$$\frac{\partial^2 \phi}{\partial z^2} = \frac{e}{K_o}\left(n - p - N_D^+ + N_A^-\right) \tag{33}$$

$$D_n = \frac{k_B T \mu_n}{e} \ . \ D_p = \frac{k_B T \mu_p}{e} \tag{34}$$

in which n and p are the carrier densities, Nd+ and Na- the charged donor and acceptor concentrations, ϕ the electrostatic potential, μ_n and μ_p the carriers mobilities, the dielectric constant, R(z) and G(z) the recombination and generation terms. In heterogeneous systems, these equations can be modified to deal with sudden changes in the carrier density. We also introduce the quasi Fermi levels :

$$F_n = k_B T \,\mathrm{Log}\, \frac{n}{N_c} - e\phi + E_c \qquad F_p = k_B T \,\mathrm{Log}\, \frac{p}{N_v} - e\phi + E_v \tag{35}$$

A reasonable assumption is to assert continuity of F_n and F_p at the interfaces ; we have :

$$\frac{n(Z_o - \varepsilon)}{n(Z_o + \varepsilon)} = \exp\left(-\frac{E_c(Z_o - \varepsilon) - E_c(Z_o + \varepsilon)}{k_B T}\right) \tag{36}$$

Consideration of these continuous Fermi levels allows for the calculation of n and p via the change of variable :

$$n^* = \frac{n}{N_c} e^{-E_c/k_B T} \qquad p^* = \frac{p}{N_v} e^{+E_v/k_B T} \tag{37}$$

in which N_c and E_c are piecewise constants. So, the change (37) in (31) can only affect R(z) and G(z). In a unipolar material e.g. n-type the equation (31) can be integrated once. This is the Scharffetter-Gummel method [19] often successful in the derivation of approximate current density values :

$$J_z^n = e \int_0^z (R - G)\, dz' + J_o^n$$

$$n(z) = n(o)\, e^{e(\phi - \phi_o)/k_B T} + \int_0^z e^{e(\phi(Z) - \phi(Z')/k_B T} \frac{e.[J_z^n]}{eD_n} dZ' \tag{38}$$

A classical solution of the system consists in the derivation first of an approximate expression of ϕ, then to R and G, and then to have direct solutions of the remaining ordinary differential first order equations :

$$J_o = \frac{n(L)e^{-e\phi_L/k_B T} - n(o)e^{-e\phi_o/k_B T}}{\int_0^L e^{-e\phi_L/k_B T} \frac{dz}{D_n}} - \frac{\int_0^L e^{-e\phi/k_B T} \frac{dz}{D_n} \int_0^z R(Z')dZ'}{\int_0^L e^{-e\phi_L/k_B T} \frac{dz}{D_n}} \tag{39}$$

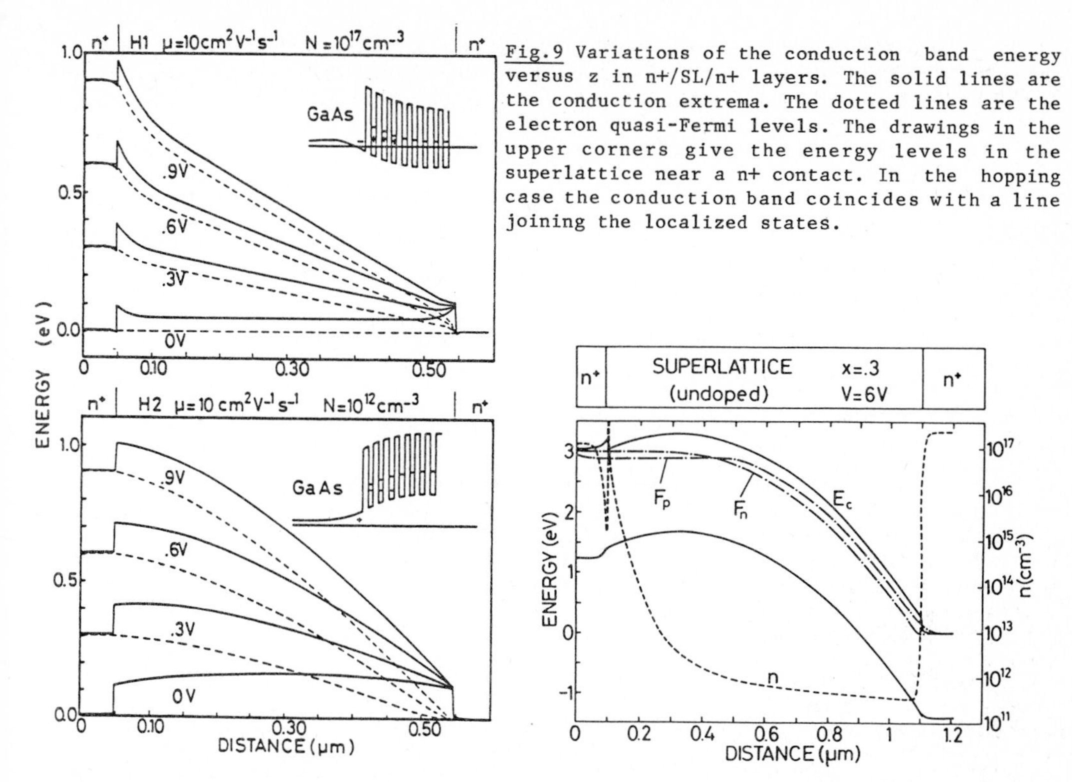

Fig.9 Variations of the conduction band energy versus z in n+/SL/n+ layers. The solid lines are the conduction extrema. The dotted lines are the electron quasi-Fermi levels. The drawings in the upper corners give the energy levels in the superlattice near a n+ contact. In the hopping case the conduction band coincides with a line joining the localized states.

Fig.10 Variation of the conduction band energy versus z for a compensated SL layer. The following parameters have been set in the model : Nd=5.10 cm-3 in N+ layers ; Na=7.10 cm-3 in SL ; Ec-Ea=.95e.v. as the position of the compensation center ; uz=10 cm^2 V-1 s-1 as the hopping mobility.

Numerical calculations are necessary when the system becomes too complex for application of (39). We have applied standard finite difference algorithms with the changes (37) to obtain current-voltages as well as band diagrams corresponding to any experimental situation [20]. In the next section we describe typical results of our modelisation.

2.2. Effective medium description of transport perpendicular to superlattices

In this case it is still possible to apply the transformation (38) to get good approximation of the current. Three different situations may occur :

- The superlattice is n-doped.

- The superlattice is not intentionally doped but uncompensated.

- The superlattice is compensated with deep centers.

The two first cases are illustrated in fig. 9 and the third in fig. 10. In the first case, the main drop in the electric field is regular within the superlattice. It is clearly seen that the current density is :

$$J_z \cong eN_D \, \mu_z \, \frac{V}{L} \qquad (40)$$

That expression will be called hypothesis H1. In fig. 9b the space charge effect is important and the electric field inhomogeneous in SL layer. So we can attribute this effect to the ideal injection law of Lampert [25]

$$J_z = \frac{9}{8} K_o \mu_z \frac{V^2}{L^3} \qquad (41)$$

Applied to not too severely compensated layers (41) seems to be confirmed by numerical simulation when the applied voltage is higher than a certain critical value. That hypothesis will be called H2. The third case is much more difficult and in fact more close to real experiences done with unintentionally doped materials. A phenomenological law is then found at moderate voltages :

$$J_z = J_o \exp \frac{eV}{\eta k_B T} \qquad (42)$$

with η 10-20. Going now to detailed analysis we may show that the number of band carriers in the example of fig. 10 would be very low, of the order of 10^{10} cm^{-3}, giving a total number of electrons of 1000 in a given sample of 10^{-4} cm^2 area. As such low densities are not physically significant, we have to find other more probable conduction processes to fit experimental data.

3. Experimental techniques

3.1. Current-voltage analysis

At room temperature, two types of I(V) curves are generally observed. A first type corresponds to $I \sim V^n$ with $1 < n < 2$. A second type is fitted by an exponential behaviour : $I \sim \exp(eV/nkT)$ in which $n \sim 20$. We are now convinced that the second kind of characteristic is due to a strong compensation of the SL layer by some kind of deep center. As discussed in the preceding part, the transition from a pure injection regime to a barrier regime occurs when the density of compensating centers is higher than 10^{10} cm^{-3}. Samples are prepared by molecular beam epitaxial techniques, the aluminium concentration being designed to minimize the contact impedance. Results for a given set of superlattices with different a and b have been given in ref. [17] and [18]. Here we give only typical results in each case. In fig. 11 we show an I(V) curve for a doped superlattice. Such a I(V) behaviour can be related to the injection in moderately doped SL layer. As we ignore the exact amount of compensating centers , two different hypotheses have been made. These hypotheses have been made in the part 2.2, say H1 and H2. For H1 the carrier density equals the donor concentration, so the mobility is over-estimated. For H2 the carrier density is very low, so the mobility is over-estimated. For H2 the carrier density is very low, so the mobility is under-estimated. Then, for each sample an experimental bar is obtained for μ_z. The resulting experimental bars for the set of samples of table (1) are given in fig. 13. It is clearly seen that the mobility values are in good agreement with the hopping predict. We have also tried to fit the data of undoped SL. Although the fit of fig. 12 seems satisfactory, many questions still remain obscure in such results and their fit with the model. Such questions are :

- The fit is much more sensitive to a small change of N_A than to μ_z, so it leads to a poor accuracy in μ_z estimation.
- A detailed analysis of the carrier density chart shows a non-physical carrier mean density in the miniband (or hopping ground state level near the GaAs conduction band).
- We also estimate that the impurity to impurity hopping at the deep levels may be the dominant process. In that case the dark currents in the structure do not involve the superlattice and could be close to that of a compensated AlGaAs layer.

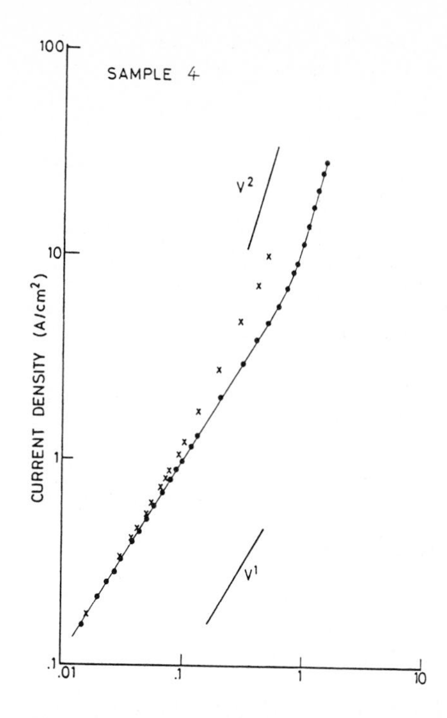

Fig.11 Current-voltage characteristics for doped sample. Results are given for two different contact areas.

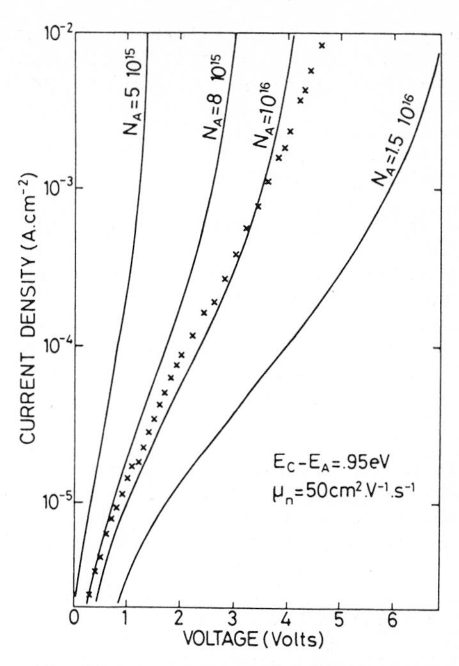

Fig.12 Current-voltage characteristics for a compensated SL layer. Different concentrations are tried for the modelling (solid lines), the x being a given datum at room temperature. The calculated curves correspond to Ec-Ea = .95 e.v.

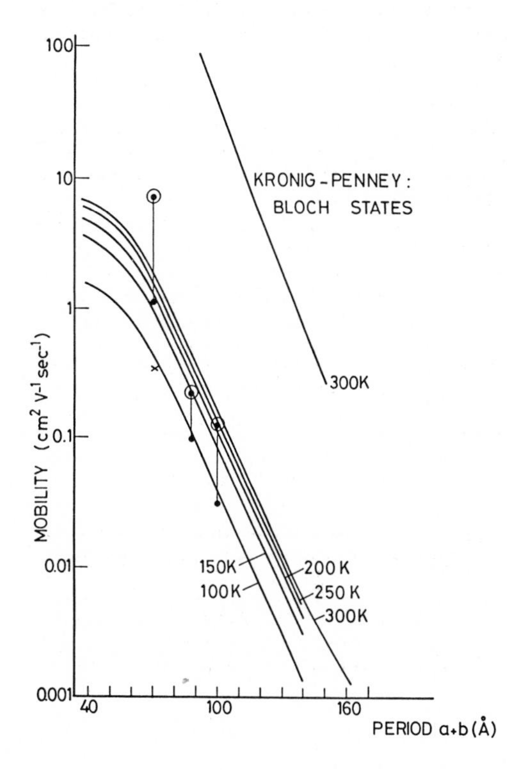

Fig.13 Estimated perpendicular mobilities as functions of the superlattice period, as deduced from H1 and H2 hypothesis. The different samples parameters are given in table 1. Here the points correspond to H1, the circles to H2 and the crosses in circles to high-frequency result of [17].

Table I Geometric and essential parameters of the studied samples

Sample ref.	a [A°]	b [A°]	x	N [cm^{-3}]	E_1 [eV]
1	50	50	.3	10^{17}	.091
2	45	45	.3	10^{17}	.102
3	40	40	.3	10^{17}	.118
4	34	34	.3	10^{17}	.140
5	34	34	.3	10^{18}	.140
6	22.5	22.5	.3	10^{17}	.15-.18

a : estimated well width from X-rays analysis

b : estimated barrier width from X-rays in Angströms

x : aluminium percentage in Angströms

N : donor concentration aimed in cm^{-3}

E_1: ground state position in the MQW from the GaAs conduction minimum

We could conclude that, in doped superlattices, coherent Bloch state conduction is not possible. However, in pure samples, photo-excitation of the carriers may be more promising.

3.2. Photocurrent response to brief light pulses

The time-of-flight technique is a well known method [23] for direct measurements of carrier mobilities. Four main parameters are rather critical to the photocurrent response :

- the electron mobility, the hole mobility
- the carrier, lifetimes [21]

As these four parameters may have much more contrasted values in the superlattices than in bulk materials, we expect two following behaviours :

- The photocurrent is rather insensitive to the applied voltage. In that case, the minority carriers are trapped somewhere and the induced current is not modified by the supplement of drift given by the external voltage.
- The photocurrent is sensitive to the applied voltage. In that case, some sweepout of one kind of carriers, or both, is occurring. To illustrate this we have numerically solved the set of equations (31-34) in the case of the n+/SL/n+ structure. The example of figure 14 is obtained with slow carriers as electrons (μ_2 = $20.cm^2\ V^{-1}s^{-1}$), and large lifetimes for both types of carriers [21]. Identification of a time-of-flight effect is made in comparing typical duration of the photocurrent and its variations with the applied voltage with the time-of flight duration tv :

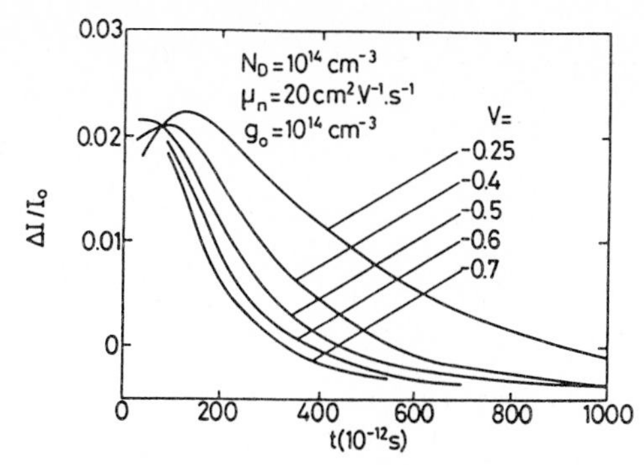

Fig.14 Numerical simulation of the time-of-flight experiment. 10 cm^{-3} carriers have been injected at one sample end with a typical absorption law in the superlattice : $n(z,t=0) = N0.\exp-(\alpha z)$. The different current responses are drawn at different applied voltages. The sign of the voltage corresponds to electrons going from the illuminated side to the other.

$$t_V = \frac{L^2}{\mu_z V} \qquad (43)$$

Moreover, the behaviour of the effect must be sensitive to the sign of the applied voltage, as a typical sort of carriers must propagate across the sample and not the other one. Preliminary results obtained with undoped superlattices give results in agreement with the I(V) data.

The point of view developed in this course has been restricted to classical aspects of the transport perpendicular to superlattices. Many other effects may be expected,namely the Stark ladder possibility. Our main purpose was to present tools which may be used for foregoing works.

APPENDIX A

DERIVATION OF COLLISION TIMES

1) Deformation potential scattering

We make use of the cylindrical coordinate system of fig. 15. Then, using relations (8) and (9), for momentum and energy conservations :

$$U^{\pm} = \frac{\hbar^2}{2m^*}(k_\perp'^2 - k_\perp^2) + \varepsilon_z(k_z') - \varepsilon_z(k_z) \pm \hbar qs$$

$$\vec{k}' = \vec{k} \pm \vec{q} + \vec{G}$$

the sum over k' is replaced by an integral :

$$\sum_{k'} = \frac{V}{(2\pi)^3}\int_{-\pi/d}^{\pi/d} dk_z' \int_0^{2\pi} d\theta' \int_0^{\infty} k_\perp' \, dk_\perp'$$

We integrate this in the order $k_\perp'$, then θ', then k'_z. The change of variable $U \leftrightarrow k_\perp'$ allows to use δ function property :

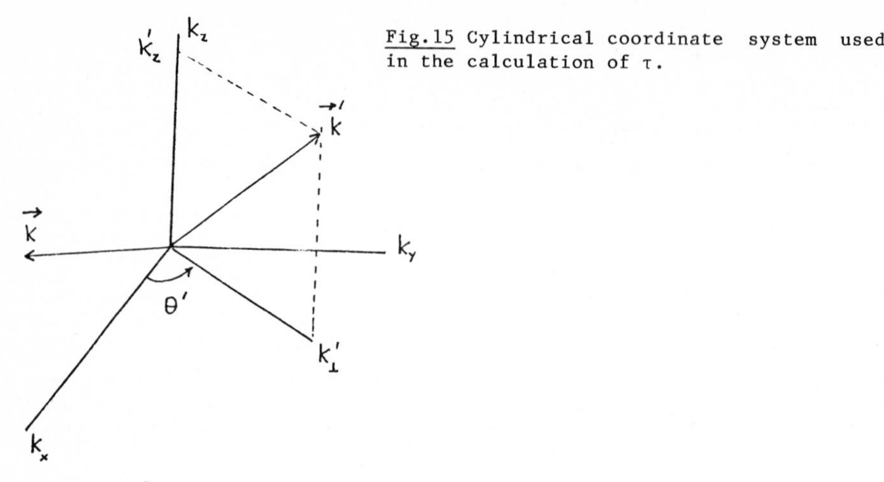

Fig.15 Cylindrical coordinate system used in the calculation of τ.

$$\frac{\partial U^{\pm}}{\partial k'_{\perp}} = \frac{\hbar^2}{m^*} k_{\perp}\left(1 \pm \frac{m^* s}{\hbar q}\left(1 - \cos\theta' \frac{k_{\perp}}{k'_{\perp}}\right)\right)$$

in which $q = |\vec{q}|$. The simplified expression (15) is obtained if we assume $m^* s \ll hq$ which is valid at moderate and high temperatures. Thus we obtain :

$$\frac{1}{\tau_a} = \frac{E_1^2 m^*}{8\pi^2 \rho s \hbar^2} \int_{-\pi/d}^{\pi/d} dk'_z \int_0^{2\pi} d\theta' \Big[q^+ (N_{q^+} + 1)\, I^+(k,k'q^+)(1 - \cos(k,k'^+)) + q^-(N_{q^-})\, I^-(k,k',q^-)(1-\cos(k,k'^-))\Big]$$

Several simplifications give the result (15):

- $N_q \simeq \frac{k_B T}{\hbar q s} \gg 1$

- $m^* s \ll \hbar q$

- $\cos(k,k') \simeq \cos\theta' \left(1 \pm \frac{k_z k'_z}{k_{\perp} k'_{\perp}}\right)$

Restricting the result to N processes (G=0) we find :

$$\frac{1}{\tau_a^N} = \frac{1}{2\pi} \frac{E^2 m\, k_B T}{\rho s \hbar^3} \begin{cases} 2k_1 \;<\; > \varepsilon(k) < 2\Delta \\ \frac{2\pi}{d} \;<\; > \varepsilon(k) \geqslant 2\Delta \end{cases}$$

in which 2Δ is the miniband width. With use of $\varepsilon = \frac{h^2 k_1^2}{2M}$ the expression (18) is straightforward.

2) Polar optical scattering

Starting from (13) we define :

$$\gamma = \frac{e^2}{8\pi^2}\left(\frac{1}{K_\infty} - \frac{1}{K_o}\right)\omega_o$$

and, with the same coordinate system than above :

$$\frac{1}{\tau} = \gamma \int_{o}^{2\pi} \int_{-\pi/d}^{\pi/d} dk'_z \int_{o}^{\infty} k'_{\perp} \, dk'_{\perp} \left[\frac{(N_q+1)\delta(U^+)I^+}{q^2} + \frac{N_q \delta(U^-)I^-}{q^2} \right]$$

Now the Jacobian of the change U <--> $k'_{\perp}$ is more simple :

$$\frac{\partial U^{\pm}}{\partial k'_{\perp}} = \hbar^2 \frac{k'}{m^*}$$

Then it comes :

$$\frac{1}{\tau} = \gamma_1 \sum_G \int_o^{2\pi} d\theta \left[\int_{-k_1^+}^{k_1^+} dk'_2 \frac{I_G(N(\omega_o)+1)}{q_G^+} + (-) \right]$$

in which $k'^{\pm}_1$ are the extrema of k'_2 corresponding to

$$\varepsilon_z(k_1) = \varepsilon \mp \hbar\omega_o$$

which may be within the segment $[-\pi/d, +\pi/d]$. Then the following relations for $q^{\pm}$ are found :

$$q_G^{\pm 2} = 2k'^2_{\perp} + (k'_2 - k_2 - G)^2 \mp \frac{2m^*\omega_o}{\hbar} \quad \ldots$$

$$+ \frac{2m^*}{\hbar^2}[\varepsilon_z(k'_z) - \varepsilon_z(k_z)] - 2k_{\perp}\cos\theta' \sqrt{\left|k_{\perp}^2 \mp \frac{2m^*\omega_o}{\hbar} + \Delta K_z^2\right|}$$

in which

$$\Delta K_z^2 = \frac{2m^*}{\hbar^2}[\varepsilon_z(k'_z) - \varepsilon_z(k_z)]$$

the integration in θ is analytical :

$$\int_o^{2\pi} \frac{d\theta}{a - b\cos\theta} = \frac{2\pi}{\sqrt{a^2 - b^2}} \qquad (|a| > |b|)$$

So we obtain the following expresions for J_n :

$$\mathbb{J}_n^{\pm} = \int_{-k_1^{\pm}}^{k_1^{\pm}} \frac{dv}{\sqrt{(1-K^{\pm})^2 + 2(1+K^{\pm})(u-v\mp g_n)^2 + (u-v\mp g_n)^4}}$$

with the scaled variables :

$$K^{\pm} = 1 \mp \Omega^* + \varepsilon_2(\lambda u) - \varepsilon_z(\lambda v) \quad ; \quad u = \frac{k_z}{k_{\perp}} \quad ; \quad v = \frac{k'_z}{k_{\perp}}$$

$$\Delta^* = \frac{\Delta}{\varepsilon_{\perp}} \quad ; \quad \frac{\hbar\omega_o}{\varepsilon_{\perp}} = \Omega^* \quad ; \quad \varepsilon_{\perp} = \frac{\hbar^2 k^2}{2m^*} \quad ; \quad g_n = \frac{2n\pi}{dk_{\perp}}$$

then the integral in k' must be made.

APPENDIX B

DERIVATION OF THE HOPPING TRANSITION PROBABILITIES

A square well between za and zb of depth W in energy give the following eigenstates [22] :

$$z < z_A \qquad \Psi = C_1\, e^{\chi(z-z_A)}$$

$$z_A \leqslant z \leqslant z_B \qquad \Psi = C \sin(K(z-z_A) + \delta)$$

$$z > z_B \qquad \Psi = C_2\, e^{-\chi(z-z_B)}$$

with

$$C_1 = C \sin \delta$$

$$C_2 = C \sin(Ka+\delta)$$

$$= K \cot(\delta)$$

and the dispersion relation :

$$K_n a + 2.a \tan\left(\frac{K_n}{\chi_n}\right) = n\pi$$

Now χ and C are related to the eigenvalue ε_n and W by :

$$\chi_n = \left[\frac{2m^*}{\hbar^2}(W - \varepsilon_n)\right]^{1/2} \qquad \left(K_n = \frac{(2m^*\varepsilon_n)^{1/2}}{\hbar}\right)$$

$$C_n = \left[\frac{a}{2} + \frac{K_n^2}{\chi_n(K_n^2 + \chi_n^2)}\right]^{-1}$$

Two different tasks must be made : i) the derivation of I_{11} ii) the summation on q_z in (25).

i) $$I_{11} = |\langle 1,L|e^{iqz}|1,R\rangle|^2$$

$$I_{11} = \frac{4\, e^{-2\chi_1 b}}{\left(\frac{a}{2}\,\frac{\chi_1^2+K_1^2}{K_1^2} + \frac{1}{\chi_1}\right)^2}\left[(\sin q_z b/2)/q_z \cdot\cdot + \frac{1}{(\chi_1^2+K_1^2-q_2^2)+4\chi_1^2 q_2^2}\left[\begin{array}{l}2\chi_1(\chi_1^2+K_1^2)\cos q_z\, b/2\\ + q_z(K_1^2-3\chi_1^2-q_z^2)\sin q_z\, b/2\end{array}\right]\right]^2$$

ii) the transition probabilities are easily summed in polar coordinates in the plane $(k'_x k'_y)$:

$$W^{\pm} = \frac{\gamma}{(2\pi)^3}\int_0^{2\pi} d\theta \int_0^{\infty} k'_\perp dk'_\perp \int_{-\infty}^{+\infty} dq_z\, \frac{(N_q + \frac{1}{2} \pm \frac{1}{2})}{q_\perp^2 + q_z^2}\,\delta(U^{\pm})\, I_{11}^2$$

Then the sum in $k_\perp$ gives the following result :

$$J_z = \frac{e^2 Ed}{8\pi^2}\left(\frac{m^*}{\hbar^2}\right)^2 \omega_o\left(\frac{1}{K_\infty} - \frac{1}{K_o}\right) N_o \int_0^{\infty} d\varepsilon I^- G^- + (N_o+1)\int_{\hbar\omega_o}^{\infty} d\varepsilon\, I^+ G^+$$

$$I^{\pm} = \int_{\infty}^{\infty} dqz\, \frac{I_{11}(q_z)}{\sqrt{\frac{4m^* q_z^2\varepsilon}{\hbar^2} + \left(q_z^2 \mp \frac{2m\omega_o}{\hbar}\right)^2}} \quad ; \quad G^{\pm} = -f'_o(\varepsilon)\; 1 - f_o(\varepsilon \pm \hbar\omega_o) - f_o(\varepsilon) f'_o(\varepsilon \pm \hbar\omega_o)$$

These integrals must be computed numerically.

References

1. L. Esaki, L.L. Chang: Proceedings of the International Conference of Semiconductors (1972)
2. L. Esaki, R. Tsu: IBM J. Res. Dev. 14, (1970), 61
3. J. Zak: Physics Letters, 76A, (1980), 287
4. P. Lebwohl, R. Tsu: J. Appl. Phys., 41, (1970), 6
5. A. Shik: Sov. Phys. Semicond., 7, (1973), 187
6. G. Shmelev, G.M. Chaikovskii, I.A. Chan Min Chon: Phys. Stat. Solidi 73, (1976), 811
7. I.M. Dykman, P.M. Tomchuk: Phys. Stat. Solidi, B76, (1976), 385
8. G. Bastard: Phys. Rev. B24, (1981), 5693
9. G. Bastard: Phys. Rev. B25, (1982), 7584
10. R. Lassnig: Phys. Rev. B30, 12, (1984), 7132
11. J.F. Palmier, A. Chomette: J. de Physique, 43, (1982), 381
12. B. Jusserand, D. Paquet, A. Regreny: Phys. Rev. B30, 10, (1984), 6245
13. J. Sapriel et al.: Phys. Rev. B4, (1983), 2007
14. K. Hess: Appl. Phys. Lett., 35, 6, (1979), 483
15. A. Chomette, J.F. Palmier: Solid State Comm., 43, (1982), 3
16. R. Tsu, G. Dohler: Phys. Rev. B12, 2, (1975), 680
17. D. Calecki, J.F. Palmier, A. Chomette: J. of Physics C 17, (1984), 5017
18. J.F. Palmier, H. Le Person, C. Minot, A. Chomette, A. Regreny, D. Calecki: Superlattices and microstructures 1, 1 (1985)
19. D.L. Scharfetter, H.K. Gummel: IEEE ED16 (1969), 64
20. E. Caquot et al.: ESSDERC 1984
21. B. Sermage et al.: to appear in J. of Appl. Phys.
22. L.D. Landau, E.M. Lifshitz: Quantum Mechanics
23. W.E. Spear: J. of Non Cryst. Solids, 1, (1969), 197
24. B. Vinter: A.P.L., Sept. 84
25. A.M. Lampert: Semiconductors and Semimetals, 6, edited by Willardson and Beer, Academic Press, (1970)
26. B. Lambert et al.: Int. Conf. GaAs and Rel. Compounds (1984)
27. C. Minot et al.: Seventh Trieste Int. Symp. "Hopping Transport"

Doping Superlattices

J.C. Maan

Max-Planck-Institut für Festkörperforschung, Hochfeld Magnetlabor,
166 X, F-38042 Grenoble Cedex, France

A doping superlattice is a layered structure with thin alternating p and n doped layers. The essential physical properties of such a system (quasi-two-dimensionality, metastability of carrier densities and tunability of the effective bandgap) are explained with simple models.

1. Introduction

In the famous paper by Esaki and Tsu [1] where the idea of superlattices was presented for the first time, the authors considered two types of superlattices, one based on a periodic variation of the composition of the layers, the other on a variation of the doping in a homogeneous material. The first idea was actively pursued from the very beginning by various groups, the second was developed theoretically in great detail by Döhler [2,3] in the early seventies, but it was not until 1981 that the first sample was grown (Ploog, Fischer and Künzel [4]). Since then a number of experiments have established the basic properties of this system. For an extensive review of the experiments and theory of doping superlattices see Ploog and Döhler [5].

A doping superlattice, schematically shown in Fig. 1, consists of a sequence of alternately p and n doped layers. By alloying Sn balls through all the layers, one

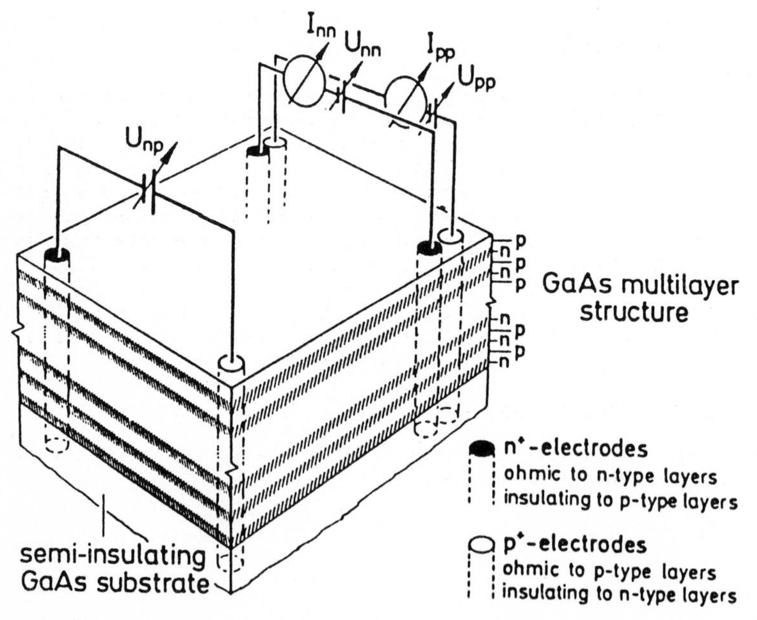

Fig.1 A doping superlattice with selective electrodes to the constituent n- and p-doped layers.

establishes an ohmic contact to all n-layers which is blocking to all p-layers. Similarly, using Sn/Zn balls, ohmic contacts to all p-layers are made, which block the n layers. In this way a voltage U_{np} can be applied between all n and all p layers, and the conductivity of the n and p layers can be measured separately. In the same way as in a pn junction, electrons are transferred from the donors in the n layers to the acceptors in the p layers, which leads to a net positive charge at the n and a net negative charge at the p side of the junctions. The resulting space-charge electric field will bend the valence and the conduction bands up to the point at which the Fermi energy is equal at both sides of the junction. This condition determines, in a given material and for a given doping density, the width of the depletion region. A doping superlattice can be considered as a series of pn junctions, but with n-type and p-type layer thicknesses comparable to the depletion layer thickness. A schematic illustration of the band energy diagram of a doping superlattice is shown in Fig. 2(top). This figure shows the two basic properties of a doping superlattice. Electrons and holes are confined in small layers by the space charge potential ("Quasi Two-Dimensional character 2D") but are spatially separated from each other by the depletion layer ("Indirect gap in real space"). The most important consequence of these two properties is that optically or electrically generated excess electrons and holes will be confined *and* spatially separated, which implies that deviations from thermal equilibrium are very long lived. ("Metastability")

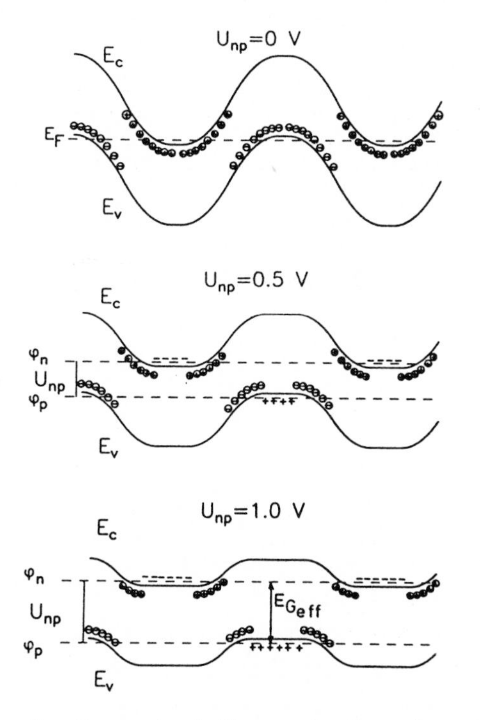

Fig.2 Schematic real space energy diagram of a doping superlattice. The parameters in the ground state are such that $N_D d_n = N_A d_p$ and the depletion layer widths $x_{n,p}$ are half the layer thickness in the ground state (top). Optically or electrically induced carriers are accumulated in the n and the p layers, reducing the depletion layer thickness and decreasing the effective gap (center and bottom).

2. Energy band diagram in thermal equilibrium

In a very simplified manner the energy band diagram of a pn junction at x=0 can be derived from the solution of Poisson's equation [6] :

$$\frac{d^2 V}{dx^2} = \frac{-qN_i}{\varepsilon} \tag{1}$$

with $N_i = N_D$ the net donor concentration in the n region for $x > 0$ and $N_i = N_A$ the net acceptor concentration for $x < 0$ and q the elementary charge. Furthermore charge neutrality requires that :

$$N_D x_n = N_A x_p \tag{2}$$

with $x_{n,p}$ the width of the depletion layer in the n and p regions respectively. Integrating (1) with boundary conditions such that at $x=x_n$ and $x=-x_p$ the electric field is zero, and that at $x=x_n$ the potential equals $-(\phi_n-E_c)/q$, with E_c the conduction band edge and ϕ_n the Fermi energy of the electrons in the n-layer and for the p-layers at $x=-x_p$ $V = \{E_g + (E_v-\phi_p)\}/q$ (E_v is the valence band edge and ϕ_p the hole Fermi energy) leads to

$$V = \frac{qN_D}{2\varepsilon}(x-x_n)^2 - \frac{(\phi n-E_c)}{q} \qquad x > 0$$

$$V = \frac{qN_A}{2\varepsilon}(x+x_p)^2 + \frac{E_G+(E_v-\phi_p)}{q} \qquad x < 0 \tag{3}$$

for the potential profiles. The depletion width in the ground state x_n is given by

$$x_n = \left\{\frac{2\varepsilon V_{bi} N_D}{q^2 N_A(N_A+N_D)}\right\}^{1/2} \tag{4}$$

with V_{bi} the built-in potential energy defined as :

$$V_{bi} = \{(E_G + (E_V - d_p) + (\phi_n - E_c)\} \tag{5}$$

and x_p by the same expression with N_D and N_A interchanged. For a doping layer thickness $d_{n,p} < 2x_{n,p}$ the potential profile is parabolic and the subband energies derived from the solution of Schrodinger equation lead to the well-known harmonic oscillator with eigenvalues :

$$\hbar\omega_{n,p} = \left\{\frac{q^2 N_{D,A}}{m_{c,v}\varepsilon}\right\}^{1/2} \quad \text{and } E_N = (N+1/2)\hbar\omega_i \tag{6}$$

in the n and the p layers where m_c and m_v are the effective masses of the conduction and the valence band. As an example, for GaAs with $N_D = N_A = 10^{18}cm^{-3}$, $\hbar\omega_n = 38meV$ for electrons. This value is much larger than the broadening due to scattering which one can derive from the mobility at these doping concentrations ($\sim$ 2000 cm^2/Vs) and which corresponds to 8 meV. I.e. a sample with these parameters behaves dynamically as a 2D system. This 2D character has been demonstrated in optical, transport and tunnelling experiments [7-11].

In the preceding section, electrons and holes have been treated in a similar manner. In a realistic sample, however, due to the high value of the acceptor binding energy ($\sim$ 30meV) it is more plausible that the conduction in the p-layers will be impurity band conduction. The conclusions about the 2D character of the system are therefore valid only for the electrons.

3. Deviations from thermal equilibrium

When light with photon energies higher than the bandgap is incident on a doping superlattice, electrons and holes are created which are immediately separated by the internal electric field, thus accumulating the electrons in the n and the holes in the p layers. These excess electrons and holes will neutralize part of the ionized impurities, reduce the depletion layer thickness and lead to a photovoltage between the p and the n layers. Equivalently, one can apply a voltage U_{np} between the n and the p-layers and inject carriers in these layers. The non-equilibrium situation is schematically shown in Fig. 2. To obtain the band bending, the boundary conditions for the potential in the solution of (1) have to be modified, which leads to the replacement of V_{bi} with $V_{bi} - qU_{np}$ in the preceding equations. The depletion width $x_{n,p}'$ in the non-equilibrium situation is then given by :

$$x_n' = \left\{ \frac{2\varepsilon N_D(V_{bi}-qU_{np})}{q^2 N_A(N_A+N_D)} \right\}^{1/2} \tag{7}$$

It is useful to define a threshold voltage as the voltage at which either $2x_n = d_n$ or $2x_p = d_p$ depending on which of the two gives the highest value. Below this voltage, either the n or the p layers, whichever one of the two depending on the doping concentration and the layer thickness of the sample, will be completely isolating and U_{np} loses its meaning. Assuming that the p-layers determine the threshold voltage, and using (7) and $n,p^{(2)} = N_{D,A}(d_{n,p} - 2x_{n,p}')$, the following relation between U_{np} and $n^{(2)}$ holds :

$$n^{(2)} = N_D d_n \left[1- \frac{N_A}{N_D} \left\{ \frac{V_{bi}-qU_{np}}{V_{bi}-qU_{th}} \right\}^{1/2} \right] \tag{8}$$

The relation between $p^{(2)}$ and U_{np} can be derived in the same manner. In the case of electrical excitation U_{np} is the applied voltage which leads to a change in $n^{(2)}$ and, in the case of optical excitation, it is the photovoltage, as a consequence of a change in $n^{(2)}$. The modulation of the 2D carrier densities and thereby of the conductivity in the n and the p layer has been experimentally demonstrated [12,13].

The induced carriers neutralize the central part of the potential, which becomes approximately flat and the depletion region becomes smaller (7) but the potential remains parabolic. It is clear that the subband spacing will decrease and not remain equidistant anymore. The subband structure in such cases has to be calculated, with more detailed arguments, c.f. Ruden and Dohler [14].

The implicit assumption in the derivation of (8) is that the recombination current I_{np} is small enough, so that the change in the carrier densities is quasi static. This assumption is especially valid for doping superlattices, because of the confinement of electrons and holes in spatially separated layers. Several processes have to be considered fòr I_{np} : tunnelling and avalanche breakdown under reverse bias and thermionic emission, direct optical recombination and Auger recombination with forward bias. The tunnel and avalanche breakdown current are very small as long as the depletion zone is relatively large, which implies moderate doping levels ($<$ few times 10^{18} cm^{-3} in GaAs) and negative U_{np} values of the order or less then E_G/q. Thermionic emission is negligible as long as the barrier height is small compared to kT i.e. $(E_G-qU_{np}) < kT$. Recombination processes are dependent on the overlap between electron and hole wavefunctions. The enhancement of the lifetimes of excess carriers in a doping superlattice with respect to bulk lifetimes is proportional to the square of the overlap matrix element between the electron and the hole subbands. As explained earlier if $2x_n=d_n$ the electronic states can be described by an harmonic oscillator. In this case the ground state wavefunctions are given by

$$\psi_i(x) = \pi^{-1/4} \alpha_i^{1/2} \exp(-x^2\alpha_i^2/2) \quad \text{with} \tag{9}$$

$$\alpha_i = (m_i\omega_i/\hbar)^{1/2}$$

and ω_i as defined in (6). The lifetime enhancement can be evaluated from the overlap matrix element squared and is given by

$$\frac{\tau_{ds}}{\tau_{bulk}} = \frac{2\alpha_c\alpha_v}{(\alpha_c^2+a_v^2)} \exp\left[- \frac{d^2}{4} \frac{\alpha_c^2\,\alpha_v^2}{(\alpha_c^2+\alpha_v^2)} \right] \tag{10}$$

For GaAs and with $N_D = N_A = 10^{18}$ and $d_n = d_p = 40$nm this factor in $\sim 10^{13}$ which corresponds to a lifetime of the order of seconds. Such long lifetimes are indeed observed in optical and electrical excitation [15]. The preceding estimate of the lifetime is valid as long as the wavefunction is mainly localized in the center of the layer. With increasing densities the depletion widths are reduced. Therefore the wavefunction overlap will increase and the lifetimes decrease rapidly.

It can readily be imagined that the property of metastability, which is specific for doping superlattices, can be used for several practical purposes, of which a few will be mentioned.

- The conductivity of the p and the n layers can be varied through a change in the carrier density with an external voltage U_{np}. In this respect the system functions in a manner similar to a JFET (Junction Field Effect Transistor).
- Furthermore, one can put the n and the p channel in a conducting state with a positive U_{np} and these channels will remain highly conducting for several seconds, while the U_{np} may be floating as long as the n and the p contacts are not shorted. Shorting the n and p contacts immediately restores the low conductivity state (memory function).
- Similarly, the system will work as an integrating detector, because every absorbed photon will create an electron and hole which recombine very slowly. One can detect the radiation either as a photovoltage U_{np} or as a conductivity change in the p or the n layers. By shorting U_{np} the detector can be reset rapidly to the initial state.

4. Tunable bandgap

It is clear from the preceding sections that the effective bandgap, defined as $E_{eff} = E_G - V_{bi} + qU_{np}$ can be varied in a wide range, through the variation of the carrier densities $n^{(2)}$ and $p^{(2)}$, by either optical or electrical excitation. The physical meaning of this effective gap is that its energy corresponds to the lowest energy for optical absorption and the highest for emission. As with the estimation of the recombination lifetimes, the probability of absorption or emission is proportional to the square of the overlap matrix element between the electron and hole subbands. The value of this matrix element is very small for large values of the depletion length x_n' and x_p', but increases rapidly as these lengths decrease, which is the case as E_{eff} approaches E_G as can be seen from eq. 7. Therefore a doping superlattice shows optical absorption and emission below the fundamental bandgap. This phenomenon is nothing other than the Franz-Keldysh effect, which describes the shift of the fundamental absorption edge in semiconductors and insulators under the influence of an electric field. The electric field in this case is caused by the space charge and can be very strong ($\sim 1^5$V/cm) leading to an appreciable shift in the absorption or emission edge. Experimentally, photoluminescence has been observed at an energy of 1.2 eV in GaAs doping superlattices [16,17,7]. At low temperatures this represents a shift of 320 meV with respect to the GaAs bandgap. Since the effective gap depends on the carrier densities $n,p^{(2)}$ the emission frequency can be tuned through a variation of these densities. In the case of photoluminescence, this implies a strong dependence of the emission frequency on the intensity of the exciting radiation, because this intensity will determine the steady state populations $n^{(2)}$ and $p^{(2)}$. Furthermore, one can vary the luminescence frequency at a given excitation intensity by changing U_{np} with external contacts [18]. Also, electroluminescence can be directly observed, in which case the recombining carriers are exclusively generated by the U_{np} voltage. In all cases the luminescence intensity will increase rapidly as its radiation energy approaches the gap, because of the strong increase in the overlap matrix element. It is important to note that the observation of tunable emission requires thin depletion layers and small values of V_{bi} whereas the opposite is required to observe the metastable character. The lifetimes in the former type of sample are much shorter [17].

5. Conclusions

The essential properties of doping superlattices have been described. The models which have been used have been very elementary. However, these kinds of simple descriptions have been shown to be accurate enough to describe experimental results. It should be realized that doping superlattices are inherently disordered systems, so that the observation of refined quantum effects should not be expected. In this respect the system is far inferior to the well known GaAs/GaAlAs heterojunctions or quantum wells. Instead, doping superlattices have some other specific unusual properties which may lead to some useful devices. The most remarkable general property is the large variability of several physical parameters such as carrier density, optical absorption coefficient, excess carrier lifetime, etc., with external excitation. In recent years all these effects have been experimentally established, and it may be expected that future research will concentrate more on their exploitation.

References

1. L. Esaki, R. Tsu : IBM, J. Res. Dev., 14, 61 (1970).
2. G.H. Döhler : Phys. Stat. Sol. (b), 52, 79 (1972).
3. G.H. Döhler : Phys. Sol. (b), 52, 533 (1972).
4. K. Ploog, A. Fisher, H. Künzel : L. Electrochem. Soc. 128, 400 (1981).
5. K. Ploog, G.H. Döler : Adv. Phys. 32, 286 (1983).
6. S.M. Sze : Physics of Semiconductor Devices, (Wiley, New York 1981).
7. G.H. Döhler, H. Künzel, D. Olego, K. Ploog, P. Ruden, H.J. Stolz, G. Abstreiter : Phys. Rev. Lett. 47, 864 (1981).
8. Ch. Zeller, B. Vinter, G. Abstreiter, K. Ploog : Phys. Rev. B 26 (1982).
9. J.C. Maan, Th. Englert, Ch. Uihlein, H. Künzel, A. Fisher, K. Ploog : Solid State Comm. 47, 383 (1983).
10. Th. Englert, J.C. Maan, G. Remenyi, H. Künzel, L. Ploog, A. Fisher : Surf. Sci. 142, 68 (1984).
11. Ch. Zeller, G. Abstreiter, K. Ploog : Surf. Sci. 142, 456 (1983).
12. K. Ploog, H. Künzel, J. Knecht, A. Fischer, G.H. Döhler : Appl. Phys. Lett. 38, 870 (1981).
13. H. Künzel, G.H. Döhler, A. Fischer, K. Ploog : Appl. Phys. Lett. 38, 171 (1983).
14. P. Ruden, G.H. Döhler : Phys. Rev. B27, 3538, 3547 (1983).
15. G.H. Döhler, H. Künzel, K. Ploog : Phys. Rev. B25, 2616 (1982).
16. H. Jung, G.H. Döhler, H. Künzel, K. Ploog, P. Ruden, H.J. Stolz : Solid State Commun. 43, 291 (1982).
17. W. Rehm, H. Künzel, G.H. Döhler, K. Ploog, P. Ruden : Physica 117B & 118B, 732 (1983).
18. H. Künzel, G.H. Döhler, P. Ruden, K. Ploog : Appl. Phys. Lett. 41, 852 (1982).

Electronic Properties of InAs-GaSb and Related AlSb Superlattices

L.L. Chang

IBM T.J. Watson Research Center, Yorktown Heights, NY 10598, USA

Abstract

We describe the InAs-GaSb superlattices and quantum wells with emphasis on their transport and optical properties. We include related structures of InAs-AlSb and GaSb-AlSb which are of interest in their own right.

Studies in semiconductor superlattices and quantum wells have advanced greatly in recent years, covering wide ranges of material systems and structural configurations [1]. The focus has been on the electronic properties of two-dimensional electrons associated with subbands from electric quantization [2]. Aside from GaAs-GaAlAs, the InAs-GaSb system represents perhaps the most extensively studied [3,4]. The lattice constants of the two components are relatively close. Their energy bands are aligned in such a way that electrons and holes are confined in spatially separated regions, and both carriers may exist to give rise to a semimetallic behavior. For this reason, this superlattice is sometimes referred to as Type II, in contrast to the Type I structure typified by GaAs-GaAlAs where both electrons and holes are confined in GaAs.

In this work, we will describe the transport and optical properties of InAs-GaSb heterostructures, including both superlattices and quantum wells. In addition, we will touch upon related structures involving AlSb to the extent that they exhibit new features. Shown schematically in Fig. 1 are the lattice constants and energy gaps of these materials. Here, we have used the low-temperature values at the Γ-point for the gaps. The bandedge alignment in InAs-AlSb, like that in InAs-GaSb, puts it also in the Type II category, while GaSb-AlSb is similar to GaAs-GaAlAs [5,6]. For comparison, we include GaAs and AlAs in the figure. The bandedge relationship between the two is extrapolated from that obtained recently for the

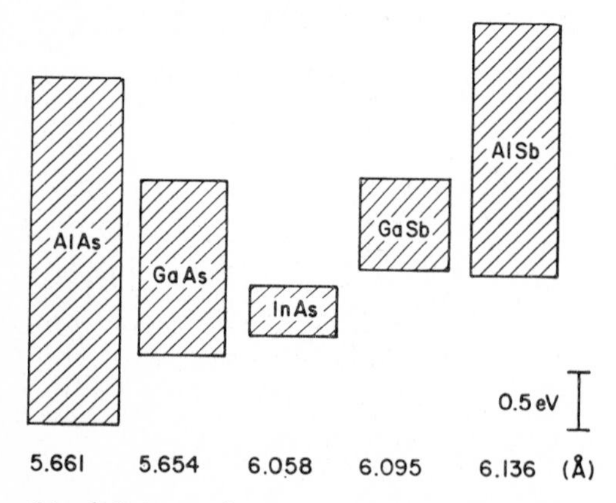

Fig.1 Schematic energy gaps and lattice constants of the various semiconductors under consideration.

GaAlAs alloys [7,8]. For GaAs and InAs, some uncertainties exist concerning the value of the valence band offset [9,10].

1. Indium Arsenide-Gallium Antimonide

The InAs-GaSb system is a special case of the more general alloy system with GaAs, i.e., InGaAs-GaSbAs. By increasing the GaAs composition, the valence bandedge of GaSbAs can be moved from above to below the conduction bandedge of InGaAs. The first experimental evidence of this behavior came from simple transport measurements across the heterojunctions with different alloys [11]. The current-voltage characteristics are either ohmic or rectifying, depending on whether the carriers are accumulated or depleted at the interface. In addition, double-barriers fabricated with a single well of InGaAs sandwiched between GaSbAs, exhibited resonant tunneling [4]. These preliminary experiments have demonstrated that InGaAs-GaSbAs is uniquely interesting,and that it is possible to achieve energy quantization in this system. Subsequently, optical absorption spectra were taken in superlattices with different alloy compositions and thicknesses [12]. For InAs-GaSb, the critical parameter of energy separation, the difference of the valence bandedge of GaSb above the conduction bandedge of InAs, was found to be about 0.15eV.

The investigation in superlattices in this alloy system has since been focussed on the binary compounds, for they represent the interesting case of bandedge separation and can be made with relative ease. The calculated quantum states or subbands as a function of the layer thickness is shown in Fig. 2. This is based on the envelope-wavefunction method in the two-band model, using equal layers of InAs and GaSb and the energy parameter of 0.15 eV as mentioned [13]. The results, similar to those from the tight-binding calculation [3], predict the crossing of the ground subbands of electrons (E_{1e}) and heavy holes (E_{1hh}) at about 85 Å. Theoretically, this is strictly valid only by neglecting the wavevectors parallel to the layers ; their inclusion may give rise to substantial coupling between the heavy and the light holes [14].

The InAs-GaSb system differs from GaAs-GaAlAs in two major aspects. One is the spatial separation of carriers as mentioned earlier, since electron states are formed in InAs with hole states in GaSb. This has a number of consequences on the absorption characteristics. The intensity is relatively weak, contributed only by the overlapping part of the wavefunctions of electrons and holes near the interfaces. The wavefunctions of these two carriers are not necessarily in phase for subbands with the same energy-indices,so that transitions involving differently indexed subbands may be equally probable. These points are borne out from experimental results shown in Fig. 3 where arrows indicate calculated transition energies, taking into account the strength of the overlapping integrals [15]. In all cases, the absorption edges are given by the ground state of electrons and

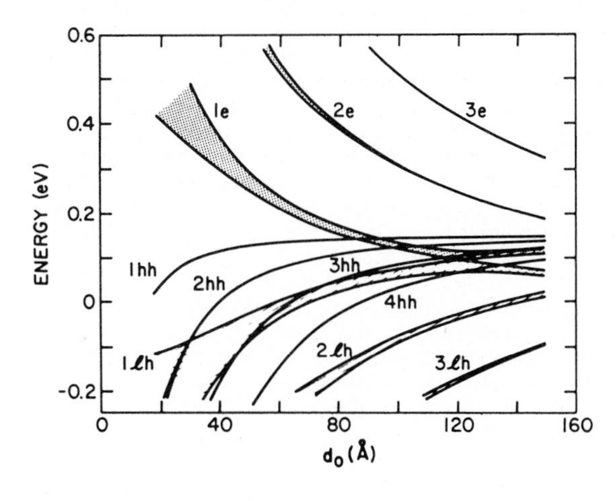

Fig.2 Calculated subband energies by the envelope wavefunction method for the InAs-GaSb superlattice with equal layer thickness d_0. Indicated are subband indices (n) for electrons (e), heavy (hh), and light (lh) holes.

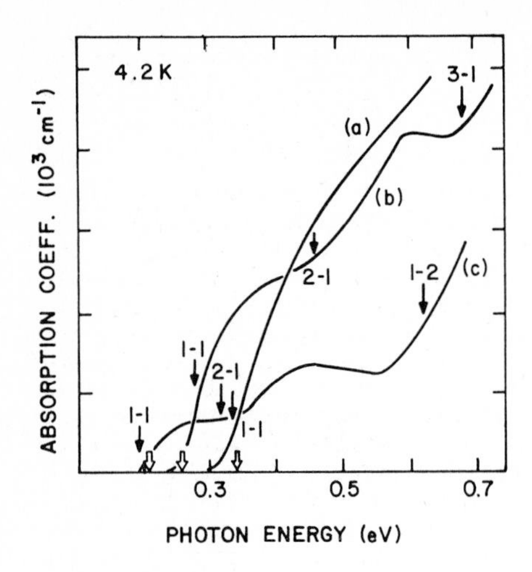

Fig.3 Absorption spectra of InAs-GaSb superlattices (a) 21-21 Å, (b) 27-27 Å, and (c) 37-37 Å at 4.2K. Indicated by arrows are calculated subband transitions from heavy holes to electrons. The block arrows mark the absorption edges obtained from magnetoabsorption.

heavy holes, while light-hole contributions are unimportant. Also evident from Fig.3 is the absence of excitonic effect, another consequence of carrier separation [15]. Many of the features in absorption were also observed in luminescence in the InAs-GaSb system [17].

The other major difference of InAs-GaSb from GaAs-GaAlAs is that the former exhibits the semimetallic behavior. As can be seen in Fig. 2, the superlattice gap, defined as $E_{1e} - E_{1hh}$, becomes zero and then negative with an increase in the layer thickness of InAs. In the semimetallic regime, electrons in the valence band of GaSb are transferred to the conduction band of InAs, creating in the ideal case an equal number of electrons and holes. This results in a sharp increase in the carrier concentration at the transition point, as first demonstrated from Hall measurements [18]. More detailed experiments of Shubnikov-de-Haas oscillations in magneto-resistance have been used to determine the electron density and the Fermi energy [19], as shown in Fig. 4. The value of E_f first increases as electron transfer starts. The space charges thus created lead to band bendings, which tend to push toward each other the subbands of electrons and holes. This effect, together with the occupation of multiple subbands as their energy spacings are narrowed with increasing thickness, reduces E_f. Eventually, the space-charge potentials localize the carriers near the interfaces, and the superlattice behaves in essence as a series of isolated heterojunctions.

Since the Shubnikov-de Haas effect, in general, reflects the extremal cross sections in the Fermi surface, it is very powerful to probe low-dimensional electron systems. That the angular dependence of the oscillatory period for a two-dimensional system should follow a cosine function of the tilting angle of the field has been used routinely to demonstrate two-dimensionality. In a superlattice where the wells are coupled, the electrons become somewhat three-dimensional with a finite subband width. Two extremal Fermi cross-sections arise, a maximum at the zone center and a minimum at the zone edge. Experimentally, two sets of oscillations were observed in this case, and their periods themselves oscillated with the tilting angle, as expected, when the extended zone of the superlattice was traversed [20]. Also observed was the enhancement of the g-value at high fields for electron spins in InAs.

Of great recent interest is the quantum Hall effect, first observed in Si-inversion layers [21] and subsequently in GaAs-GaAlAs heterojunctions [22]. Over certain ranges of the field, the magneto-resistance becomes vanishingly small, and the Hall resistance exhibits plateaus whose values depend only on fundamental constants and the occupation of the Landau levels. The results [23] for a single InAs well between GaSb are shown in Fig. 5. In addition to the usual features as

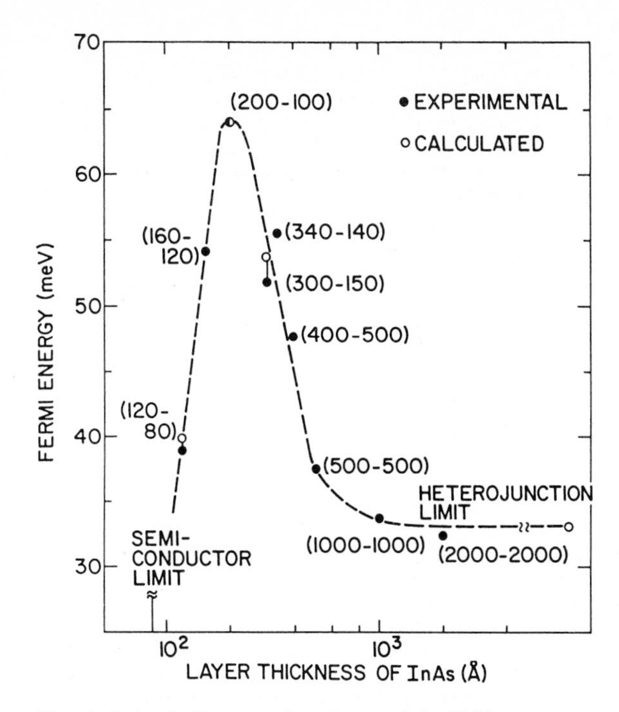

Fig.4 InAs-GaSb superlattices with different layer thickness in angstroms in parentheses to illustrate the Fermi energy variation in the semimetallic regime.

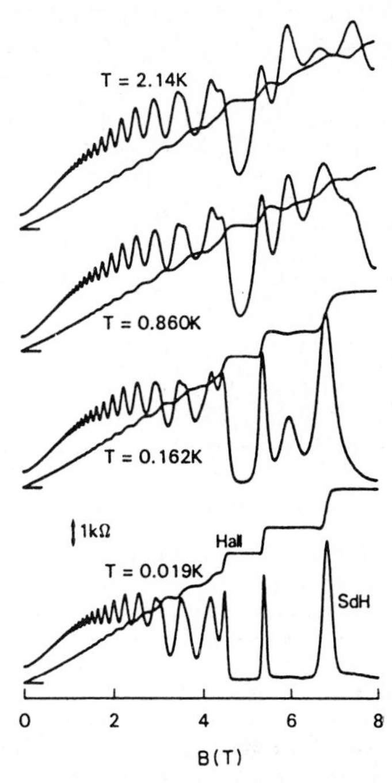

Fig.5 Magneto and Hall resistance at different temperatures for a single quantum well of InAs of 150 Å between GaSb. The sample has an electron concentration of $7.4\times10^{11}cm^{-2}$ and a mobility of $1.7\times10^{5}cm^{2}/V.sec$.

seen at the lowest temperature, extra peaks in magneto-resistance appear as the temperature is increased. These peaks, in comparison with those associated with electrons, are much broader and have a much stronger temperature dependence. The observation, which cannot be simply due to holes, is attributed to electron-hole interactions in a way which is not understood [23,24]. Such phenomenon was not observed in structures of thin wells with low electron mobilities. They exhibit, instead, a negative magneto-resistance at low fields, contrary to the present situation. In all cases, the carrier concentrations are too high to reach the extreme quantum limit with available fields to explore the fractional quantum Hall regime [25].

While the oscillatory density of states is probed through electron scattering at the Fermi energy in magneto-resistance, it can be studied conveniently also by magneto-absorption involving transitions with photoexcitation. Two types of transitions are generally of interest. One is inter-subband absorption arising from Landau levels between the valence and conduction bands ; the energies of absorption peaks provide information about the subband structure. The other is cyclotron resonance, sometimes referred to as intra-subband absorption, between Landau levels across the Fermi energy within the same band of electrons or holes. The variation of absorption energies with fields in this case gives a direct determination of the effective mass. Information can also be obtained from cyclotron interactions with phonons and impurities.

Measurements of magneto-absorption in the semiconducting regime have yielded values of the absorption edges,as marked in block arrows in Fig. 3, which are

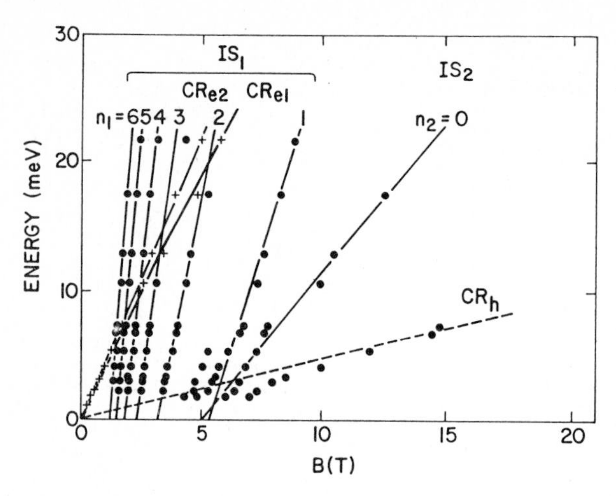

Fig.6 Energies of magneto-absorption in an InAs (1000 Å)-GaSb (1000 Å) superlattice in the heterojunction limit. Indicated are inter-subband transitions (IS_1 and IS_2) and cyclotron resonances of electrons (CR_{e1} and CR_{e2}) and holes (CR_h).

consistent with those from simple optical absorption. But most extensive work was done on semimetallic superlattices, showing a number of interesting observations. Figure 6 plots the results in the standard fashion for a thick superlattice in the heterojunction limit [26]. Two branches of inter-subband transitions are identified : IS_1 corresponds to the ground states E_{1e}-E_{1hh} ; and IS_2 to the same hole subband but the second electron subband E_{2e}. Extrapolated energies to zero field of both branches are negative, a direct demonstration of the semimetallic nature, and the difference between the two gives a direct measure of the energy spacing of the electron subbands. In addition, three cyclotron transitions (CR) are marked. Two are associated with electrons of the two subbands, a consequence of the nonparabolic conduction band of InAs. The third one is attributed to heavy holes in GaSb.

Similar absorption experiments were performed on superlattices with different configurations [27], and with the field applied in directions away from the surface normal [28]. For structures with a significant subband width, this width can be measured by observing the difference in absorption energies between transitions at the center and edge of the superlattice zone. In the case of off-normal fields, mixed electric and magnetic quantization occurs. Subbands and Landau levels must be considered in an integrated fashion. The latter dominates at high, parallel magnetic fields to yield well-defined transitions, which correspond to the cyclotron resonance under proper conditions [28].

2. Indium Arsenide-Aluminum Antimonide

The semiconductor AlSb is an indirect material with its minimum energy gap at the X-point about 0.8 eV below that indicated in Fig. 1. From the heteroepitaxial point of view, InAs-AlSb represents the most undesirable among the three systems under consideration. Its lattice mismatch of 1.3% is twice as large as that in InAs-GaSb, and it does not share a common element as in the case of GaSb-AlSb. In addition, AlSb is hygroscopic,which creates problems in processing [5,29]. Our interest in this system originated from the proposal of polytype superlattices involving three materials [30]. We carried out investigations of InAs-AlSb in parallel with InAs-GaSb, anticipating the similarity between GaSb and AlSb.

Figure 7 shows the electron concentrations and mobilities as a function of the layer thickness in single wells of InAs sandwiched between AlSb [5]. Both

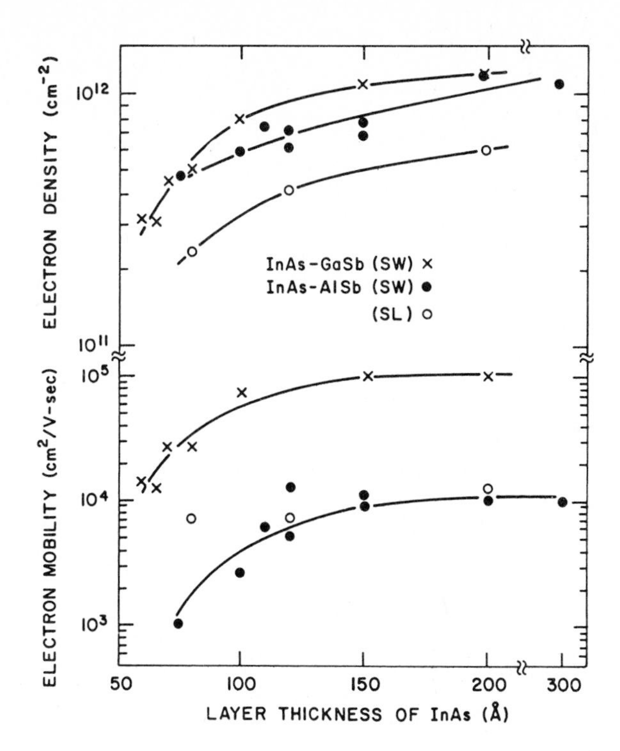

Fig.7 Electron concentrations and mobilities at 77K of InAs-AlSb single wells and superlattices, and InAs-GaSb single wells.

quantities are seen to increase with the thickness and tend to saturate. The rise in electron concentration is believed to originate basically from the same transfer mechanism as that in InAs-GaSb whose results from single wells are included in the figure for comparison. The rise in mobility is due predominantly to a reduction in interface scattering rather than the effect of carrier screening, as the mobilities in superlattices, also included in Fig. 7, are much higher in the thin regions. That the interface quality is improved with the superlattice is consistent with the experience in GaAs-GaAlAs. The overall quality of InAs-AlSb is inferior to InAs-GaSb, as evidenced from the order-of-magnitude difference in their saturated mobility values. In both cases, the presence of holes provides an additional process for the scattering of electrons.

It is noted from Fig. 7 that the electron concentrations for a given InAs layer thickness are smaller in the superlattices than in the single wells. While the thin AlSb layers in the superlattice may limit the electron transfer, it is likely that the difference results mainly from the influence of strain to be described below for GaSb-AlSb. In the present case, the effect of bilateral dilation of InAs, because of its smaller lattice, is expected to be more significant in single wells, causing a greater bandgap shrinkage. This lowers the conduction bandedge of InAs, with the consequence of increasing the above-mentioned energy separation, and thus the electron concentration. Indeed, similar differences have also been found in InAs-GaSb. The critical thickness for the onset of the semimetallic behavior, as is apparent in Fig. 7, is smaller than the 85 Å obtained from superlattice structures. Another point of interest is that the electron concentrations of InAs-AlSb, scattered as they are, appear to be lower than those of InAs-GaSb. This may be taken as evidence that the valence bandedge of AlSb lies below that of GaSb. A rough estimate of this value is consistent with the results of GaSb-AlSb in the following section.

Only preliminary transport measurements under magnetic fields have been performed. They demonstrated the two-dimensional nature of the electron system, and exhibited enhanced spin g-factors and well-defined Hall plateaus. Recently, the local order in the system was examined from extended x-ray absorption fine structure (EXAFS) [31]. The results are consistent with a tetragonal distortion of the InAs lattice with the local environment in the superlattice being as well ordered as in the bulk. An anisotropy appears to exist in the electron mean path, which is shorter in the direction perpendicular to the layers.

3. Gallium Antimonide-Aluminum Antimonide

With a mere replacement of As by Sb in GaAs-AlAs, the GaSb-AlSb superlattice can be readily made. The lattice mismatch, about 0.7 %, is similar to that in InAs-GaSb. A number of studies have been reported concerning the optical properties of the system, including both luminescence [32-34] and absorption [6,35]. The lack of efforts in transport measurements is due to the fact that this system, unlike the previous two involving electrons in InAs, deals with holes in the valence band.

Earlier experiments have demonstrated the formation of the superlattice both metallurgically from x-ray diffraction and electronically from photoluminescence and electroreflectance [33]. The temperature dependence of the luminescent energy was found to follow that of the GaSb energy gap, indicating the confinement of both electron and hole subbands in this material. The subbands at the Γ-point were shown to dominate the process, which persisted even for thin GaSb layers when conduction band minima at other points in the Brillouin zone became lower in energy because of their larger effective masses [34]. Absorption measurements, subsequently , provided accurate values of the transition energies and revealed the two-dimensional density of states [35]. From the intensities, the Type I nature of the superlattice was confirmed. In addition, the derived energy gap of GaSb was found to be invariably smaller than the bulk value, providing initial evidence of the effect of strain.

Figure 8 shows the results of absorption experiments on improved samples , which illustrate not only the gap shrinkage but also the excitonic absorption peaks and, more remarkably, the reversal in energy of the ground subbands of heavy and light holes [6]. The lattice mismatch causes the GaSb to stretch and the AlSb to contract in the plane of the layers. For GaSb where the strain effect is of primary concern, this biaxial dilation results in a hydrostatic tensile stress which shrinks the gap and a uniaxial compressive stress in the perpendicular direction which splits the valence band. The ground state of the light holes is lowered in (hole) energy with respect to that of the heavy holes and, under favorable conditions of the layer

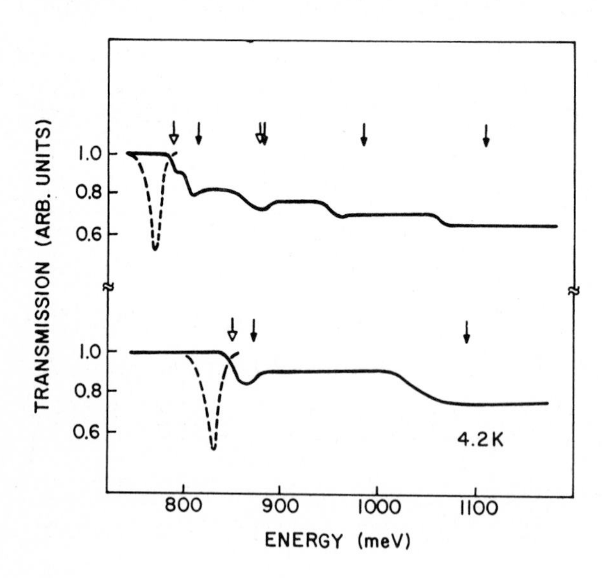

Fig.8 Transmission spectra of GaSb-AlSb superlattices of 181-452 Å (upper) and 84-419 Å (lower) at 4.2K. Solid and open arrows indicate calculated transitions from heavy and light holes, respectively, to electrons. Dotted curves represent luminescence.

thickness, becomes the ground state of the valence band in the system. Quantitative fit to the experimental data, as seen in Fig. 8, led to the valence band offset between GaSb and AlSb of about 40 meV for heavy holes and somewhat higher for light holes under the strained conditions.

Recently, Raman scattering, a technique which has been used extensively to probe low-dimensional electrons in GaAs-GaAlAs [36], was applied to the present system. In one experiment [37], the LO phonon energy in GaSb was found to shift downward from the bulk value, consistent with the effect of lattice strain. In the other experiment, resonant enhancement of GaSb LO-phonons was observed [38]. The resonant energies, increasing in the case of superlattices, were interpreted as due to increases of the gap at the L-point in GaSb.

Acknowledgments

We acknowledge important contributions to this work from many of our colleagues and collaborators. The research is partially sponsored by the US Army Research Office.

References

1. L.L. Chang: J. Vac. Sci. Technol. B1, 120 (1983) ; *in Molecular Beam Epitaxy and Heterostructures*, ed. by L.L. Chang and K. Ploog (Nihoff, Dordrecht, 1985), Chapter 13
2. T. Ando, A.B. Fowler and F. Stern: Rev. Modern Phys. 54, 437 (1982)
3. G.A. Sai-Halasz, L. Esaki, and W.A. Harrison: Phys. Rev. B13, 2812 (1978)
4. L.L. Chang and L. Esaki: Surf. Sci. 98, 70 (1980)
5. C.A. Chang, L.L. Chang, E.E. Mendez, M.S. Christie, and L. Esaki: J. Vac. Sci. Technol. 214 (1984) ; Surf. Sci. 142, 598 (1984)
6. P. Voisin, C. Delalande, M. Voos, L.L. Chang, A. Segmüller, C.A. Chang, and L. Esaki: Phys. Rev. B30, 2276 (1984)
7. R.C. Miller, A.C. Gossard, D.A. Kleinman, and O. Munteanu: Phys. Rev. B29, 3740 (1984)
8. W.I. Wang, E.E. Mendez, and F. Stern: Appl. Phys. Lett. 45, 639 (1984)
9. S.P. Kowalczyk, W.J. Schaffer, E.A. Kraut, and R.W. Grant: J. Vac. Sci. Technol. 20, 705 (1982)
10. G. Margaritondo, A.D. Katnani, N.G. Stoffel, R.R. Daniels, and T.X. Zhao: Solid State Commun. 43, 163 (1982)
11. H. Sakaki, L.L. Chang, R. Ludeke, C.A. Chang, G.A. Sai-Halasz, and L. Esaki: Appl. Phys. Lett. 31, 211 (1977)
12. G.A. Sai-Halasz, L.L. Chang, J.M. Welter, C.A. Chang, and L. Esaki: Solid State Commun. 27, 935 (1978)
13. G. Bastard: Phys. Rev. B24, 5693 (1981) ; in *Molecular Beam Epitaxy and Heterostructures*, ed. by L.L. Chang and K. Ploog (Nihoff, Dordrecht, 1985), Chapter II
14. M. Altarelli: Physica 117-118BC, 744 (1983)
15. L.L. Chang: G.A. Sai-Halasz, L. Esaki, and R.L. Aggarwal, J. Vac. Sci. Technol. 19, 589 (1981)
16. G. Bastard, E.E. Mendez, L.L. Chang, and L. Esaki: B26, 1974 (1982)
17. P. Voisin, G. Bastard, C.E.T. Goncalves da Silva, M. Voos, L.L. Chang, and L. Esaki: Solid State Commun. 39, 79 (1981)
18. L.L. Chang, N.J. Kawai, G.A. Sai-Halasz, R. Ludeke, and L. Esaki: Appl. Phys. Lett. 35, 939 (1979)
19. L.L. Chang, N.J. Kawai, E.E. Mendez, C.A. Chang, and L. Esaki: Appl. Phys. Lett. 38, 30 (1981)
20. L.L. Chang, E.E. Mendez, N.J. Kawai, and L. Esaki: Surf. Sci. 113, 306 (1982)
21. K. von Klitzing, G. Dorda, and M. Pepper: Phys. Rev. Lett. 45, 494 (1980)
22. D.C. Tsui and A.C. Gossard: Appl. Phys. Lett. 38, 550 (1981)
23. S. Washburn, R.A. Webb, E.E. Mendez, L.L. Chang, and L. Esaki: Phys. Rev. B31, 1198 (1985) ; Phys. Rev. B29, 3752 (1984)
24. E.E. Mendez, L.L. Chang, C.A. Chang, L.F. Alexander, and L. Esaki: Surf. Sci. 142, 215 (1984)

25. D.C. Tsui, H.L. Störmer, and A.C. Gossard: Phys. Rev. Lett. 48, 1559 (1982)
26. Y. Guldner, J.P. Vieren, P. Voisin, M. Voos, J.C. Maan, L.L. Chang, and L. Esaki: Solid State Commun. 41, 755 (1982) ; Phys. Rev. Lett. 45, 1719 (1980)
27. J.C. Maan, Y. Guldner, J.P. Vieren, P. Voisin, M. Voos, L.L. Chang, and L. Esaki: Solid State Commun. 39, 683 (1981)
28. J.C. Maan, Ch. Uihlein, L.L. Chang, and L. Esaki: Solid State Commun. 44, 653 (1982)
29. C.A. Chang, H. Takaoka, L.L. Chang, and L. Esaki: Appl. Phys. Lett. 40, 983 (1982)
30. L. Esaki, L.L. Chang, and E.E. Mendez: Jpn J. Appl. Phys. 20, L529 (1981)
31. E. Canova, A.I. Goldman, S.C. Woronick, Y.H. Kao, and L.L. Chang: submitted for publication
32. M. Naganuma, Y. Suzuki and H. Okamoto: in *Proc. Int. Sym. GaAs and Related Compounds*, ed. by T. Sugano, Oiso, 1981 (Inst. of Phys. University of Reading, Berkshire, 1981), p. 125
33. E.E. Mendez, C.A. Chang, H. Takaota, L.L. Chang, and L. Esaki: J. Vac. Sci. Technol. B1, 152 (1983)
34. G. Griffiths, K. Mohammed, S. Subbana, H. Kroemer, and J. Merz: Appl. Phys. Letter. 43, 1059 (1983)
35. P. Voisin, G. Bastard, M. Voos, E.E. Mendez, C.A. Chang, L.L. Chang, and L. Esaki: J. Vac. Sci. Technol. B1, 409 (1983)
36. G. Abstreiter and K. Plog: Phys. Rev. Lett. 42, 1308 (1979) ; G. Abstreiter, in *Molecular Beam Epitaxy and Heterostructures*, ed. by L.L. Chang and K. Ploog (Nihoff, Dordrecht, 1985), Chapter 12
37. B. Jusserand, P. Voisin, M. Voos, L.L. Chang, E.E. Mendez and L. Esaki: submitted for publication
38. C. Tejedor, J.M. Calleja, F. Meseguer, E.E. Mendez, C.A. Chang, and L. Esaki: presented at the *17th Int. Conf. Phys. Semicond.*, San Francisco, 1984. (To appear in Proceedings)

Strained Superlattices

J.Y. Marzin

Laboratoire de Bagneux*, C.N.E.T., 196 rue de Paris, F-92220 Bagneux, France
*Laboratoire associé au C.N.R.S. (L.A. 250)

We discuss some features of strained superlattices, built with III-V semiconductors : we describe their structure and we show how to include the strain effects in an envelope function-type model of the band structure. These properties are illustrated by experimental results (mainly optical data) obtained on such strained systems.

After Esaki and Tsu proposed the idea of superlattices [1], the first to be realized were strained GaAsP-GaAs superlattices [2]. With the development of Molecular Beam Epitaxy and Organo Metallic Vapor Phase Epitaxy, "unstrained" systems, where the small lattice mismatches between the constitutive materials are easily accommodated by elastic strains, have been mainly investigated. In recent years, the "strained" systems have regained interest, because their use broadens the choice of epitaxial materials on a given substrate. A large number of new such systems [3-6], built with III-V compounds with large lattice mismatches of the order of 1%, have been investigated. For these systems, the strain effects on the band structure have to be taken into account. We will first study some general features of these strained superlattices built with III-V compounds, before we discuss some experimental results obtained on two strained systems : InGaAs-GaAs and GaSb-AlSb both grown on GaAs substrates.

1. Strain analysis

A strained superlattice is a periodic structure consisting of alternate layers of two materials (1) and (2), with lattice parameters a1 and a2, (we will take a1 > a2, for example), and thicknessès L1 and L2 small enough [7] so that the lattice mismatch a1-a2/a2 is elastically accommodated. The two materials inside such superlattice are strained in order to match their in-plane parameter $a_\perp$. Throughout the paper Z will be the growth axis.

The strained superlattices are in general grown either on graded buffer layer or on a substrate of material (1) or (2). In the former case, the buffer layer is built to have a parameter at the interface with the superlattice that matches its equilibrium parameter $\bar{a}$. If we assume that the superlattice remains planar and quadratically strained, a can be easily evaluated by writing the equilibrium conditions of the superlattice boundaries (Fig. 1 (a)), and is equal to L1.a1+L2.a2/L1+L2. The material (1) layers are under biaxial compression, and the others in biaxial tension. In the latter case, where, for example, the superlattice is grown on a material (2) substrate, the whole structure is elastically strained to take an in-plane $a_\perp$ parameter, which is nearly equal to a2. In the superlattice, the (2) layers are essentially unstrained, while the others are, in our case a1 > a2, in biaxial tension. As can be seen in Fig. 1(b), the strain induced forces tend to bend the structure. The equilibrium radius of curvature can be estimated[8] by

$$R = \frac{a2.Ls^2}{(\bar{a}-a_2)Ls1}$$

where Ls and Ls1 are the substrate and superlattice thicknesses, with Ls1 << Ls. In

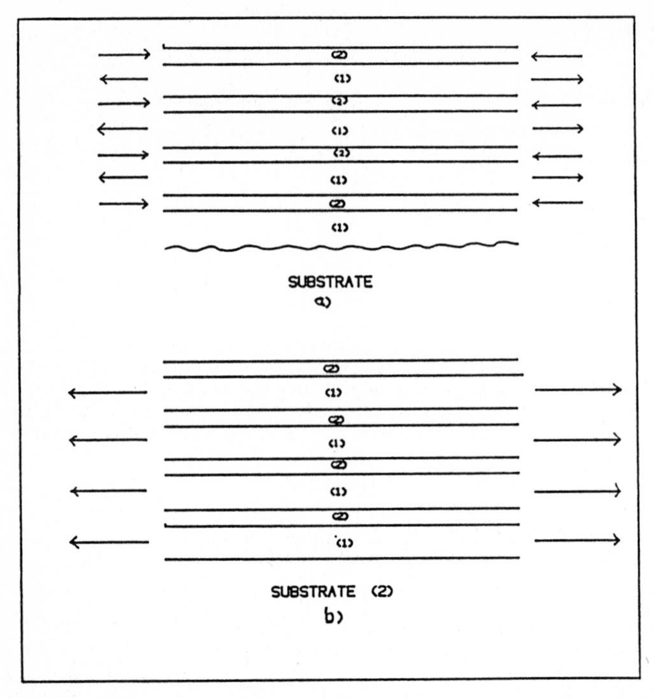

Fig.1 Schematic diagram of two extreme strain configurations in a superlattice built with materials (1) and (2), a1 > a2 :
(a) the superlattice has taken its in plane equilibrium parameter, (1) is under compression and (2) under tension, there is no resultant force on the surface of the superlattice.
(b) The superlattice is strained on the substrate. The planar configuration drawn here where the in plane parameter is the mean parameter of the whole structure, substrate included, (and is slightly larger than a2), is not the equilibrium : there are forces on the substrate (not represented) and the superlattice surfaces and this system of forces tends to bend the structure.

the systems we will describe here, the effects of this curvature remain small enough to be neglected. They should be included, on the contrary, for example in a detailed study of the widths of the X-ray satellites.

The second option (no buffer layer) avoids the existence of a poor quality buffer layer, allows to vary the thickness of one sublayer alone without modifying the strains, but leads to an additionnal limitation of the overall superlattice thickness, beyond which it will tend to take its equilibrium parameter.

In both cases, the strains are constant in each sublayer, if we neglect curvature as well as edge effects, and the only non-vanishing strains are

$$\varepsilon_{xx}^{(1)},\ \varepsilon_{yy}^{(1)},\ \varepsilon_{zz}^{(1)},\ \varepsilon_{xx}^{(2)},\ \varepsilon_{yy}^{(2)},\ \varepsilon_{zz}^{(2)}$$

with

$$\varepsilon_{xx}^{(1,2)} = \varepsilon_{yy}^{(1,2)} = \frac{a_{\perp} - a_{(1,2)}}{a(1,2)} \qquad \varepsilon_{zz}^{(1,2)} = \frac{a_z^{(1,2} - a_{(1,2)}}{a(1,2)}$$

They are linked with the stresses by the usual relation

$$\sigma_{ii}^{(1,2)} = C_{ij}^{(1,2)} \varepsilon_{jj}^{(1,2)}$$

where the Cij's are the bulk elastic constants, in standard notation.

The strains in the Z direction are obtained by writing that the (x,y) surfaces are not stressed :

$$\sigma_{zz}^{(1,2)} = 2C_{12}^{(1,2)} \varepsilon_{xx}^{(1,2)} + C_{11}^{(1,2)} \varepsilon_{zz}^{(1,2)} = 0$$

so

$$\varepsilon_{zz}^{(1,2)} = -2 \frac{C_{12}^{(1,2)}}{C_{11}^{(1,2)}} \varepsilon_{xx}^{(1,2)}$$

The biaxial stresses, for a layer in compression, are equivalent to the sum of a hydrostatic pressure and a uniaxial tension along the Z axis. Both of them, for a typical lattice mismatch of 1% are of the order of 10 kbars. Thus, strained superlattices allow one to study the effects of large stresses, including tension, on thin layers.

In the superlattices, the existence of these strains will affect the cristallographic properties as well as the band structure.

2. X-ray diffraction patterns

The X-ray diffraction pattern analysis appears to be a very powerful tool for the study of strained structures [9]. A typical spectrum obtained on a InGaAs-GaAs superlattice is shown in Fig. 5. In this system, they allow a precise determination of the In composition, of the sublayers thicknesses, L1 and L2, and also give an estimation of the overall superlattice strain state with respect to the substrate [9]. These profiles are obtained by double diffraction in the vicinity of the GaAs 004 diffraction peak. The reflecting planes are parallel to the surface. The precision in the determination of the sample parameters mainly results from the following :

if a1 and a2 are different enough, and for rather small L1 and L2 (of the order of 100 Å), the diffraction pattern of one period of the superlattice should look like Fig. 2 (Curve C_1), i.e. mainly two lines centered on Bragg positions for a1z and a2z, which are the lattice parameters in the Z direction, largely broadened by a small size effect in a $\sin^2 x/x^2$ (curves C2,C3) shape, eventually slightly interfering. If several periods are stacked, then the new periodicity will give an additional Bragg condition in order to have non-destructive interference between the two envelope-shaped patterns diffracted by different periods. This gives the final spectrum, also shown in Fig. 2a. This diffraction pattern gives immediately the superlattice period from the spacing of the satellites. Their intensity, which is modulated by the one-period envelope, allows a determination of the individual sublayers thicknesses, and of a1z and a2z. In the experimental spectra, the line corresponding to the 004 substrate diffraction is also observed. If the whole superlattice is relaxed with respect to the substrate, the sublayers still keeping the same $a_\perp$, although now different from the substrate parameter, then the in plane additional strains will be the same for all sublayers. If the elastic constants ratio 2.C12/C11 is the same in the two materials, the a1z and a2z will be modified by this relaxation of the same amount -2 C12/C11 Δ $a_\perp$. Therefore, the superlattice diffraction pattern will essentially shift with respect to the substrate line, without being modified, so that their relative position will give an indication of the superlattice strain state.

More formally, the intensity calculated in a one-dimension model is given by

$$I(\theta)\alpha \left| \sum_{m=0}^{n-1} e^{i4m\pi(L_1+L_2)\frac{\sin\theta}{\lambda}} \left[\sum_{n=0}^{n_1-1} e^{2i\pi n a_{1z}\frac{\sin\theta}{\lambda}} (F_A^{(1)}+F_B^{(1)} e^{i\pi a_{1z}\frac{\sin\theta}{\lambda}}) \right.\right.$$

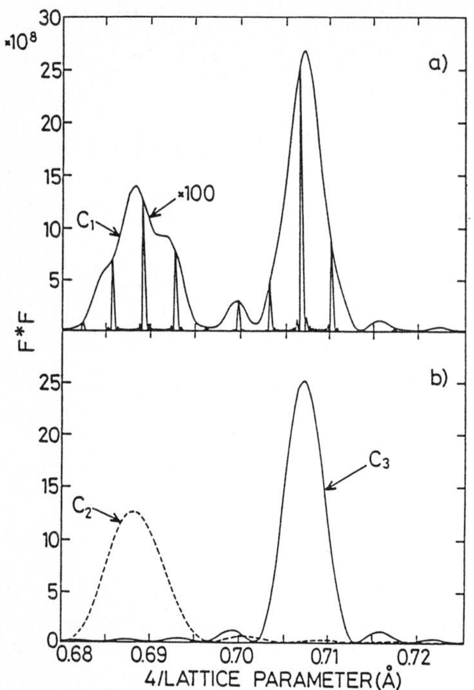

Fig.2 a) Calculated profile on a N=10 periods strained superlattice (one period = 60 monolayers f GaAs and 30 of $In_{.3}Ga_{.7}$ As). The envelope C1 is the calculated profile for one period, with a N^2 = 100 intensity correction.
b) profiles obtained for single $(GaAs)_{60}$ (C3) and $(In_{.3}Ga_{.7}As)_{30}$ (C2) sublayers. Comparing a) and b) shows that each set of peaks refers almost exactly to each material.

$$+\sum_{n=0}^{n_2-1} e^{2i\pi(na_{2z}+n_1a_{1z})\frac{\sin\theta}{\lambda}}\left(F_A^{(2)}+F_B^{(2)}e^{i\pi a_{2z}\frac{\sin\theta}{\lambda}}\right)]|^2$$

N number of periods

where λ is the X-ray wavelength, $F_A^{(1)}, F_B^{(1)}, F_A^{(2)}, F_B^{(2)}$ the structure factors of the constituent atoms in the two materials and the n_i's given by $L_i = n_i.a_i/2$. For strained superlattices, contrarily to what is observed in unstrained ones (GaAs-GaAlAs), the diffraction profiles intensities are due more to the differences in the lattice parameters than to the differences in the structure factors of successive layers. This leads, in the strained case, to much stronger satellites which can be more easily used to determine the characteristics of the studied samples.

3. Electronic band structure

In this part, we will restrict ourselves to superlattices where the relevant states can be built with near-Brillouin zone center Γ6, Γ8, Γ7 states of the hosts materials. In order to calculate the energies of the transitions observed in the optical experiments, we need a description of the superlattice which takes into account the effects of the strains. For this purpose, we will first focus on what happens in a strained bulk material. We will neglect the influence of the lack of inversion symmetry in the III-V compounds, as it is expected to lead to small effects.

From group theory, the expression of the strain + spin orbit hamiltonian, operating on the valence bands zone center states of Td semiconductors is the following [10], for a quadratic deformation of axis Z, in a $|J, m_J\rangle$ basis, with J "quantized" along Z :

$$H_{ST} = \begin{array}{c} \\ \end{array}\begin{array}{ccc} |3/2, 3/2\rangle_z & |3/2, 1/2\rangle_z & |1/2, 1/2\rangle_z \\ \end{array}$$

$$H_{ST} = \begin{bmatrix} -\delta E'_H - \frac{1}{2}\delta E_s & 0 & 0 \\ 0 & -\delta E'_H + \frac{1}{2}\delta E_s & \frac{1}{\sqrt{2}}\delta E'_s \\ 0 & \frac{1}{\sqrt{2}}\delta E'_s & -\Delta_o - \delta E''_H \end{bmatrix} \quad (1)$$

and an identical expression on the negative m_J functions.

Where

$$\delta E''_H \simeq \delta E'_H = c(\varepsilon_{xx} + \varepsilon_{yy} + \varepsilon_{zz})$$
$$\delta E'_s \simeq \delta E_s = 2b\ (\varepsilon_{zz} - \varepsilon_{xx}) \quad (2)$$

The lowering of the symmetry results in a splitting at the Γ point of the J = 3/2 valence band states equal to

$$\delta E_s = 2b\ (\varepsilon_{zz} - \varepsilon_{xx}) \quad (3)$$

in the small strain limit. The $m_J = \pm\ 3/2$ "heavy" hole states remain uncoupled to the light particle ones, while the light hole and split-off bands are coupled by the strain. The eigen states can be noted $\phi_{3/2}$, $\phi^+_{1/2}$ and $\phi^-_{1/2}$ ($\phi^-_{1/2}$ extrapolating to $|1/2, 1/2\rangle_z$ at zero stress). The Γ_6 conduction band is only affected by the hydrostatic part of the strain, which gives finally the following strain-induced variations of the (Γ) valence to conduction bands transitions energies, in the first order in $\delta E_s/\Delta_o$

$$\phi_{3/2} \rightarrow CB \qquad E(\bar{\bar{\varepsilon}}) - Eg(o) = \delta E_H + \frac{\delta E_s}{2}$$

$$\phi^+_{1/2} \rightarrow CB \qquad E(\bar{\bar{\varepsilon}}) - Eg(o) = \delta E_H - \frac{\delta E_s}{2} - \frac{1}{2}\frac{(\delta E_s)^2}{\Delta o} \qquad \text{with}$$

$$\phi^-_{1/2} \rightarrow CB \qquad E(\bar{\bar{\varepsilon}}) - Eg(o) = \delta E_H + \Delta o + \frac{1}{2}\frac{(\delta E_s)^2}{\Delta o} \quad (4)$$

$$\delta E_H = a(\varepsilon_{xx} + \varepsilon_{yy} + \varepsilon_{zz})$$

a and b are the deformation potentials and in GaAs a = 8.4 eV, b = -2eV. Fig. 3 represents the evolution of these valence to conduction bands optical transitions energies, in a thick InGaAs layer strained on a InP substrate, as a function of the In content in the alloy. One can note the asymmetrical behaviour of the band gap, as well as the effect of the coupling of the two $m_J = 1/2$ valence bands. We want to

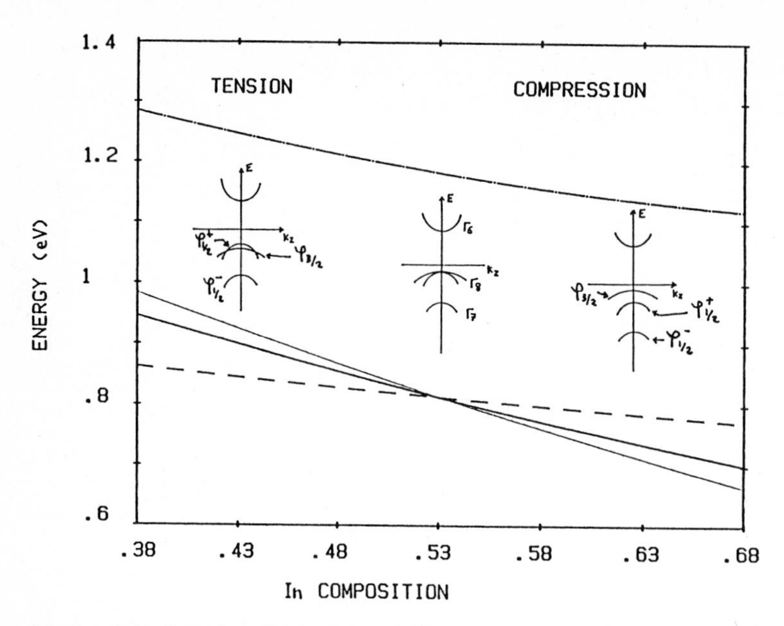

Fig.3 Calculated energies for the valence to conduction bands extrema transitions in a thick $In_xGa_{1-x}As$ layer strained on an InP substrate, as a function of the In content in the alloy, around the lattice matching x = .53 composition.

—— unstrained band gap, —— $\phi^{+}_{3/2} \rightarrow$ CB ----- $\phi^{+}_{1/2} \rightarrow$ CB –·–·– $\phi^{-}_{1/2} \rightarrow$ CB in the strained case.
The insets schematize the band configurations when the layer is under tension, unstrained and under compression, from left to right.

emphasize here that for a given application, the sign of the strain is important. For example, to get a low gap layer on a substrate, for a given unstrained material gap, the strained layer gap will be lower if it can be chosen with a smaller lattice parameter than the substrate, so that it will be in tension inside the structure. Otherwise, the strain effects will shrink the band gap difference. In general, in the III-V compounds, the band gap decreases with increasing lattice parameter, so that the strain-induced variation of the band gap will constitute a limiting factor.

The selection rules for these transitions can be obtained from group theory and in the dipolar approximation, if the light polarisation is linear, the only symmetry-forbidden transition is $\rho_{3/2} \rightarrow$ CB, for a light polarisation along Z. This result is the same as in an unstrained superlattice, which overall symmetry is also D2d.

In the superlattice, the effective masses of the $\phi_{3/2}$ and $\phi^{+}_{1/2}$ valence band will be important parameters. They can be obtained easily, when the J = 3/2 bands can be considered as decoupled from the J = 1/2 one, ie for a large spin orbit splitting Δ_o, and $\delta E_s << \Delta_o$. Then, for small k values, we can write the hamiltonian restricted to these two twofold bands as the sum of a strain hamiltonian, obtained from (1) and of a Luttinger [11,12] one, as

$$
\begin{array}{cccc}
|3/2,3/2>_z & |3/2,1/2>_z & |3/2,-1/2>_z & |3/2,-3/2>_z
\end{array}
$$

$$
\begin{bmatrix}
\frac{-\hbar^2}{2m_o}[(\gamma_1+\gamma_2)k_\perp^2+(\gamma_1-2\gamma_2)k_z^2] - \delta E_H - \frac{\delta Es}{2}, & B, & C, & 0 \\
B^*, & \frac{-\hbar^2}{2m_o}[(\gamma_1-\gamma_2)k_\perp^2+(\gamma_1+2\gamma_2)k_z^2, -\delta E_H - \frac{\delta Es}{2} & 0, & C \\
C^*, & 0, & \frac{-\hbar^2}{2m_o}[(\gamma_1-\gamma_2)k_\perp^2+(\gamma_1+2\gamma_2)k_z^2] - \delta E_H + \frac{\delta E_s}{2}, & -B \\
0, & C^*, & -B^*, & \frac{-\hbar^2}{2m_o}[(\gamma_1+\gamma_2)k_\perp^2+(\gamma_1-\gamma_2)k_z^2] - \delta E_H - \frac{\delta E_s}{2}
\end{bmatrix} \quad (5)
$$

$$B = 2\sqrt{3}\,k_z(k_x-ik_y)\frac{\hbar^2}{2m_o} \qquad C = \sqrt{3}\,[\gamma_2(k_x^2-k_y^2)-2i\gamma_3k_xk_y]\frac{\hbar^2}{2m_o}$$

$$k_\perp^2 = k_x^2 + k_y^2$$

where the γi's are the Luttinger parameters.

For $k_\perp = 0$,(5) is diagonal, so that the effective masses along k are not affected by the strain. For $k_z = 0, k_\perp \neq 0$ and zero stress, after diagonalisation of (5), we find, as expected, the same effective masses in X and Y directions as in the Z one, because the non-zero C terms couple $|3/2, \pm 3/2>_z$ and $|3/2, \mp 1/2>_z$ states. This will not remain true for finite stresses, since, although the C terms are not changed, the δE_s diagonal terms tend to decouple, for large strains and small ks, the $m_J = 3/2$ band from the other. In this large stress limit, the effective transverse masses are simply determined from the diagonal k dependant terms in (5) and :

$$
\left.\begin{aligned}
\frac{m_o}{m_z} &= 2\gamma_2 - \gamma_1 \\
\frac{m_o}{m_\perp} &= -(\gamma_1+\gamma_2)
\end{aligned}\right\} m_J = \pm 3/2
\qquad
\left.\begin{aligned}
\frac{m_o}{m_z} &= -2\gamma_2-\gamma_1 \\
\frac{m_o}{m_\perp} &= \gamma_2-\gamma_1
\end{aligned}\right\} m_J = \pm 1/2 \quad (6)
$$

The expressions for the longitudinal masses do not depend on the stress, in this frame.

The light holes masses are slightly modified if one takes into account the coupling of the $|3/2, \pm 1/2>_z$ and $|1/2, \pm 1/2>_z$ bands.

This phenomenon is often called the mass reversal, because the in-plane mass of the "heavy" holes becomes lighter than the in-plane mass for the "light" holes. It should happen as soon as a perturbation splits strongly the valence band without adding additional coupling, as it is the case in unstrained superlattices.

As the masses in the Z direction do not depend on the band splitting, it will not modify the confinement energies in the superlattice, but the excitons binding energies [13], the magnitude of the joint density of state, and consequently the absorption coefficient [14].

In order to calculate the band structure of the superlattice, we will use the envelope functions formalism developed by G. Bastard [15]. As the strain does not couple heavy and light hole particles, the heavy hole problem can be, in this approximation, still treated separately for $k_\perp = 0$. We will only include the strain effects in the approximation we used above for the bulk material, where we have assumed $H_{ST} = H_{ST}$ ($\vec{k} = 0$).

In the cases where non-parabolicity can be neglected, and for strains effects small enough compared to the spin-orbit splitting, so that additional coupling between light and split-off bands is also negligible, the superlattice can be viewed as shown in Fig. 4. Electrons, heavy and light holes motions in the Z direction are quantized in the wells arising when matching the corresponding band extrema of the strained bulk materials. This picture reveals the numerous possible band configurations that may exist according to the strain states and band offsets. In the general case where one material is in compression and the other in extension, the holes can be confined in the same layers as the electrons (type I), or the electrons in one material and the holes in the other (type II). An original situation shown in Fig.4 (c) can also occur where the system is of type I for heavy (or light) holes, and of type II for the light (resp. heavy) ones, and this could be called a mixed type superlattice.

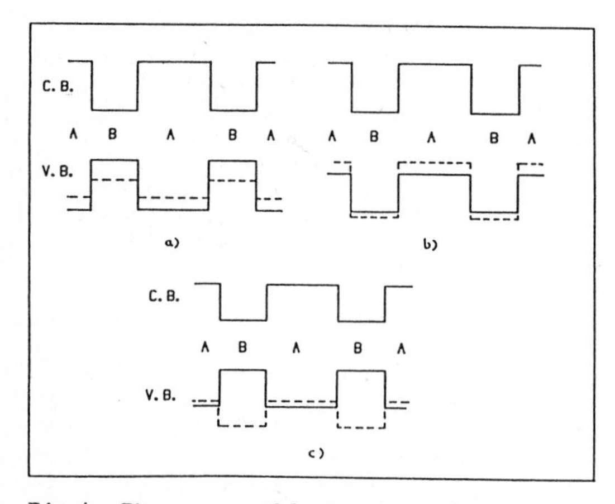

Fig.4 Three possible band configurations for a strained superlattice built with material A in tension and B in compression :
a) type I for heavy and light holes
b) type II for heavy and light holes
c) type I for heavy holes and II for the light ones.

In the envelope function formalism, the superlattice wave functions are written as follows :

$$\psi_{q,k_\perp=0} = \sum_{n=1}^{8} u n_o(\vec{r})\, f_n(z) \qquad f_n(z+L_A+L_B) = e^{iq(L_A+L_B)} f_n(z)$$

where the ui's are the 8 zone center wave functions for the 4 considered bands, which are assumed to be the same in each bulk material which will be noted as A and B.

These bulk materials are described by Kane hamiltonians, where we have included the strain effects. For $k_\perp = 0$, in A (resp. B) material, it reduces to two identical 4.4 components :

$$
\begin{array}{cccc}
S|1/2,1/2>_z & |3/2,3/2>_z & |3/2,1/2>_z & |1/2,1/2>_z
\end{array}
$$

$$
\begin{bmatrix}
0 & 0 & -\sqrt{\frac{2}{3}}Phk_z & \frac{1}{\sqrt{3}}\,Ph\,k_z \\
0 & \begin{matrix}-E_A-\delta E_H^A \\ -\frac{1}{2}\,\delta E_s^A\end{matrix} & 0 & 0 \\
-\sqrt{\frac{2}{3}}\,Ph\,k_z & 0 & \begin{matrix}-E_A-\delta E_H^A \\ +\frac{1}{2}\delta E_s^A\end{matrix} & \frac{1}{\sqrt{2}}\,\delta E_s^A \\
\frac{1}{\sqrt{3}}\,Ph\,k_z & 0 & \frac{1}{\sqrt{2}}\,\delta E_s^A & \begin{matrix}-\Delta_A-E_A \\ -\delta E_H^A\end{matrix}
\end{bmatrix} \tag{7}
$$

E_A and Δ_A are the unstrained bulk bandgap and spin orbit splitting, P is the Kane matrix element, assumed to be the same in A and B and

$$
\delta E_H^A = a(\varepsilon_{xx}^A + \varepsilon_{yy}^A + \varepsilon_{zz}^A) \qquad \delta E_s^A = 2b\,(\varepsilon_{zz}^A - \varepsilon_{xx}^A)
$$

It gives the following dispersion relations

$$
\begin{aligned}
&-E[(-E_A + \tfrac{1}{2}\,\delta E_s^A - \delta E_H^A - E)(-\Delta_A - E_A - \delta E_H^A - E) - \tfrac{1}{2}\,(\delta E_s^A)^2\,] \\
&\quad = \frac{P^2}{3}\,\hbar^2\,k_A^2\,[2(-\Delta_A - E_A - \delta E_H^A - E) + 2\delta E_s^A + (-E_A + \tfrac{1}{2}\,\delta E_s^A - \delta E_H^A - E)]
\end{aligned} \tag{8}
$$

which can be written formally as

$$
E = \frac{\hbar^2 k_A^2}{2\alpha_A(E)} \tag{9}
$$

The eigenstates in the superlattice are found to satisfy the corresponding effective mass differential system :

$$
\begin{bmatrix}
V_s(z) & -\sqrt{\frac{2}{3}}\,PPz & 0 & \frac{P}{\sqrt{3}}\,Pz \\
-\sqrt{\frac{2}{3}}\,PPz & -E_A+V_p(z) & 0 & \frac{1}{\sqrt{2}}\,\delta\,E_s(z) \\
0 & 0 & -E_A+V'_p(z) & 0 \\
\frac{P}{\sqrt{3}}\,Pz & \frac{1}{\sqrt{2}}\,\delta E_s(2) & 0 & \begin{matrix}-E_A - B_A \\ -V\delta(z)\end{matrix}
\end{bmatrix}
\begin{bmatrix} f_1 \\ f_2 \\ f_3 \\ f_4 \end{bmatrix}
= E \begin{bmatrix} f_1 \\ f_2 \\ f_3 \\ f_4 \end{bmatrix} \tag{10}
$$

where f_1, f_2, f_3, f_4 are the envelope functions associated to $S|1/2,1/2>_Z$, $|3/2,1/2>_Z$ $|3/2,3/2>_Z$ and $|1/2,1/2>_Z$, respectively

and

	z in A	z in B
$V_s(z) =$	δE_H^A	$\delta E_H^B + V_s$
$V'_p(z) =$	$-\frac{1}{2}\delta E_s^A$	$-\frac{1}{2}\delta E_s^B + V_p$
$V_p(z)$	$+\frac{1}{2}\delta E_s^A$	$+\frac{1}{2}\delta E_s^B + V_p$
$V\delta(z) =$	0	$V\delta$
$\delta E_s(z) =$	δE_s^A	δE_s^B

$$E_A + V_s = E_B + V_p \qquad\qquad E_A + D_A + V_s = E_B + D_B + V\delta$$

By substitution we obtain one equation on f_1

$$V_s(z)f_1 + P^2 P_z \left[\frac{1}{\mu(z,E)}\right] P_z f_1 = E f_1 , \tag{11}$$

with $\mu(z,E)$ given in each material by (9),

which together with the continuities of f_1 and of the probability current

$$\frac{1}{\mu(z,E)} \frac{\partial}{\partial z} f_1$$

at the interfaces give the eigen energies in the superlattice, with $k_\perp = 0$, as the solution of

$$\cos(qd) = \cos(k_A L_A)\cos(k_B L_B) - \frac{1}{2}\left(\xi + \frac{1}{\xi}\right)\sin(k_A L_A)\sin k_B L_B$$

$$\xi = \frac{k_B \alpha_A(E)}{k_A \alpha_B(E)}$$

together with the dispersion relations (8). As in the unstrained heterostructure case, only one unknown parameter appears in the band structure calculation, which is one cf the band offsets V_s, V_p or V_δ. In this section, we have shown that it is easy to include the strain effects in a description which takes into account non parabolicity. This is necessary to calculate precisely the excited states in the superlattice. However, the more important result remains that the strains will deeply modify the band structure, mainly because the confining superpotentials for heavy and light holes become different. This will be now illustrated by some data obtained in optical experiments on strained superlattices.

4. $In_xGa_{1-x}As$-GaAs on GaAs

$In_xGa_{1-x}As$ alloy has been used by several laboratories as a constituent in strained structures on GaAs [3-4], for there is no small gap III-V semiconductor lattice matched to GaAs. The two approaches described in the first part of this paper, i.e. with or without graded buffer layer, have been reported. Both of them led to the realization of devices such as photodetectors [16], lasers [17,18], or field effect transistors [19], and the crystallographic [9,20] and electronic [21-25] properties of these systems have been thoroughly studied. We chose to present here optical data we have obtained on $In_{.15}Ga_{.85}As$-GaAs superlattices strained on GaAs. As already mentioned, then

i) the GaAs layers are not strained

ii) the strains in the $In_xGa_{1-x}As$ layers are independent of the sublayers thicknesses and are fixed by the In content in the alloy. For x = 15, $\Delta a/a$ = 1%; the $In_xGa_{1-x}As$ layers are in compression, so that the higher states in energy in the ternary valence band are the $\phi3/2$ states. The unstrained band gap is Eg = 1.295 eV at 77K, and strain effects given by (4) lead to :

$$E\ (CB)\ -E\ (\phi_{3/2}) = 1.352\ eV$$

$$E\ (\phi_{3/2})\ -E\ (\phi^{+}_{1/2}) = 72\ meV$$

The spin orbit splitting is equal to 360 meV. As the $In_xGa_{1-x}As$ is in compression, the effect of the coupling between the m_J = 1/2 valence band states is to increase the zone center energy of the $\phi^{+}_{1/2}$ band by 10 meV.

A series of samples, consisting of 10 periods of alternating GaAs layers (thickness 200 Å) and $In_xGa_{1-x}As$ (In content.15) layers which thickness d has been varied, were grown by M.B.E.. The X-ray diffraction pattern obtained on one typical sample is shown on Fig. 5. They reveal a good structural quality of the samples, and, together with growth parameters, allowed a precise determination of the period, sublayer thicknesses and of the In content x. This last parameter, when determined from X-ray results, depends on the strain state of the superlattice. In many samples, this strain state can be readily obtained from the position of the GaAs related envelope with respect to the substrate diffraction line. In some samples, however, especially those with large d where some relaxation of the superlattice as a whole and the substrate occurs, as it can be shown by T.E.M.; the In composition and strain state should be more reliably obtained from a simultaneous fit of the X-ray results together with optical data. In the rather large d limit, d > 100 Å, the optical band gap hardly depends on the value of the only unknown parameter giving the band offsets, but is very sensitive to x and to the strain state, so that this transition can be used.

Fig. 6 shows the optical transmission data obtained at 77K, for 2 samples with different d. The excitonic absorption lines reveal the high quality of these samples. There is strong evidence of strain effects : the first transition in sample A, for example, is shifted by 105 meV, with respect to the unstrained band

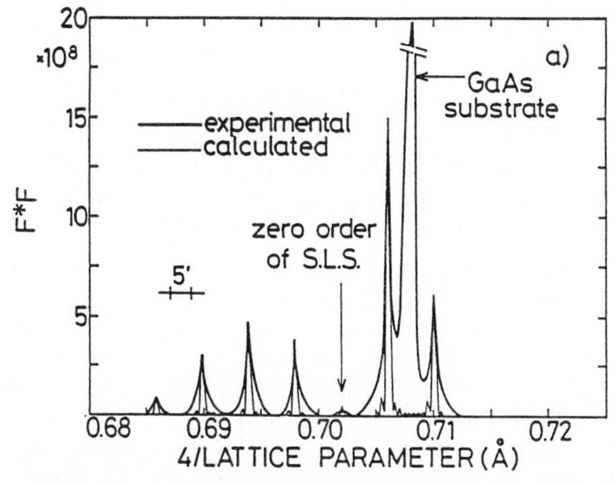

Fig.5 Typical X-ray double diffraction profile obtained on InGaAs-GaAs 10 periods superlattice. The corresponding calculated profile is also shown. (The calculation is somehow different from what is indicated in the text for this sample, see Ref. 9 for further details).

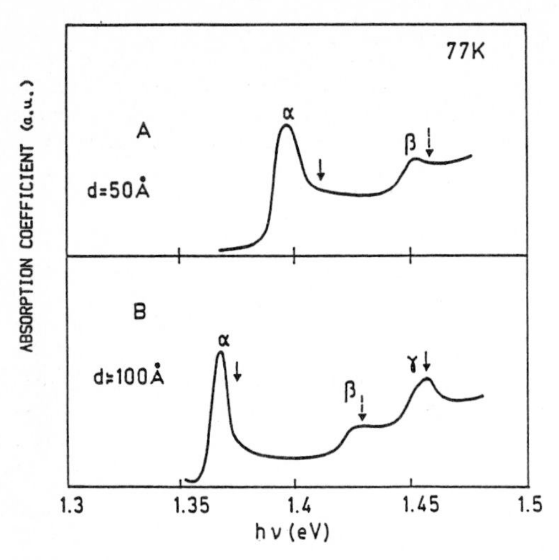

Fig.6 Absorption coefficient deduced from single beam transmission experiments, at 77K, for two $In_{.15}Ga_{.85}As$-GaAs 10 periods strained superlattices. Full and dashed arrows indicate the calculated subband to subband transitions involving heavy and light holes respectively.

gap, while the confinement shift is expected to be of the order of 70 meV only. In an unstrained structure, we would also expect the two first transitions HH1-E1 and LH1-E1 to be separated by some 20 meV, in the same sample, which is not observed. The possible valence band configurations all lead to a first transition associated with heavy holes, and, as it is shown in another paper of this series by P. Voisin, the strength of the absorption shows that the heavy holes and electrons are both confined in the ternary layers. The assignment of the other observed transitions is more difficult, due to the strain effects. As the α and γ transitions have similar shapes, they are likely to arise from n = 1 and n = 2 heavy holes excitonic absorptions, whereas β transitions, also similar in shape, could be related with light holes. These assignments were checked by polarized excitation of the photoluminescence experiments. In this system, because the substrate is transparent in the energy region of the superlattice transitions, it is possible to excite the superlattice photoluminescence through the substrate, with a dye laser focused on the edge of the sample. This configuration allowed us to set the exciting light polarization either in the plane of the layers, which is the case for usual "surface" excitation, or on the Z axis. In this last polarization, as already mentioned, the heavy holes associated transitions are forbidden, while the light ones are not, which was verified for the assignment given above in the studied samples.

The fit of the energies of the observed transitions was done using the model presented in the previous section, for these samples which were shown to be well strained on the substrate and whose characteristics were deduced from their X-ray patterns. The only adjustable parameter was the conduction band discontinuity. The best fit, also shown in fig.6, leads to a mixed type band configuration, as shown in Fig. 4. For the light holes, the InGaAs layers constitute the barriers, whose height is of the order of 20 meV. The observed light holes related transitions can be either due to transitions between states confined inside the GaAs, but strongly leaking in the InGaAs layers, because of the small barrier height (and eventually thickness) and the n=1 electron states, or to transitions between diffusion states nearby the energy of the light holes band maximum in the InGaAs and n=1 electron states, or both, depending on the ternary material layer thicknesses.

The system, which appears to exhibit an original band configuration, should also be interesting for its transport properties, for the built-in strains should enhance the mass reversal, leading to a higher lying "heavy" hole band having a small in-plane effective mass. This fact has already been observed [26] and can have consequences in the lasers application, because of the reducing of the joint density of states.

5. GaAb-AlSb

The optical properties of this system, grown by M.B.E. [27], have been studied by P. Voisin et al. [28]. The studied samples have been grown on a thick AlSb buffer layer, which presents a large mismatch with the GaAs substrate, so that it relaxes to its own parameter. The buffer-layer plus superlattice structure is expected to take its equilibrium parameter. Fig. 7 shows the X-ray diffraction pattern of one sample. Again, it can be used to obtain the sample parameters, in this case the period, sublayers thicknesses and strain state of the superlattice.

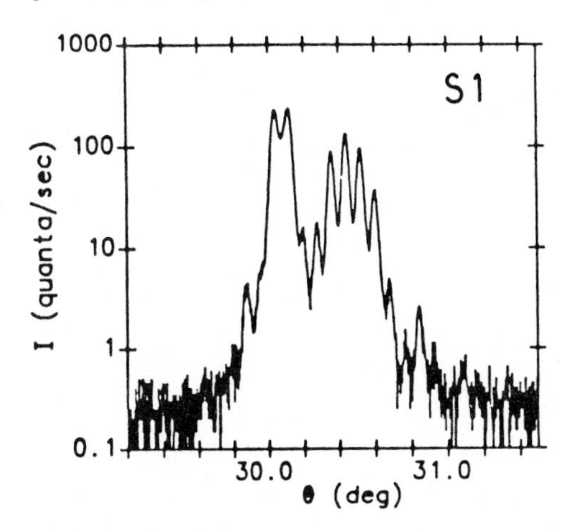

Fig.7 X-ray double diffraction profile for a GaSb-AlSb superlattice whose characteristics are given in Fig.8 caption.

The situation is rather different compared to the InGaAs-GaAs system

i) the constituents are binary semiconductors, so that their properties are well known and there is no composition to determine. It has allowed in this case an evaluation of the strain state of these superlattices by the observation of the strained induced shifts of the energy of the GaSb longitudinal optical phonon [29].

ii) the strain state depends on the relative thickness of the sublayers, and, as the dependence of the band offsets versus the built-in strain is unknown, all samples have to be analysed independently.

iii) the GaSb smaller gap layers are in tension, so that the higher energy valence band in the strained GaSb is the "light" holes one. Furthermore, as the spin orbit splitting in GaSb is large, and as the strain is a tension, the two $m_J = 1/2$ valence band states can be considered as decoupled in Γ point.

The low temperature absorption spectrum of Fig. 8 is again characteristic of a type I system, because of the absorption strength. It displays a step-like shape, which is typical of 2D systems, together with excitonic lines showing the good quality of the samples. The strain effect is striking in this spectrum, for the first transition occurs at 0.81 eV, which is equal to the bulk GaSb band gap, while the confinement energy HH1+E1, for this well thickness should be of the order of 25 meV.

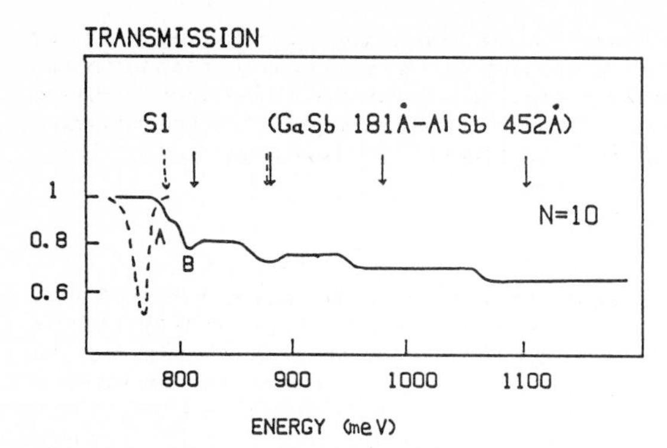

Fig.8 Transmission (full line) and luminescence (dashed line) of a GaSb-AlSb superlattice. Full and dashed arrows indicate the calculated transitions involving heavy and light holes, respectively.

Moreover, there is a shoulder in the absorption edge, which is interpreted as the first light hole excitonic transition, pushed by the strain effects at lower energy than the first heavy hole one.

As the AlSb layers are much thicker than the GaSb ones, they remain essentially unstrained. AlSb is a large gap semiconductor, which has a common Sb anion with GaSb. The valence band offsets should then be rather small, so that there will be very deep wells for the electrons. The number of HHn → En and LHn → En transitions which are observed will be determined by the number of holes levels, and this has been used in Ref.28 to determine the offsets for each sample.

The strain analysis for this sample gives the following relations for the band edges positions in the GaSb :

$$\varepsilon_{xx} = 0.46\ 10^{-2}$$

$$E(CB) - E\ (\phi^{+}_{1/2}) = 0.758\ \mathrm{eV}$$

$$E(\phi_{3/2}) - E\ (\phi^{+}_{1/2}) = -\ 30\ \mathrm{meV}$$

the offset obeys $40\ \mathrm{meV} \leqslant E^{GaSb}_{(\phi 3/2)} - E^{AlSb}_{(\phi 3/2)} \leqslant 60\mathrm{meV}$

Here, the small gap layer is under tension, and the stress effects enlarge the gap difference.

The system is of type I for heavy and light holes, and the calculated transitions are indicated by arrows in Fig. 8.

In this system, the relative positions of the first light and heavy holes subbands is the result of a competition between the stress effects and the confinement energies : for the same stresses, there can be either a light hole band gap, for large GaSb layers thicknesses, or a heavy hole gap for small ones.

6. Other strained systems

$In_xGa_{1-x}As$-$In_yGa_{1-y}As$ on InP

This two-ternary layer system can be easily grown by M.B.E. with a mean composition of .53, so that the superlattice, lattice matched to InP, is at its equilibrium parameter (for equal sublayers thicknesses, x+y = 1.06), so there is no need for a graded buffer layer. These superlattices, with x = .42 and y = .64, have been shown to have band gaps corresponding to wavelengths as large as 2m, at 300K.

They are potentially interesting for emission or detection devices operating at such wavelengths. For sublayers of very small thicknesses (typically some monolayers), the limit of this system can be reached, where the superlattice constitutes an InAs-GaAs pseudo-alloy. In this extreme case, the 7% lattice mismatch between the constituents is expected to lead to large stress effects.

$GaAs_{1-x}P_x$-GaP on GaP, $GaAs_{1-x}P_x$-GaAs on GaAs

Although these systems have been both theoretically [30] and experimentally [4,6] studied, we will only mention them. Because they involve indirect gap materials, their analysis requires, in general, more sophisticated treatments than the effective-mass approach, although it can still be used.

InAs-GaSb

This system is often mentioned because, in this type II system, the InAs conduction band minimum lies 150 meV below, the GaSb valence band maximum. It results that for small sublayers thicknesses, the superlattice is a semiconductor, while, for large periods, a $k_\perp = 0$ analysis leads to a semimetallic structure. In fact, as it was shown by M. Altarelli [31], there is an anticrossing between E1 and HH1, which are strongly repelling each other for $k_\perp \neq 0$. As already pointed out by Voisin, the discrepancy between experiments [32] and theory can be due to strain effects : there is a rather large lattice mismatch of .62 % between InAs and GaSb, leading to stress effects of some tens of meV, which are larger than the predicted band gap in the anticrossing region. These strain effects should then change the actual position of the anticrossing.

GaAs-GaAlAs on GaAs

In general, in this most studied system, the strain is small enough so that, for usual superlattice thicknesses, the whole structure takes an in-plane lattice parameter which is equal to the GaAs one : the barriers are strained with $\varepsilon_{xx} = 1.3\ 10^{-3}.x$, while the wells remain unstrained. The effects of these small strains on the energy levels, only due to variations of the barrier heights in the superlattice, are negligible. This is the main reason why they are never observed in the experiments where the substrate is kept intact, as luminescence or excitation spectroscopy. On the contrary, in absorption experiments, where it is necessary to remove part of the substrate, the local equilibrium parameter of the remaining structure can be noticeably different from the substrate's. In these experiments, strain effects are indeed observed, and they can entail several meV shifts of the transitions [33]. As in GaSb system, GaAs is under tension, so that when the confinement energies are small, the light hole exciton can lie at a lower energy than the heavy hole one. Often, in such cases, the analysis of the results is very difficult, because the strain state of the structure is not known. It should be also the case when the superlattice is too thick, in a regime where at least partial relaxation is expected.

7. Conclusion

We have discussed some aspects of the superlattices built with III-V compounds. We have assumed throughout the paper a simple strain state, which would not hold if any stress relaxation occurs inside the superlattice. The study of these stress relaxation conditions may be important for the applications : they are already known to entail very short lifetimes for c.w. strained superlattice lasers, for example. In our treatment of the strain effects on the band structure of superlattices, we did not include a discussion of the evolution under strain of the other extrema of the conduction band, which would not modify our analysis in InGaAs-GaAs or GaSb-AlSb, but should be kept in mind for larger strains cases. In other systems, involving II-VI (HgTe-CdTe [34]), IV-IV (Ge-Si), or IV-VI ($Pb_xSn_{1-x}Te$-PbTe [35]) semiconductors, materials with large lattice mismatches have been used to realize superlattices. The strain effects, especially for the small gap systems, have to be considered, although their analysis can be more complicated than in III-V compounds, because of the band structure of the hosts materials.

Acknowledgements

I wish to thank my colleagues M. Quillec, L. Goldstein, M.N. Charasse, B. Sermage and D. Paquet, as well as G. Bastard and M. Voos, from the E.N.S., for the discussions we had and their fruitful comments on this work. I am also grateful to P. Voisin, with whom I had the pleasure to work on the properties of these strained systems.

References

1. L. Esaki and R. Tsu: IBM J. Res. Dev. 14, 61 (1970)
2. J.W. Matthews and A.E. Blakeslee: J. Cryst. Growth 27, 118 (1974)
3. L. Goldstein, M. Quillec, E.V.K. Rao, P. Henoc, J.M. Masson and J.Y. Marzin: Journal de Physique 12 C5, 201 (1982)
4. P.L. Gourley and R.M. Biefeld: Appl. Phys. Lett. 45, 749 (1984)
5. I.J. Fritz, L.R. Dawson, G.C. Osbourn, P.L. Gourley and R.M. Biefeld: 10 International Symposium on GaAs and related compounds, Albuquerque, New Mexico, Sept.82, Inst. Phys. Conf. Ser. 65, 241 (1982)
6. R.M. Biefeld and P.L. Gourley: Appl. Phys. Lett. 41, 172 (1982)
7. J.H. Van der Merwe: J. Appl. Phys. 34, 117 (1963)
8. F.K. Reinhart and R.A. Logan: J. Appl. Phys. 44, 3171 (1973)
9. M. Quillec, L. Goldstein, G. Le Roux, J. Burgeat and J. Primot: J. Appl. Phys. 55, 2904 (1984)
10. F.H. Pollack: Surf. Science 37, 863 (1973) and references therein
11. J.M. Luttinger: Phys. Rev. 102, 1030 (1956)
12. G.E. Pikus and G.L. Bir: Fiz. Tverd. Tela 1, 1642 (1959), reproduced in Soviet Phys. Solid State 1, 1502 (1959)
13. R.L. Greene, K.K. Bajaj and D.E. Phelps: Phys. Rev. B29, 1807 (1984)
14. P. Voisin: in this school
15. G. Bastard: Phys. Rev. B24, 5693 (1981), S.R. White and L.J. Sham, Phys. Rev. Lett. 47, 879 (1981)
16. D.R. Myers, T.E. Zipperian, R.M. Biefeld and J.J. Wiczer: International Electronic Device Meeting 83, 700 (1983)
17. W.D. Laidig, D.R. Blanks and J.F. Scherzina: J. Appl. Phys. 56, 1791 (1984)
18. M.J. Ludowise, W.T. Dietze, C.R. Lewis, M.D. Camras, N. Holonyak, B.K. Fuller and M.A. Nixon: Appl. Phys. Lett. 42, 487 (1983)
19. T.E. Zipperian, L.R. Dawson, G.C. Osbourn and I.J. Fritz: International Electronic Device Meeting 83, 696 (1983)
20. S.T. Picraux, L.R. Dawson, G.C. Osbourn and W.K. Chu: Appl. Phys. Lett. 43, 930 (1983)
21. J.Y. Marzin and E.V.K. Rao: Appl. Phys. Lett. 43, 560 (1983)
22. M.D. Camras, J.M. Brown, N. Holonyak, M.A. Nixon, R.W. Kaliski, M.J. Ludowise, W.T. Dietze and C.R. Lewis: J. Appl. Phys. 54, 6183 (1983)
23. M. Nakayama, K. Kubota, H. Kato and N. Sano: Solid State Comm. 51, 343 (1984)
24. W.D. Laidig, P.J. Caldwell, Y.F. Lin and C.K. Peng: Appl. Phys. Lett. 44, 653 (1984)
25. I.J. Fritz, L.R. Dawson and T.E. Zipperian: Appl. Phys. Lett. 43, 846 (1983)
26. J.E. Schirber, I.J. Fritz and L.R. Dawson: Appl. Phys. Lett. 46, 187 (1985)
27. C.A. Chang, H. Takaoka, L.L. Chang and L. Esaki: Appl. Phys. Lett. 40, 983 (1982)
28. P. Voisin, C. Delalande, M. Voos, L.L. Chang, A. Segmuller, C.A. Chang and L. Esaki: Phys. Rev. B 30, 2276 (1984)
29. B. Jusserand, P. Voisin, M. Voos, L.L. Chang, E.E. Mendez and L. Esaki: to be published in Appl. Phys. Lett.
30. G.C. Osbourn: J. Appl. Phys. 53, 1586 (1982)
31. M. Altarelli: Phys. Rev. B28, 842 (1983)
32. J.C. Maan, Y. Guldner, J.P. Vieren, P. Voisin, M. Voos, L.L. Chang and L. Esaki: Solid State Comm. 39, 683 (1981) and references therein
33. R. Dingle and W. Wiegmann: J. Appl. Phys. 46, 4312 (1975)
34. P. Voisin: Springer Series in Solid State Science Vol. 53, p. 192 (1984), Y. Guldner, same book p. 200
35. E.J. Fantner and G. Bauer: Springer Series in Solid State Science vol. 53, p. 207

HgTe-CdTe Superlattices

Y. Guldner

Groupe de Physique des Solides de l'Ecole Normale Supérieure,
24, rue Lhomond, F-75231 Paris Cedex 05, France

HgTe-CdTe superlattices, recently grown by molecular beam epitaxy, exhibit far more diverse characteristics than the conventional GaAs-Al_x $Ga_{1-x}As$ heterostructures because of the specific bulk band structure of HgTe and CdTe. The superlattices subband structure can be calculated in the framework of the envelope function model, and the calculations are confirmed by magneto-optical experiments. Finally, this new system presents interesting potential applications as infrared detectors.

Introduction

Following the work of ESAKI and TSU in 1970 [1], there has been strong interest in semiconductor superlattices (SL) fabricated from III-V compounds, such as GaAs-$Al_xGa_{1-x}As$ heterostructures. Recently, a new SL system, involving II-VI materials, i.e. HgTe-CdTe SL, has been grown by molecular beam epitaxy (MBE) by FAURIE, MILLION and PIAGUET [2]. Both HgTe and CdTe crystallize in the zinc-blende lattice and they are closely matched in lattice constant (6.46 Å in HgTe and 6.48 Å in CdTe). This lattice-matching within 0.3% makes them good candidates for the growth of heterostructures, and high quality SL's with a small interdiffusion between HgTe and CdTe layers have been obtained at low temperature ($\sim$ 200°C) by MBE [2,3].

This system presents a great physical interest, because of the specific bulk band structure of HgTe and CdTe and of the band line-up of the two host materials (see fig.1). CdTe is a wide gap semiconductor ($\sim$ 1.6eV at low temperature) with a direct gap at the zone center (Γ point). At k = 0 the conduction band edge has S-type symmetry (Γ_6) whereas the upper valence band edge (Γ_8) is fourfold degenerate (P symmetry, J = 3/2). The spin-orbit split off Γ_7 band (P symmetry, J = 1/2) lies below the Γ_8 states with $\Delta = E_{\Gamma 8} - E_{\Gamma 7} \sim$ 1eV. HgTe is a *zero-gap semiconductor* due to the inversion of the relative positions of Γ_6 and Γ_8 edges. What was the Γ_8 light hole band in CdTe forms a conduction band in HgTe and the Γ_6 conduction band in CdTe becomes a light hole band in HgTe. The ground valence band is the Γ_8 heavy hole band, so that the Γ_8 states represent both the top of the valence band and the bottom of the conduction band, yielding to a zero-gap configuration (Fig.1). The spin-orbit separation Δ is also $\sim$ 1eV in HgTe. The evidence for the inverted structure of HgTe was mainly provided by magneto-optical measurements [4,5,6].

The band structure of the SL's depends on the offset between the various band edges at the HgTe-CdTe interface. From Harrison's common anion argument [7], one can infer that the offset Λ between the Γ_8 band edges of HgTe and CdTe is small. As shown later from the experimental data, $\Lambda \sim$ 40 meV [8]. This value of Λ implies that HgTe layers are potential wells for heavy holes, whereas the situation for light particles (electrons or light holes) is more complicated : the bands which most significantly contribute to the light particle SL states are the Γ_8 conduction band in HgTe and the Γ_8 light valence band in CdTe. These two bands have *opposite* curvatures and the *same* Γ_8 symmetry. This mass-reversal for the light particles at each of the HgTe-CdTe interface is a unique property of the HgTe-CdTe SL's. This contrasts with the case of GaAs-Al_x $Ga_{1-x}As$ SL where the SL states are mainly formed from bands in GaAs and AlGaAs displaying the same curvature.

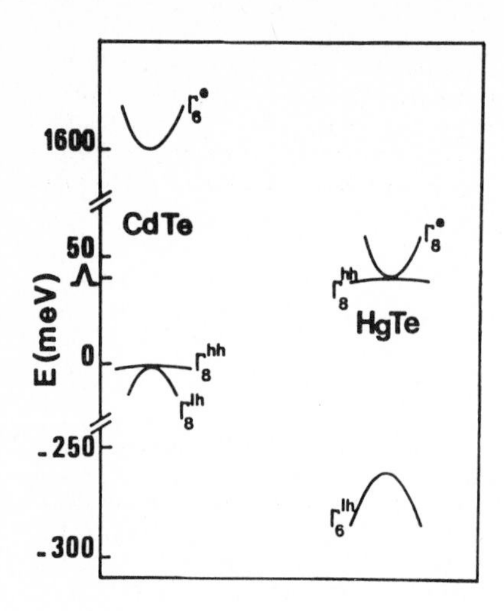

Fig.1 Band structure of bulk HgTe and CdTe. The lh, hh, and e indices refer to light holes, heavy holes, and electrons, respectively.

1. SL Subband structure calculations

1.1 Light particle subbands

The simplest description of the SL band structure is obtained in the framework of the envelope function approximation [9]. The band structure of both HgTe and CdTe near the Γ point is described in this approach by the Kane model [10] which takes into account the non-parabolicity of the Γ_6 and light Γ_8 bands. This non-parabolicity is important in HgTe where the energy separation $E_{\Gamma 8} - E_{\Gamma 6}$ is small.

If the energies E are measured from the top of the Γ_8 valence band of CdTe, the dispersion relation for the Γ_6 and light Γ_8 bands are

$$E\,(E-\varepsilon_2) = \frac{2}{3}\, p^2K^2\, (k_2^2 + k_\perp^2) \qquad \text{in CdTe}, \qquad (1)$$

$$(E-\Lambda)(E-\Lambda + |\varepsilon_1|) = \frac{2}{3}\, p^2k^2\, (k_1^2 + k_\perp^2) \qquad \text{in HgTe}. \qquad (2)$$

In (1) and (2), ε_1 and ε_2 are the interaction gap $E_{\Gamma 6}-E_{\Gamma 8}$ in HgTe and CdTe (Table 1), Γ is the valence band offset at the HgTe-CdTe interface and P is the Kane matrix element, which is assumed to be the same for HgTe and CdTe. Here k_1 and k_2 are the wave vectors along the z axis and $\vec{k}_\perp=(k_x,k_y)$. The influence of the Γ_7 band is neglected because the spin orbit separation is large in both materials.

In the envelope function approximation, the SL wave functions can be written :

$$\psi = \sum_{i=1}^{6} e^{ik\vec{r}} \quad f_i(z)\, u_i\,(\vec{r},z)$$

where z is along the SL axis, $\vec{r}_\perp$ and $\vec{k}_\perp$ describe the SL in-plane motion, i runs over the Γ_6 and Γ_8 band edges. The u_i functions are the Γ_6 and Γ_8 periodic parts of

Table I Interaction gap of HgTe and CdTe at several temperatures [18].

T	4K	77K	300K
ε_1 (meV)	-302.5	-261	-122
ε_2 (meV)	1600	1550	1425

the Bloch functions of HgTe and CdTe which are assumed to be *identical* in both material. The transverse wave vector $\vec{k}_\perp$ is a constant of motion and the SL hamiltonian is a a 6x6 matrix acting only on the envelope functions f_i. Taking into account the SL periodicity d, applying the Bloch theorem and the appropriate matching conditions at the interface [9] leads to the approximate dispersion relation of the SL light particle subbands, which is a simple Kronig-Penney expression

$$\cos qd = \cos k_1 d_1 \cos k_2 d_2$$

$$- \frac{1}{2} \left[(\xi + \xi^{-1}) + \frac{k_\perp^2}{4k_1k_2} (r + r^{-1} - 2) \right] \sin k_1 d_1 \sin k_2 d_2 , \qquad (3)$$

where q is the SL wave vector ($-\pi/d \leqslant q < +\pi/d$), d_1 and d_2 are the thickness of HgTe and CdTe layers respectively,

$$\xi = \frac{k_2}{k_1} \frac{E - \Lambda}{E} \quad \text{and} \quad r = \frac{E - \Lambda}{E} .$$

The wave vectors k_1 and k_2 at a given energy E are obtained from relations (1) and (2) and can be real or imaginary, depending on whether E corresponds to propagating or evanescent states in the host materials. For a given energy E, a SL state of wave vector q exists if the right-hand-side of (3) lies in the range (-1, +1).

1.2 Heavy hole subbands

The Kane simple model does not describe the heavy hole band of the host materials. At $k = 0$, the SL heavy hole subbands are obtained from (3) with $\xi = k_1/k_2$. k_1 and k_2 are given by

$$- \frac{\hbar^2 k_1^2}{2m_{hh}} = E - \Lambda ,$$

$$- \frac{\hbar^2 k_2^2}{2m_{hh}} = E ,$$

where m_{hh} is the heavy hole mass of the host materials ($\sim 0.3 m_o$).

Results

Using the previous model and the values of ε_1 and ε_2 given in Table 1, the band structure of HgTe-CdTe SL can be calculated as a function of the layer thickness. The Kane matrix element P is obtained from the magneto-optical measurements in bulk HgTe [5,6] by using the relation

$$\frac{2P^2}{3|\varepsilon_1|} = \frac{1}{2m^*} ,$$

where m* is the electron cyclotron mass in HgTe.

The valence band offset is taken to be $\Gamma = 40$ meV [8]. Fig.2 gives the energy and the width at $\vec{k}_\perp = 0$ of the ground heavy hole (HH_m), light hole (h_m) and conduction (E_m) subbands as a function of layer thickness for SLs with $d_1 = d_2$ (Fig.2a) and $d_1 = 4d_2$ (Fig.2b). The calculations are made at 4K. It should be noted that d_1 essentially controls the SL band gap (i.e., $E_g = E_1 - HH_1$) while d_2 governs the width of the subbands (i.e., the electron effective mass along the SL axis).

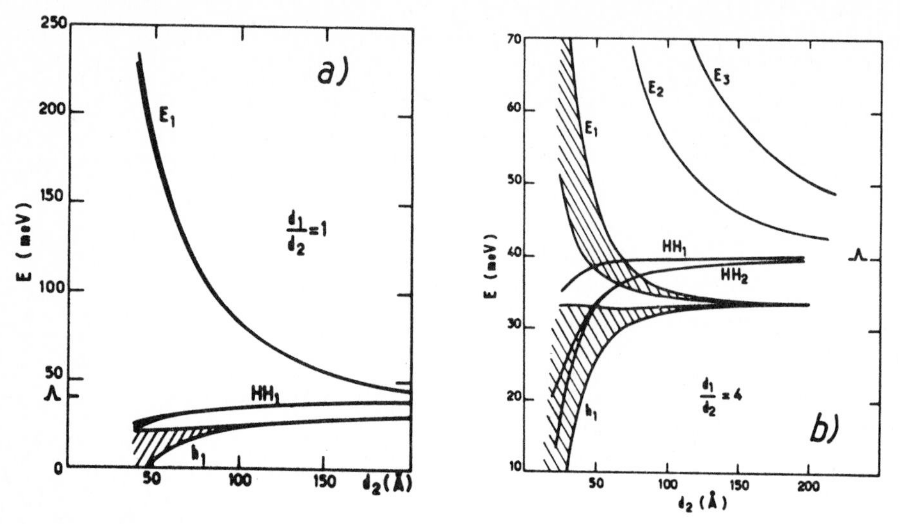

Fig.2 Energy and width of different subbands in HgTe-CdTe SL as a function of the CdTe thickness d_2.
a) $d_1 = d_2$; b) $d_1/d_2 = 4$

The ground light particle subband h_1 lies in the forbidden energy gap (0, Λ) corresponding to evanescent states in both kinds of layers *(k_1 and k_2 imaginary).* The SL wave function associated to h_1 is found to peak at the interfaces instead of peaking at the center of the layers, so that h_1 corresponds to interface states [11]. Such interface states are a consequence of matching bulk states which belong to two bands, displaying the same symmetry and opposite curvatures.

In Fig.2b, it is shown that the cross-over of E_1 and HH_1 at $q = 0$ (i.e., $E_g = 0$) occurs for $d_2 \sim 45$ Å and $d_1 \sim 180$ Å. For $d_2 > 50$ Å, E_1 states drops in the energy gap (0, Λ) and become interface states.

2. Experimental results

Interesting information on the SL band structure can be obtained from far-infrared (FIR) magneto-absorption experiments [8,12]. When a strong magnetic field B is applied perpendicular to the layers ($\theta = 0$), the ground subbands (HH_1, E_1) are split into landau levels $HH_1(n)$, $E_1(n)$. At low temperature ($T < 4K$), the FIR transmission signal being recorded at fixed photon energies as a function of B presents pronounced minima which correspond to the resonant optical transitions between the different Landau levels. The FIR sources are generally FIR laser or carcinotrons. Fig. 3 shows typical transmission spectra for several FIR wavelengths obtained in a SL sample which consists of one hundred periods of (180 Å) HgTe - (44 Å) HgTe - (44 Å) CdTe grown onto CdTe at 200°C. B is applied perpendicular to the layers ($\theta = 0$). The energy positions of the transmission minima are plotted as a function of B on Fig. 4a. The observed transitions extrapolate to an energy $E \sim 0$ at $B = 0$ but they cannot be due either to electron cyclotron resonance (because the SL is found to be p-type for $T < 20K$), or to hole cyclotron resonance because they would lead to hole masses much too small. Fig. 5 shows the calculated band structure of this SL along the SL wave vector q using $\Lambda = 40$ meV and for $B = 0$. The ground heavy hole subband HH_1 is almost flat along q at 2 meV below the Γ_8 HgTe band edge. E_1 has a mixed electron (HgTe) and light hole (CdTe) character, which explains its energy position and the zero-gap nature of this sample. The magneto-optical transitions (Fig. 4a) are attributed to interband transitions from Landau levels of HH_1 up to Landau levels of E_1 occurring at $q = 0$. The investigation of the θ-dependence of the magnetic field position of the transitions shows a variation like $(\cos\theta)^{-1}$ as evidenced on Fig. 4b for $\theta = 45°$. This variation corresponds to a

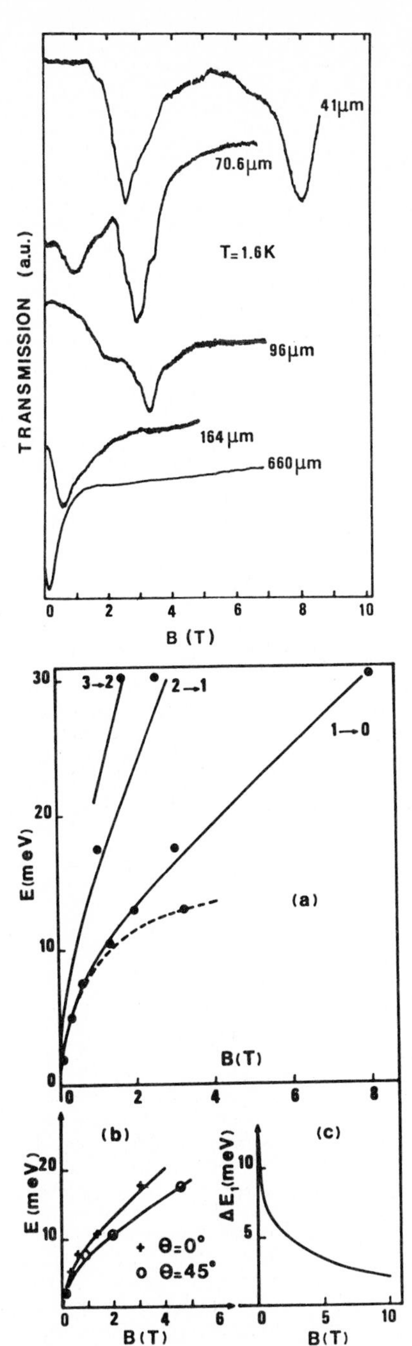

Fig.3 Typical transmission spectra obtained as a function of the magnetic field B for several infrared wavelengths in a (180 Å) HgTe - (44 Å) CdTe SL.

Fig.4 a) Energy position of the transmission minima as a function of B observed in a (180 Å) HgTe - (44 Å) CdTe SL. The solid lines are theoretical fits.
b) Transmission minima (o,+) corresponding to the transition 1 → 0 for two values of θ. The solid line for θ = 0 is the calculated energy and that for θ = 45° corresponds to a perfect two-dimensional behavior (cos θ law).
c) Calculated width ΔE_1 of the E_1 (0) Landau level versus B.

two-dimensional character induced by the magnetic field. To interpret quantitatively these results, one has to calculate the SL Landau level energies. The light particle Landau levels are simply obtained by replacing, in relation (3), $k_\perp^2$ by $(2n+1)eB/\hbar$ where n = 0,1... is the Landau level index. The heavy holes magnetic levels are

$$HH_1(n) = HH_1 - (n + 1/2)\, heB/m_{hh} .$$

In this model, the spin effects are neglected [13].

The selection rules for the $HH_1(n) \rightarrow E_1(n')$ transitions are taken to be $n'-n = \pm 1$ as for the interband $\Gamma_8 \rightarrow \Gamma_8$ transitions in bulk HgTe. The calculated transition energies using $\Gamma n = -1$ are shown in Fig.4a (solid lines). For example, the curve labelled $1 \rightarrow 0$ corresponds to $HH_1(1) \rightarrow E_1(0)$. A good agreement is obtained between theory and experiment for $\Lambda = (40 \pm 10)$ meV. The experimental data could be interpreted equally well with the selection rule $\Delta n = + 1$ except for the transition $1 \rightarrow 0$. In fact, the transitions $n-1 \rightarrow n$ and $n+1 \rightarrow n$ would practically coincide since most of the transition energies arise from $E_1(n)$. The transition energies are remarkably non-linear versus B as a consequence of the $\vec{k}.\vec{p}$ interaction between E_1, HH_1 and h_1. The deviation from the theoretical fit of the experimental data for the $1 \rightarrow 0$ transition around 2.5 T may be due to an interband polaron effect, the LO-phonon energy being 16 meV in HgTe. Figure 4c presents the calculated width ΔE_1 of the $E_1(0)$ Landau level as a function of B. E_1 drops rapidly with B and $E_1(0)$ becomes nearly flat along q. This behavior explains the two-dimensional character observed in this sample (Fig. 4b).

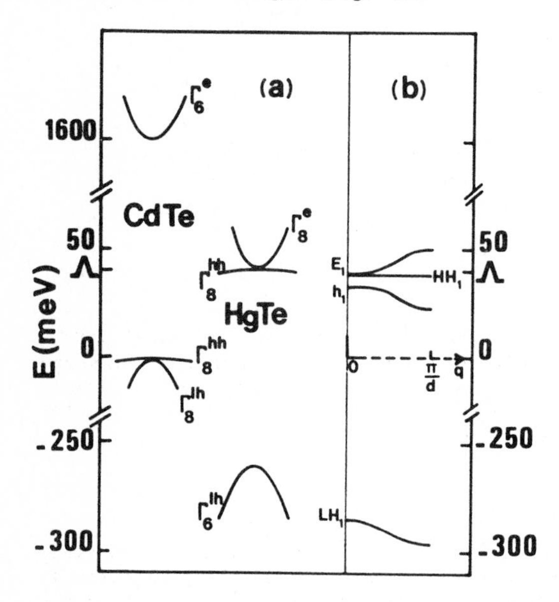

Fig.5 (a) Band structure of bulk HgTe and CdTe.
(b) Calculated band structure along the SL wave vector q of a (180 Å) HgTe - (44 Å) CdTe SL at 4K.

The calculated band structure (Fig.5b) is confirmed by the observation of interband transitions from Landau levels of LH_1, which is the topmost valence band derived from the Γ_6 HgTe states, up to Landau levels of E_1 in the photon energy range 300-400 meV. Figure 6b shows a typical magneto-transmission spectrum observed in this energy range in the Faraday configuration. The position of the transmission minima are presented in Fig.6a and the solid lines correspond to the calculated transitions slopes. One takes the same selection rules ($\Delta n = \pm 1$) as those established for $\Gamma_6 \rightarrow \Gamma_8$ magneto-optical transitions in bulk HgTe (Faraday configuration) [5]. The observed broad minima (Fig.6b) correspond to the two symmetric transitions $n \rightarrow n + 1$ and $n + 1 \rightarrow n$ which are not experimentally resolved. The agreement between theoretical and experimental slopes is rather good. The transitions converge to 344 meV at B = 0 while the corresponding energy $E_1 - LH_1$ at q = 0 is calculated to be 330 meV. The 14 meV difference can be explained by the approximations of the model, in particular the use of the two-band Kane model to describe the band structure of the host materials. It might be also explained by the 0.3% lattice mismatch between HgTe and CdTe which results in an increase of the interaction gap $|\varepsilon_1|$ in HgTe and, therefore, in an increase of $E_1 - LH_1$ [14].

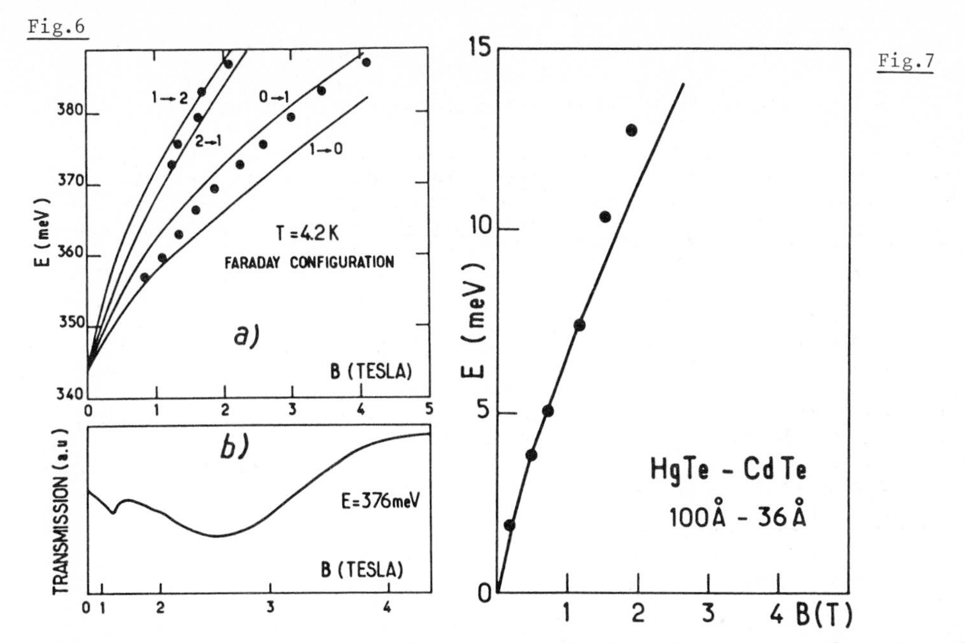

Fig.6 (a) Energy position of the transmission minima as a function of B corresponding to $LH_1 \rightarrow E_1$ transitions in a (180 Å) HgTe - (44Å) CdTe SL.
(b) Typical magneto-transmission spectrum associated to the $LH_1 \rightarrow E_1$ transitions.

Fig.7 Cyclotron resonance in a (100 Å)HgTe-(36Å) CdTe SL.

Moreover, recent magneto-optical results obtained on SLs with d_1 ranging from 40 to 100 Å are also consistent with Λ = 40 meV [15]. Fig. 7 shows the cyclotron resonance [i.e. the transition $E_1(0) \rightarrow E_1$ (1)] observed in a (100 Å) HgTe - (36 Å) CdTe n type SL. The solid line is the calculated energy and the agreement between theory and experiment is good for Λ = 40 meV.

3. HgTe-CdTe SL : an infrared detector material

It has been suggested recently that HgTe-CdTe SL could be an interesting infrared detector material [16]. Using Λ = 40 meV and the HgTe and CdTe bulk parameters at several temperatures (see Table I), one can calculate the energy gap $E_g = E_1 - HH_1$ and the cutoff wavelength λ_g (µm) = 1.24/Eg(eV) for HgTe-CdTe SL's with equally thick HgTe and CdTe layers (Fig.8) [17]. The SL energy gap is found to increase when the temperature is increased, as observed in bulk $Hg_{1-x}Cd_x$ Te alloys with similar energy gap [18]. Nevertheless, the calculated temperature variation is smaller for a SL than for the corresponding ternary alloy. The interesting cutoff wavelengths for infrared detectors (8-12 µm) should be obtained at 77K for thickness in the 50-70 Å range. Recent optical absorption experiments performed at 300K on three SLs with d_1 ranging from 40 to 100 Å [3] confirm the calculated dependence of E_g shown in Fig.8, in particular the increase of λ_g as d_1 increases. However, when similar measurements are carried out at 30K, no important change is observed in opposition to theoretical predictions. Clearly, more experimental results are needed to prove or disprove the correctness of the theoretical approach.

Conclusion

HgTe-CdTe SLs are new and important materials whose properties are only beginning to be investigated. More generally, SL's made from II-VI ternary alloys, such as $Hg_{1-x}Cd_xTe$ or the semimagnetic semiconductor $Hg_{1-x}Mn_xTe$, $Cd_{1-x}Mn_xTe$, should present

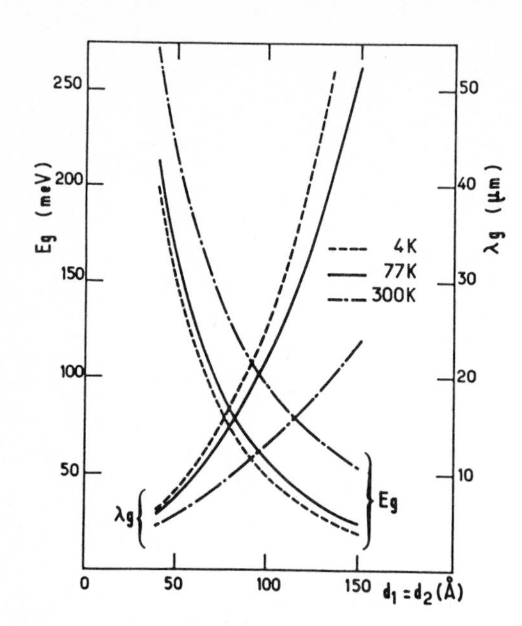

Fig.8 Energy gap and cutoff wavelength as a function of layer thickness for HgTe-CdTe SLs with $d_1 = d_2$.

a great fundamental and technical interest. In particular $Hg_xCd_{1-x}Te$-CdTe SLs look promising for avalanche detectors [19] at 1.3 and 1.5 μm which are important wavelengths for optical communications.

Acknowledgements

It is my pleasure to thank G. BASTARD, J.M. BERROIR, J.P. VIEREN, and M. VOOS from *Ecole Normale Supérieure*, J.P. FAURIE from University of Illinois, A. MILLION and J.P. GAILLARD from the Laboratoire Infrarouge (LETI-CENG, Grenoble).

References

1. L. Esaki, R. Tsu : IBM, J. Res. Develop. 14, 61 (1970).
2. J.P. Faurie, A. Million, J. Piaguet : Appl. Phys. Lett. 41, 713 (1982)
3. J.P. Faurie, M. Boukerche, S. Sivananthan, J. Reno, C. Hsu : 1st Intern. Conf. on Superlattices, Microstructures and Microdevices, Champaign-Urbana, USA (1984). Journal of Superlattices and Microstructures 1, 237 (1985)
4. S.H. Groves, R.N. Brown, C.R. Pidgeon : Phys. Rev. 161, 779 (1967)
5. Y. Guldner, C. Rigaux, M. Grynberg, A. Mycielski : Phys. Rev. B8 3875 (1973)
6. J. Tuchendler, M. Grynberg, Y. Couder, H. Thomé, R. Le Toullec : Phys. Rev. B8, 3884 (1973)
7. W. Harrison : J. Vac. Sci. Techn. 14, 1016 (1977)
8. Y. Guldner, G. Bastard, J.P. Vieren, M. Voos, J.P. Faurie, A. Million : Phys. Rev. Lett. 51, 907 (1983).
9. G. Bastard : Phys. Rev. B25, 7584 (1982)
10. E.O. Kane : J. Phys. Chem. Solids 1, 249 (1957)
11. Yia-Chung Chang, J.N. Schulman, G. Bastard, Y. Guldner, M. Voos : Phys. Rev. B31, 2557 (1985)
12. Y. Guldner, J.P. Vieren, P. Voisin, M. Voos, L. Esaki, L.L. Chang : Phys. Rev. Lett. 45, 1719 (1980)
13. A more sophisticated description of the hole Landau levels in heterostructures is given by :
    A. Fasolino, M. Altarelli : In Springer Ser. Solid-State Sci. Vol. 53, by G. Bauer, F. Kuchar, H. Heinrich (Springer, Berlin, Heidelberg (1984) p. 176
14 P. Voisin : In Springer Ser. Solid-State Sci. Vol. 53, by G. Bauer, F. Kuchar, H. Heinrich (Springer, Berlin, Heidelberg (1984) p. 192

15 J.M. Berroir, Y. Guldner, J.P. Vieren, M. Voos, J.P. Faurie : to be published
16 D.L. Smith, T.C. Mc Gill, J.N. Schulman : Appl. Phys. Lett. 43, 180 (1983)
17 Y. Guldner, G. Bastard, M. Voos : J. Appl. Phys. 57, 1403 (1985)
18 M.H. Weiler : In Semiconductors and Semimetals, Vol. 16, ed. by R.K. Willardson, A.C. Beer (Academic, New York 1981) p. 119
19. F. Capasso : J. Vac. Sci. Technol. B1, 457 (1983)
F. Capasso, W.T. Tsang, A.L. Hutchinson, G.F. Williams : Appl. Phys. Lett. 40, 38 (1982)

Part IV

Technology

Molecular Beam Epitaxy of III-V Compounds

J. Massies

Laboratoire Physique du Solide et Energie Solaire, CNRS Sophia Antipolis, F-06560 Valbonne, France

This paper does not intend to give a review on Molecular Beam Epitaxy (MBE) of III-V compounds, since several excellent review papers have already been published on this topic (see the bibliography section). According to the spirit of the school, our purpose is rather to give insights into the basic principles and the technology of MBE applied to the III-V compounds epitaxial growth, with particular emphasis on what we believe to be the crucial points in order to succeed in performing MBE.

1. Principle

The term of MBE is used to denote an evaporation process under ultra-high vacuum (UHV) conditions, involving the reaction of thermal-energy molecular (or atomic) beams with a monocrystalline subtrate surface heated at a suitable temperature to allow epitaxial growth. If the substrate is of the same nature as the deposited layer, the term of homoepitaxy is often used, while heteroepitaxy corresponds to the case of deposited layers different from the substrate.

Gunther [9] was the first to report on the growth of III-V materials using a multiple molecular beams method, known as the three-temperature technique and considered as the origin of MBE. Gunther has obtained stoichiometric films of III-V compounds by using a group V element source at temperature T_1 in order to maintain a steady pressure in the vacuum chamber, a group III element source at a higher temperature T_3 controlling the condensation rate, and a substrate held at an intermediate temperature T_2 sufficiently high to eliminate the excess condensation of group V element on the substràte (Fig. 1). This process is indeed very close to

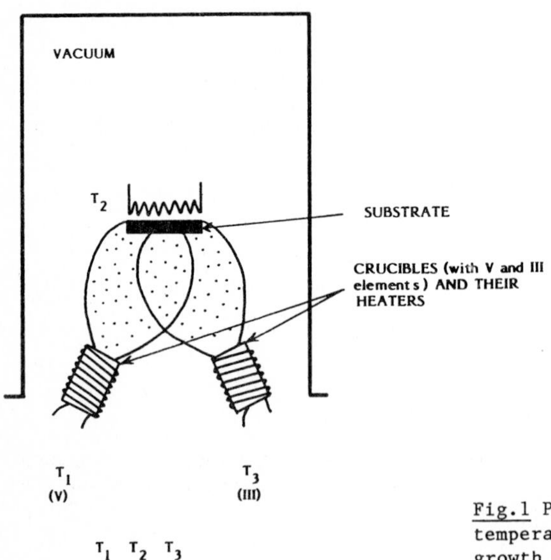

Fig.1 Principle of the three-temperature method for the growth of III-V compounds.

the one used in MBE of III-V materials. The major improvement of MBE technique with regard to the Gunther's method is the use of UHV and clean single crystal substrates.

2. Basic Considerations

2.1. Motion of molecules in rarefied gases-Molecular beams

The kinetic theory of gases postulates that gas molecules are in constant motion and that their velocities are temperature dependent [1-3]. The density of molecules having a velocity within v and v+dv is given by

$$\frac{(dn)\,v}{dV} = n\,f\,(v) \tag{1}$$

where dV is defined as the volume included between spheres of radius v and v+dv, n is the number of molecules per unit volume, and f(v) is the distribution function of the velocities.

By definition

$$\int_{v=0}^{v=\infty} n\,f\,(v)\,dV = n \qquad \text{since} \tag{2}$$

$$dV = 4\pi\,v^2\,dv \tag{3}$$

$$\int_0^\infty 4\pi\,v^2 f(v)\,dv = 1 \tag{4}$$

Maxwell has given the following expression for the distribution function

$$f(v) = a\,\exp(-bv^2) \tag{5}$$

where a and b are constants.

By using (2), (3), (4), and (5) it results that

$$4\pi\,a\int_0^\infty \exp(-bv^2)\,v^2 dv = \pi a\sqrt{\frac{\pi}{b^3}} = 1 \qquad \text{thus} \tag{6}$$

$$a = \sqrt{\frac{b^3}{\pi^3}}$$

$$\text{taking} \quad b = \frac{m}{2kT} \tag{7}$$

$$\text{or} \quad b = \frac{M}{2RT} \tag{8}$$

where m is the mass of molecule, M the molar mass ($M = mN_A$, N_A being the Avogrado's number), k is the Boltzmann's constant, R the gas constant and T the absolute temperature.

Introducing (7) in (5) and (8) in (5) it results that

$$f(v) = \left[\frac{m}{2\pi kT}\right]^{3/2} \exp\left(-\frac{m}{2kT}v^2\right) \tag{9}$$

or

$$f(v) = \left[\frac{M}{2\pi RT}\right]^{3/2} \exp\left(-\frac{M}{2RT}v^2\right) \tag{10}$$

Thus the number of molecules dn having speeds between v and v+dv is given by substituting (10) in (1)

$$(dn)_v = \frac{4\pi}{\sqrt{\pi}}\left[\frac{M}{2RT}\right]^{3/2} v^2 \exp\left(-\frac{M}{2RT}v^2\right) dv \tag{11}$$

(11) represents the Maxwell-Boltzmann distribution law.

The value of $f(v) = 1/n\ dn/dv$ is zero for $v = 0$ and $v = \infty$ and has its maximum for

$$v_p = \left[\frac{2kT}{m}\right]^{1/2}$$

or

$$v_p = \left[\frac{2RT}{M}\right]^{1/2} = 1.289 \times 10^4 \left[\frac{T}{M}\right]^{1/2} \text{ cm/sec} \tag{12}$$

which is known as the most probable velocity. This velocity is different from the arithmetic average v_{av} which results from

$$v_{av} = \frac{1}{n} \int_0^\infty v(dn)_v \tag{13}$$

with (11), (13) gives

$$v_{av} = \frac{4}{\sqrt{\pi}}\left[\frac{M}{2RT}\right]^{3/2} \int_0^\infty v^3 \exp\left(-\frac{M}{2RT}v^2\right) dv \tag{14}$$

since

$$\int_0^\infty v^3 \exp(-bv^2)\, dv = \frac{1}{2b^2}$$

Therefore

$$v_{av} = \left[\frac{8RT}{\pi M}\right]^{1/2} = \left[\frac{8kT}{\pi m}\right]^{1/2} = 1.455 \times 10^4 \left(\frac{T}{M}\right)^{1/2} \text{ cm/sec} \tag{15}$$

the mean square velocity is obtained from

$$\overline{v^2} = \frac{1}{n} \int_0^\infty v^2 (dn)_v \tag{16}$$

By using (11) this gives

$$\overline{v^2} = \frac{4}{\sqrt{\pi}}\left[\frac{M}{2RT}\right]^{3/2} \int_0^\infty v^4 \exp\left(-\frac{M}{2RT}v^2\right) dv \tag{17}$$

since

$$\int_0^\infty v^4 \exp(-bv^2)\, dv = \frac{3}{8}\sqrt{\frac{\pi}{b^5}}$$

therefore

$$\overline{v^2} = \frac{3RT}{M} \tag{18}$$

and the root mean square velocity is

$$v_r = \left[\frac{3RT}{M}\right]^{1/2} \tag{19}$$

These different velocities are of interest as representing the average behavior of a gas. When the velocity of molecules directly influence the process under consideration (e.g. flow of gases) the arithmetic average is used, while the mean square velocity should be used in energy calculations. For example, the average kinetic energy of a molecule is

$$\overline{E} = \frac{1}{n}\int_0^\infty \frac{1}{2} mv^2\, (dn)_v = \frac{m}{2n}\int_0^\infty v^2 (dn)_v \tag{20}$$

by using (16) and (18)

$$\overline{E} = \frac{3}{2}\frac{mRT}{M} = \frac{3kT}{2} \tag{21}$$

The mean free path of molecules is also an important notion which enters into many aspects of MBE technology. It can be defined as the average distance that a molecule travels between successive collisions with other molecules in the gas phase. The mean free path λ is related to the average velocity given in (15) by

$$\lambda = v_{av}\, \tau \tag{22}$$

where τ is the average time for a given molecule between collisions. The mean free path depends on the density and diameter of the molecules σ

$$\lambda = \frac{1}{\sqrt{2}\, \pi\, n\, \sigma^2} \tag{23}$$

Using the perfect gas law

$$P = n\, k\, T \tag{24}$$

(23) becomes

$$\lambda = \frac{kT}{\sqrt{2}\, \pi\, \sigma^2 P} = 2.33 \times 10^{-20}\, \frac{T}{\sigma^2 P} \tag{25}$$

where T is in °K, σ in cm and P in Torr.

Another useful quantity is the molecular arrival rate Φ which is defined as the rate of molecular collisions per unit area on a wall. It is also related to the average velocity

$$\Phi = \frac{n\, v_{av}}{4} \tag{26}$$

by substituting from (15) and (24)

$$\Phi = P\left[\frac{1}{2\pi mkT}\right]^{1/2} \quad \text{or} \tag{27}$$

$$\Phi = 3.51 \times 10^{22}\, P \frac{1}{\sqrt{MT}} \qquad \text{molecules/cm}^2\text{sec} \tag{27'}$$

where P is in Torr.

This equation (27') is very useful because it can also be used to calculate the quantity of gas passing through small orifices in thin walls at low pressures. From (27) and (27') the mass flowing rate is then

$$G = m\,P \left[\frac{1}{2\pi mkT}\right]^{1/2} \qquad \text{or}$$

$$G = 5.83 \times 10^{-2}\, P \sqrt{\frac{M}{T}} \qquad 9/\text{cm}^2\text{sec} \tag{28'}$$

Knudsen [4] has shown that the number of molecules effusing from a unit area of an orifice of the wall per unit time within the solid angle dω (see Fig.2) is a function of the angle θ from the orifice center line

$$d\Phi = \frac{n\, v_{av} \cos\theta}{4\pi}\, d\omega \tag{29}$$

This equation represents the cosine law of molecular effusion. It shows that emission of material from a small orifice does not occur uniformly, but favors directions close to the normal of the emitting surface.

Considering (26), (27') and (29) the flux angular density distribution (flux/unit solid angle dω) of material evaporating from a crucible with an emitting surface A_e is

$$\frac{d^2\Phi}{dt\,d\omega} = 1.12 \times 10^{22}\, A_e \frac{P}{\sqrt{MT}} \cos\theta \tag{30}$$

where P is the pressure (in Torr) in the crucible.

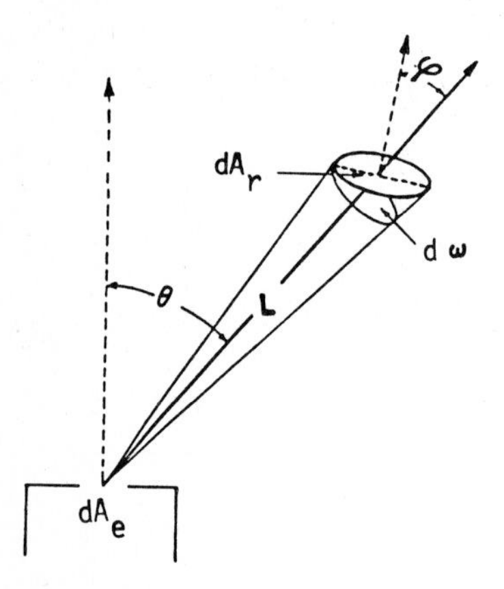

Fig.2 Molecular beam of solid angle dω corresponding to a receiving surface dA_r (see text, Eq. 31).

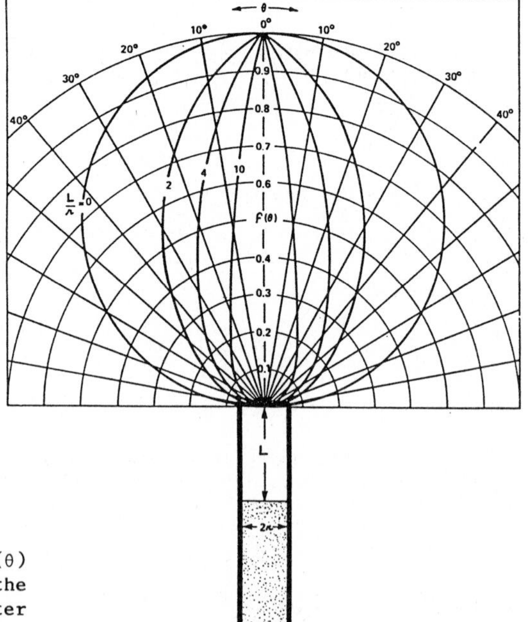

Fig.3 Dependence of the angular term f(θ) of the flux density distribution on the charge level L in a crucible of diameter 2r (after ref.6).

Consider now a small receiving surface A_r, at the distance L from the source, whose normal makes an angle ϕ with direction of the vapour stream and corresponding to an evaporant beam of solid angle $d\omega$ defined as (Fig.3)

$$d\omega = \frac{\cos\phi}{L^2} d A_r \tag{31}$$

The number of molecules striking the receiving surface is obtained by combining (30 and (31)

$$\frac{d^2 \Phi_r}{dtdA_r} = 1.12 \times 10^{22} A_e \frac{P}{\sqrt{MT}} \frac{\cos\theta \cos\phi}{L^2} \tag{32}$$

For a substrate normal to the crucible axis, the receiving flux density at$\theta=0$ is given by

$$\frac{d^2 \Phi_r}{dtdA_r} = 1.12 \times 10^{22} A_e \frac{P}{\sqrt{MT}} \frac{1}{L^2} \tag{33}$$

The number of molecules adsorbed on the substrate in this geometry ($\theta=0$) may be written

$$\frac{d^2 \Phi_{ra}}{dtdA_r} = 1.12 \times 10^{22} A_e \frac{P}{\sqrt{MT}} \frac{1}{L^2} S \tag{34}$$

where S is the sticking coefficient of the incident molecules on the substrate.

It should be recalled that the preceding equations (26 to 34) are valid only in the free molecule flow regime. In this regime inter-molecular collisions are negligible and therefore the flow may be analyzed by using simple collisionless kinetic theory, as shown above. The free molecule theory can be only applied if the mean free path λ (given by (23)) is much greater than the diameter D of the emitting orifice. This condition can be expressed by

$$\frac{\lambda}{D} \gg 1 \tag{35}$$

where λ/D is known as the Knudsen number.

A case of practical importance in MBE technology is the one concerning the use of crucibles in form of tubes instead of true Knudsen cells. In this case the free molecular regime is only obtained if

$$\lambda \gg D \tag{36}$$

and

$$\lambda \gg L \tag{37}$$

where D and L are respectively the diameter and the length of the tube.

On the other hand, the cosine law distribution (29) is only strictly valid when L = 0, i.e. when the tubular crucible is completely full. When the evaporating charge level falls in the tube, L increases and the crucible behaves like a collimator restricting the flux to progressively more sharply peaked angular distributions [6,7].

Therefore, (30) which gives the flux angular distribution for a Knudsen cell should be replaced by [6] :

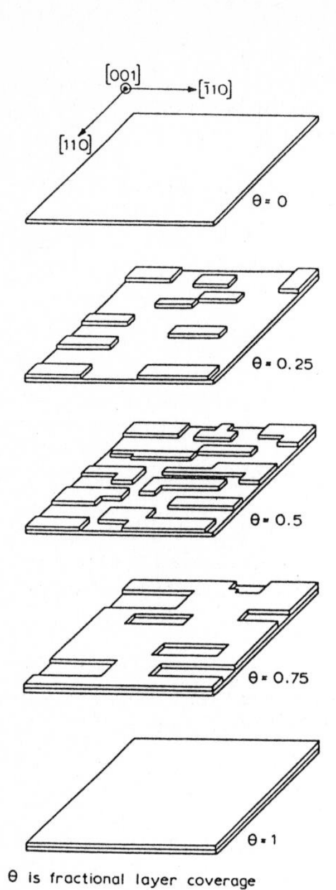

Fig.4 Schematic representation of the formation of a single complete layer on the (001) surface of GaAs in the two-dimensional mode of growth (after ref.10).

$$\frac{d^2 \Phi}{dt d\omega} = 1.12 \times 10^{22} A_e \frac{P}{\sqrt{MT}} f(\theta) \tag{38}$$

The angular term $f(\theta)$ is given in Fig. 4 (after refs 6 and 7) for different melt levels L in a crucible of diameter D = 2r.

It should be emphasized that such angular distribution is only valid for conditions of free-molecule flow. It has been shown that the flow through a sharp-edged orifice ($L/D \simeq 0$) does not become completely free-molecular until $\lambda/D \gtrsim 20$ [8]. For such high Knudsen numbers the angular distribution and the center-line intensity are in close agreement with cosine law predictions. When the Knudsen number decreases, the angular distribution becomes increasingly narrow, and the center-line intensity ($\theta = 0$) rises above that derived from the free-molecule theory. A result of high practical importance is that when tubes are used, the angular distribution becomes increasingly diffuse as the Knudsen number decreases [8]. Consequently, in practical MBE conditions which often correspond to rather low Knudsen numbers, the increasing nonuniformity of the deposited layer resulting from the increase of L values as the crucible charges are depleted may be less pronounced than indicated in Fig. 4.

Moreover, it should be kept in mind that to ensure molecular flows, true Knudsen cells which contain condensed phase and vapour at equilibrium must be used. Obviously, it is not the case for practical MBE evaporation cells, since the orifices of the crucibles are not very small compared to their size. Therefore the preceding equations are only approximate for usual MBE conditions.

Numerical Applications

Estimate of the GaAs growth rate for a given temperature of the Ga cell

The vapour pressure for a source material at the temperature T can be expressed as

$$\log P \simeq \frac{A}{T} + B \log T - C$$

where A, B and C are constants specific to the element. In the case of Ga [6]

$$\log P\,(\text{Torr}) = -\frac{11021.9}{T} + 7 \log T - 15.42$$

for

$$T = 960°C = 1233°K$$
$$P = 1.87\times10^{-3}\ \text{Torr}$$

For a type 125 RIBER cell (orifice diameter 20 mm) and a distance from the substrate of 12.5 cm (typical working distance for a RIBER system) the flux density calculated for $\theta = 0$ following (34) is 1.44×10^{15} atoms/cm^2 sec.

In the case of GaAs growth, for a typical growth temperature between 500°C to 600°C, the sticking coefficient of Ga is unity and an equivalent number of impinging arsenic atoms (provided by an excess flux or arsenic) react with gallium atoms at the growth interface to form stoichiometric material.

Seeing that a monolayer of GaAs (a = 5.6532 Å) contains

$$\frac{2}{(5.6532 \times 10^{-8})^2} = 6.258 \times 10^{14}\ \text{atoms/cm}^2$$

the growth rate is given by

$$\frac{2 \times 1.44 \times 10^{15}}{6.258 \times 10^{14}} = 4.6\ \text{ml/sec} \qquad \text{or}$$

$$\frac{5.6532}{4} \times 4.6 = 6.5\ \text{Å/sec} = 2.34\ \mu\text{m/h}$$

For a variation of the Ga cell temperature T = + 1°C (from 960°C to 961°C) the Ga flux density becomes 1.48×10^{15} atoms/cm^2 sec. This represents a variation of about 2.7 % of the growth rate.

Average kinetic energy of Ga atoms

Using (21) for T = 960°C = 1233°K
E = 0.16 eV

Mean free path of Ga atoms

For T = 960°C = 1233°, σ = 1.41 A Eq.(25) gives

λ = 77 cm

Knudsen number for the Ga cell

For a 125 RIBER type cell (diameter 20 mm, length 90 mm) containing Ga at 960°C

$$\frac{\lambda}{D} = \frac{77}{2} = 38.5$$

For a half-full cell $\frac{\lambda}{L} = \frac{77}{4.5} = 17$

Therefore the Ga flow can be still considered as close to a molecular flow.

2.2. Growth mode and growth mechanism

2.2.1. Growth mode

In order to obtain nearly perfect single crystal epitaxial layers with high electrical and optical performances, the mode of growth of the layer on the substrate should be of the Frank-Van der Merwe type, i.e. two-dimensional layer-by-layer growth mode. Actually, it has been recently demonstrated by using the so-called RHEED (for reflection high-energy electron diffraction) intensity oscillation technique that it is indeed the case in typical MBE growth of III-V materials such as GaAs, $Ga_{1-x}Al_xAs$ or $Ga_xIn_{1-x}As$ on clean GaAs substrates [10-15]. The RHEED intensity oscillations, which are generally measured on the specular beam, are interpreted in terms of variation of the surface reflectivity, which is maximum for a complete monolayer formation and minimum for half-monolayer coverage. This can be explained by the step density variation, along with the formation of a monolayer, the specular beam intensity being minimum when the step density and therefore the surface roughness are maximum, i.e. when the layer is half complete (Fig.4) [10]. It has been established that the period of the oscillation corresponds precisely to the growth of a single layer of the material (defined as a complete layer of element III, plus one complete layer of element V) (Fig.5).

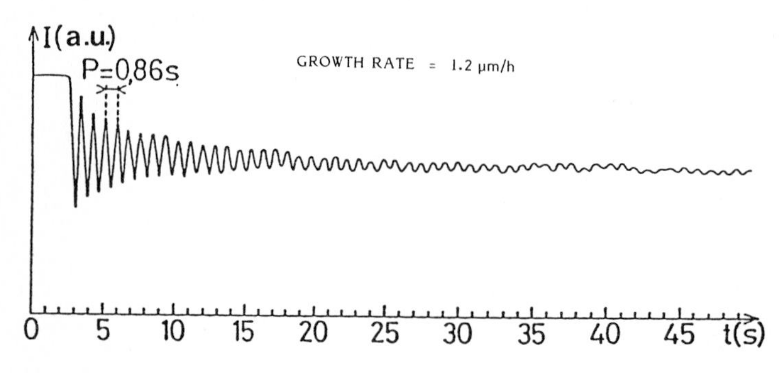

Fig.5 RHEED intensity oscillations during the MBE growth of GaAs (after ref.15).

This clearly demonstrates the two-dimensional character of the growth. However, the RHEED oscillations are gradually damped as the growth proceeds, indicating that the growth progressively deviates from the ideal two-dimensional layer by layer process, i.e. a new layer may start before the preceding one has been completed. This damping of the oscillation can be more or less pronounced, depending on the growth parameters, and therefore the RHEED oscillation can be used to optimize these parameters. On the other hand, measurement of the oscillation period provides a continuous growth rate monitor and alloy composition can also be determined by comparing changes in growth rate between a binary compound and the ternary alloy [10]. It can also be used to investigate sticking coefficients variations as a function of the growth temperature [15].

2.2.2. Kinetic aspects of growth mechanism

Molecular beam epitaxial growth is essentially governed by kinetics, contrarily to the quasi-equilibrium conditions which exist in the so-called liquid phase epitaxy or vapour phase epitaxy growth methods.

Kinetic aspects of the III-V compound MBE growth have been recently reviewed by JOYCE [16] and therefore will be only briefly summarized here. The surface kinetic of both growth and doping processes have been studied by modulated molecular beam techniques [16-20].

In the growth of III-V materials, different atomic of molecular species can be used as constituents of the incident beams. Vapours of elements III are always monoatomic, but elements V may be obtained in vapour as dimers or tetramers molecules. Tetramers (P_4, As_4, Sb_4) result from evaporation of elemental material, while dimers (P_2, As_2, Sb_2, are produced when appropriate III-V compounds are used as evaporation source. However, in the latter case, there is a significant proportion of the group III element in the dimer flux. Therefore the generally used technique to obtain pure dimers molecular flux is a thermal cracking of tetramers by the high-temperature stage of a two-zone evaporation cell.

In the case of As_2 and Ga interactions on GaAs surface, it has been shown [17,18] that in the usual temperature range of GaAs growth (500-600°C), the sticking coefficient of Ga atoms is unity, while condensation of As_2 occurs only when bonding with Ga is possible, i.e. when Ga atoms are present on the surface due to non-congruent surface dissociation or provided by an incident beam. The sticking coefficient of As_2 increases as the Ga adsorption rate, reaching unity when the Ga flux is roughly twice the As_2 flux. Therefore stoichiometric GaAs can be obtained if the flux ratio Φ_{As2}/Φ_{Ga} is larger than 0.5, the excess of incident As_2 molecules being simply eliminated by desorption. In practice, a flux ratio $\Phi_{As2}/\Phi_{Ga} > 1$ is generally used in order to largely compensate the evaporation of arsenic from the substrate due to the partial dissociation of GaAs in the usual epitaxial growth temperature range. A detailed model for the growth of GaAs from Ga and As_2 beams has been proposed by JOYCE and co-workers [16,18] (Fig. 6a). In this model, As_2

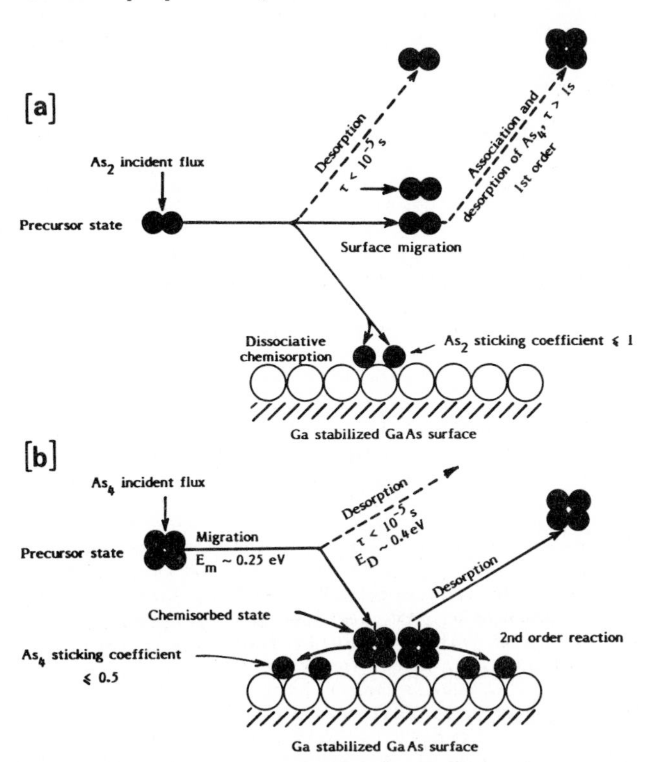

Fig.6 Model of the growth of GaAs by MBE : a) from As_2 and Ga beams b) from As_4 and Ga beams (after ref. 16).

molecules are first adsorbed into a weakly bound precursor state, mobile on the surface. Adsorbed As_2 molecules dissociate only when they encounter paired Ga lattice sites. If not they desorb as As_2 or As_4 (Fig. 6a).

Growth processes involving an incident As_4 flux are more complex (16,19). The flux ratio Φ_{As4}/Φ_{Ga} strongly influence the As_4 sticking coefficient, which has been found to be proportional to the gallium flux, when $\Phi_{Ga} << \Phi_{As4}$. In these conditions stoichiometric GaAs material is obtained provided that the substrate temperature is above 180°C (below this temperature the As_4 adsorption becomes non-dissociative) However, when $\Phi_{Ga} \leqq \Phi_{As4}$, the As_4 sticking coefficient becomes independent of the Ga flux. The crucial point is that it never exceeds 0.5. This behavior has been explained by the model schematically shown in Fig. 6b (16,19). The main feature of this model is a pairwise dissociation of As_4 molecules adsorbed on adjacent Ga atoms. The interaction of two As_4 molecules results in four As atoms incorporated into the GaAs lattice, while the remaining four atoms recombine and desorb as an As_4 molecule. Therefore the sticking coefficient of As_4 can never exceed the value of 0.5, even when the surface is entirely covered by Ga atoms. From this mechanism the important result for practical film growth is that stoichiometric GaAs can be obtained for As_4/Ga flux ratio larger than 0.5.

It is generally accepted that the kinetic models established in the case of the GaAs MBE growth are also valid for other III-V materials.

However, in the case of alloys resulting from the bonding of different group III elements to a group V element, sticking coefficients of one or several of these group III elements may be no longer unity above a certain growth temperature. In this case the composition of the growing material is not only fixed by the flux ratio of the group III elements, but depends also on the growth temperature via sticking coefficient variation (21). This behavior has been related to the thermal stability of the binary compounds of which the alloy may be considered to be composed (23), the more volatile group III elements being reduced in concentration relative to the less volatile ones (21-25).

Alloys containing different group V elements are more difficult to grow by MBE with a controlled composition, because of the considerable differences which exist in the sticking coefficients of molecules of these elements (the sticking coefficient increases along the column V from P to Sb).

Considering now the effects of element V species (dimeric or tetrameric molecules) on the epitaxial layer properties, they are still under investigation. Some available evidence of these effects has been summarized in references 16 and 26.

For mechanisms of dopant incorporation the reader is referred to references 21 to 27.

3. Technology of the MBE process of III-V compounds

3.1. MBE systems

During the first period of the development of the MBE growth technique (1970-1980), MBE systems were designed and built, at least for crucial parts, in the research laboratories active in the field. High quality device materials have been obtained by using such systems (28-32). High performance MBE machines are now commercially available from several manufacturers through the world (ISA-RIBER, Physical Electronics, Vacuum Generators, Varian ...). They are generally multichamber systems. Each of these interconnected chambers has a particular purpose, e.g introduction and preparation of the samples, UHV analyses and growth. All of these chambers have UHV type pumping systems (usually ion and titanium sublimation pumps). A schematic horizontal cross-section of the system used in the author's laboratory is given in Fig. 7 as an example of a multichamber system (the growth chamber is an ISA-RIBER 2300 equipped with ion, titanium sublimation and liquid-He

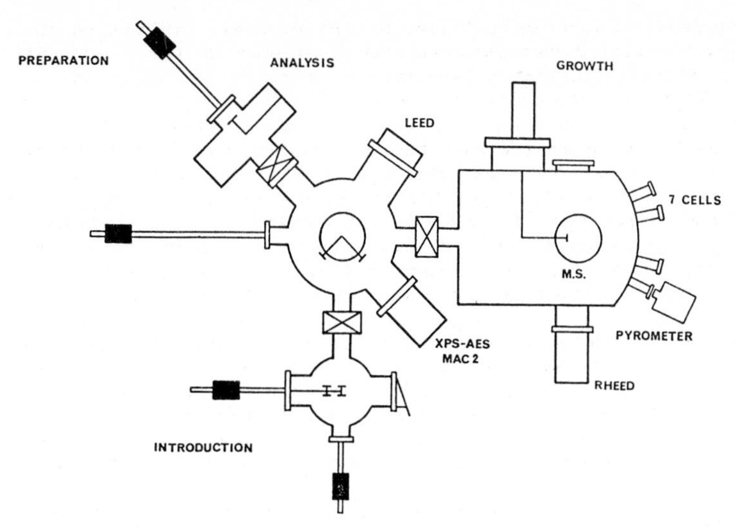

Fig.7 Schematic arrangement (top view) of a multichamber MBE system (introduction, preparation, analysis and growth chambers).

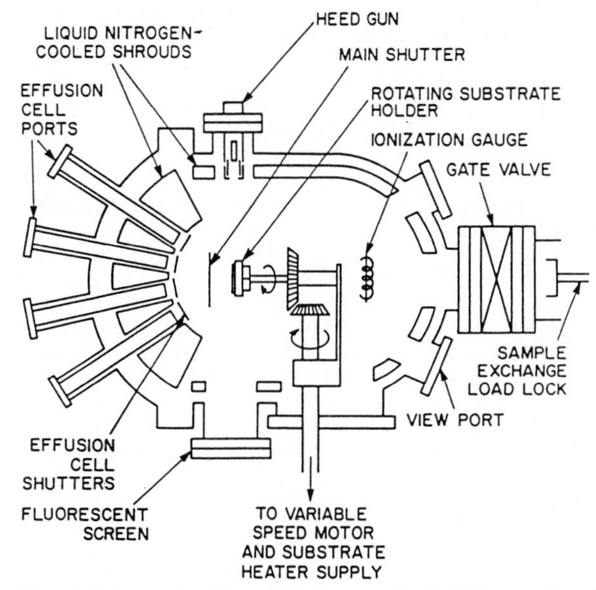

Fig.8 Schematic diagram of a typical MBE growth chamber (top view) (after [33])

pumping). Obviously, the crucial part of these systems is the growth chamber (Fig. 8, after ref. 33). A pressure less than 2×10^{10} Torr must be maintained when group III element cells are at their working temperatures. Such a low residual pressure is obtained by extensive outgassing of both the wall-chamber and the evaporation cells, and by using liquid nitrogen-cooled shrouds enclosing the entire growth area and the cells (Fig. 8).

Commercially available MBE systems can give very high quality materials [34], but they are very expensive. It should be emphasized that less expensive systems may be designed and built in laboratories having suitable mechanical manufacture facilities. The main practical advices drawn from the experience of the author and his colleagues in THOMSON-CSF Central Research Laboratory (P. ETIENNE, P. DELESCLUSE and J.F. ROCHETTE) are the following :

1. as small as possible a growth chamber, with high-speed-UHV pumping devices ;

2. source flange(s) positioned 25 to 45° below the horizontal, combining thus advantages of horizontal and vertical evaporation geometries, i.e. maximum charges of the cells and minimum contamination of the sources by falling flakes from the shutters and the walls of the vacuum chamber ;

3. flange positioned on the axis of the source flange in order to possibly use a quadrupole-mass-analyser to control the beams [35] ;

4. oven cells quasi-isolated from the growth area, with pumping on the rear;

5. extensive liquid-nitrogen shrouds (at least around the evaporation cells);

6. sample exchange load-lock system connected to the growth chamber ;

7. use of high-purity materials of low outgassing rate for the crucibles (pyrolitic boron nitride for high temperature cells, quartz for low temperature cells, e.g. arsenic cell) and their ovens (tantalum, PBN, sapphire or alumina). A typical example of a laboratory-made cell is given in Fig. 9.

3.2 Substrate preparation

A critical step in MBE is the preparation of the substrate surfaces, which must be smooth and clean on an atomic scale prior to the growth.

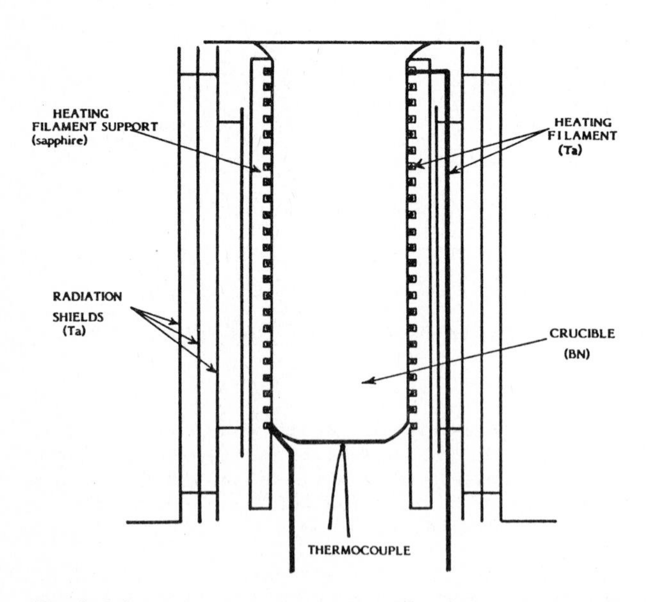

Fig.9 Schematic representation of a laboratory-made evaporation cell.

Two years ago, CHO and coworkers [28,36] have proposed a chemical etching process of GaAs substrates which has been decisive for the development of the MBE application. This procedure is still used by most groups active in MBE [33,34,37]. It mainly involves a chemical etching based on a $H_2SO_4/H_2O_2/H_2O$ solution followed by a rinse in deionized water. This last step was found to be very important in order to produce a thin oxide layer protecting the surface against contamination during subsequent manipulations of the sample [28]. Such a passivating layer is particularly attractive since it can be easily removed, after loading the sample into the MBE system, by thermal desorption. This particular point mainly explains the success of this method. Actually, even when a different etching solution is used, such as Br_2/CH_3OH, a final rinse with deionized water is often performed in order to produce an oxide layer on GaAs or InP surfaces [33,38].

In the case of GaAs substrates a typically used etching process is the following :

1. trichloroethylene degrease
2. methanol rinse
3. 18 MΩ running deionized water (RDOW) rinse
4. H_2SO_4/H_2O(5 : 1 : 1) etch for 10 min. at 60°C
5. concentrated HCl etch (5 min.)
6. 18 MΩ-RDIW.rinse

The sample is then dried in a stream of nitrogen and mounted on a molybdenum sample-holder with indium solder at ≃ 16°C. It is then quickly loaded in the introduction chamber of the MBE system.

The $H_2SO_4/H_2O_2/H_2O$ etching process can be preceded or simply replaced by chemomechanical polishing with lens paper soaked with a Br_2/CH_3OH solution (0.5-2%) [33]. This latter process is generally preferred for InP substrate preparation [32], but is also often used in the case of GaAs.

It has been recently shown by X-ray photoelectron spectroscopy [39] that the substrate surface stoichiometry is affected by these etching procedures : the $H_2SO_4/H_2O_2/H_2O$ mixture (5/1/1, 60°C) produces an arsenic-rich GaAs surface layer having an atomic ratio As/Ga close to 1.15, while the bromine-methanol mechanopolishing leads to arsenic (GaAs) or phosphorus (InP) depleted surfaces with atomic ratios As/Ga ≃ 0.7 and P/In ≃ 0.65.

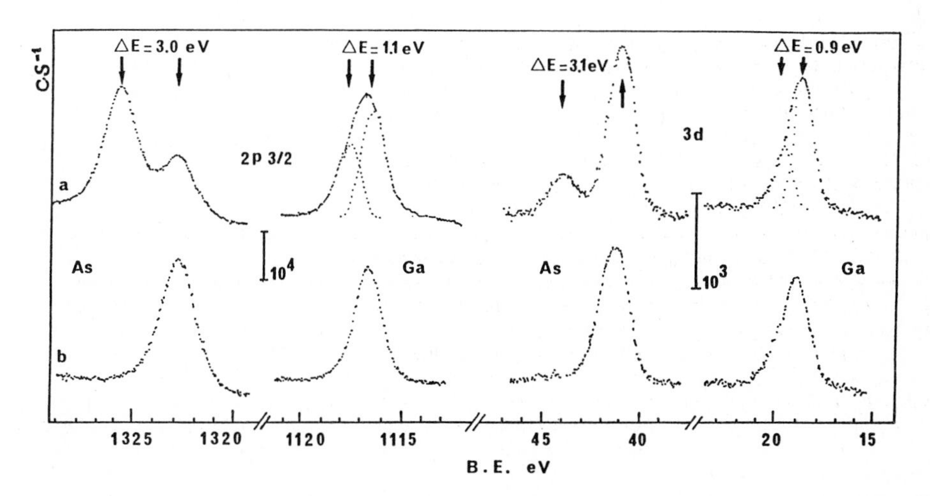

Fig.10 Photoelectron spectra recorded on a GaAs substrate after $H_2SO_4/H_2O/H_2O_2$ etching : a) sample tranferred in air after indium soldering ; b) without indium soldering and transferred under N_2 atmosphere after etching.

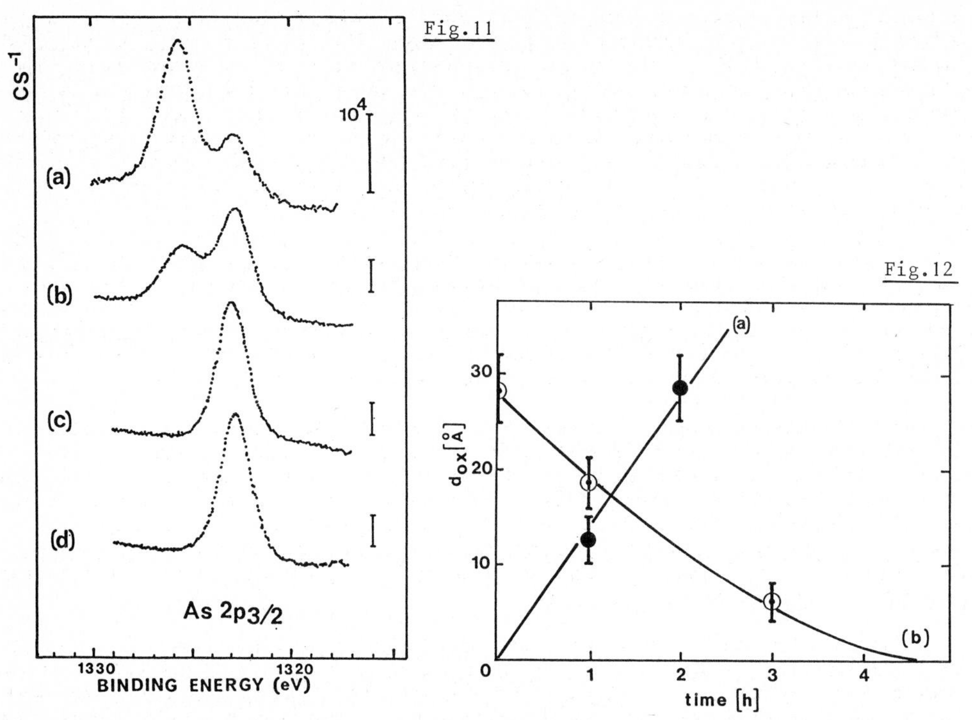

Fig.11 Effects of different preparation procedures on the As $2p_{3/2}$ level : a) as described in section 4.2 ; b) substrate mechanically fixed on the sample-holder (no indium soldering) ; c) substrate mechanically fixed and transferred under N_2 atmosphere after etching ; d) cleaning procedure without HCl etching, substrate fixed and transferred as in c) (after ref. 40).

Fig.12 Oxided layer thickness versus time for GaAs contacted with static (a) and running deionized water (b) (after ref. 41).

It has also been demonstrated that contrarily to what was generally believed, rinsing in running deionized water after etching does not produce any passivating oxide film on the substrate surface [40]. The surface oxidized phases observed after the standard preparation processes described above, using $H_2SO_4/H_2O_2/H_2O$ or Br_2/CH_3OH etching solutions, are only due to the sample manipulation in air after etching (Fig. 10) [40,41]. This oxidation process is enhanced by the sample heating for indium soldering on the sample-holder after etching (Fig. 11).

Therefore, if the substrate is mechanically mounted on the substrate holder, or if indium soldering is performed before the final etching (e.g. Br_2/CH_3OH polishing method of preparation), the surface oxidized phases can be avoided by handing the sample under an inert atmosphere after etching (fig. 10) [40]. Moreover, oxidized surface phases formed during indium soldering or transit in air of GaAs samples can be simply removed by dissolution in running deionized water (Fig. 11) [41].

This last point is of practical importance for MBE growth, since, associated with transferring and loading of the sample under dry nitrogen, it allows to obtain an oxidation-free surface, even when the sample is standardly mounted on the substrate holder with indium soldering after etching.

However, when manipulations or treatments of substrates in conditions of possible damage of the surface after etching are unavoidable, it can be useful to grow an oxide layer in controlled conditions of purity (in contrast to oxidation in air), in order to protect the surface. Such a passivating oxide film can be obtained by drying the substrate in a pure oxygen flow after etching, as proposed by CHO and TRACY [36], or by letting the substrate react in deionized water under static conditions (stagnant water) [40,41]. The latter process produces a Ga-rich oxide layer on GaAs surface, with a rate of formation of 10 to 20 A h^{-1} (an oxide layer of 20 to 40 Å can be sufficient to protect the surface against subsequent contamination). It should be emphasized that this oxide layer can be removed by running deionized water as the one obtained by air oxidation (the dissolution rate is within 5 to 10 A h^{-1} for a water flow of 0.4 l $min.^{-1}$, Fig. 12) [41].

When an oxidized layer is present on the substrate surface (intentionally or not), it must be eliminated before starting MBE growth in order to ensure a two-dimensional growth mode of the deposited material. The oxidized layer is usually removed by thermal desorption. As for the substrate chemical etching, a careful control of this step of the MBE process is undoubtedly of critical importance, in order to obtain epitaxial layer having a low defect concentration.

Above a critical temperature (e.g. 200°C for InP, 400°C for GaAs), annealing of the substrate is performed under a sufficient group V element pressure to prevent partial thermal decomposition of the surface. In the case of GaAs substrates, the oxidized arsenic phase desorbs from 150°C upwards, the complete desorption being observed around 400°C, while the oxidized gallium atoms desorb only above 550°C [40,42]. If the desorption is performed under a sufficient arsenic flux ($P_{As4} \sim 5 \times 10^{-6}$ Torr) a 2 x 4 (or C(2x8)) reconstructed surface structure is then obtained and the growth can be started.

In the case of the desorption of InP oxidized surface layer, which is complete between 450 and 500°C [33,43], the stabilization of the surface is easier to achieve with impinning arsenic molecules than with phosphorus ones. Therefore, when InP is used as substrate for epitaxial growth of materials in the (AlGaIn) As system, annealing is generally performed (from about 200°C upwards) under arsenic pressure ($P_{As4} \sim 5 \times 10^{-5}$ Torr at 500°C) in order to prevent thermal decomposition of the surface [44]. This process results in a clean As-stabilized 2x4 reconstructed surface [45] suitable for subsequent epitaxial growth. Thermal cleaning of GaSb substrates is also more easily done with arsenic (As_4) than with antimony (Sb_4) impinning molecules [46].

3.3. Molecular beams calibration and control

In the case of growth of multinary III-V alloys, vapor pressure versus temperature relationships for the source materials can be used to estimate the composition of the alloys for different crucible temperatures. However, this procedure is not generally sufficient to ensure a reproducible composition from run to run, in particular because the crucibles are not true Knudsen cells, the emitting fluxes being slightly dependent on the charge level of the large tubular crucibles generally used. Therefore, a more direct measurement of the flux densities is often performed by using an ion gauge attached to the substrate heater and which can be rotated into the normal substrate growth position. This gives a beam equivalent pressure for each element of interest for growing a given multinary alloy. Such measurements should be corrected from the sensitivity variation of the ionization gauge with the atomic number Z of the source elements. The ionization efficiency η of a given element relative to nitrogen, for which the ion gauges are calibrated, can be obtained by using an empirical expression such as [47] :

$$\eta/\eta_{N2} = [(0.4\ Z/14) + 0.6]$$

Experimental values of ionization gauge sensitivity for various gases are given in Ref. 48.

A more powerful method of comparing flux intensities is the use of a quadrupole mass spectrometer (QMS) placed in the line of sight of the effusion cells. By using an associated mass peak selector, it allows to simultaneously measure and compare the different beams used, before and during the growth. It can be directly coupled to the regulation devices of the effusion cell ovens. Owing to the rapid response and mass discrimination of the QMS, a flux feedback control of the furnace power in real time can be achieved [35]. Using a QMS system, it is therefore possible to control the ratio of two beams (e.g. Al/Ga, Al/In or Ga/In for the growth of $Al_xGa_{1-x}As$, $Al_xIn_{1-x}As$ or $Ga_xIn_{1-x}As$ alloys). However, there are several difficulties inherent in the QMS use for monitoring fluxes [6] : i) dopant fluxes are generally too low for detection by the QMS ; ii) due to the dependence of the angular distribution of the fluxes on the charge level in the crucibles, the flux ratio at the QMS detector and at the substrate may be different if the positions are not strictly equivalent ; iii) the QMS is subject to significant variation sensitivity (in particular due to the multiplier gain drift) and re-normalization of its response is regularly needed.

An accurate control of the fluxes is particularly important for the growth of III-V alloys which are lattice matched on a substrate only for very precise compositions; It is for example the case of $Ga_xIn_{1-x}As$ or $Al_xIn_{1-x}As$ alloys which are lattice matched on InP substrate for x = 0.47 and x = 0.48 respectively. A variation of 1% in the InAs mole fraction from x = 0.47 causes a $Ga_xIn_{1-x}As$/InP lattice mismatch a/a 7 x 10^{-4} [49]. It has been shown that for a/a 10^{-3}, electron mobility of this material is severely degraded [50]. Therefore, when growing such lattice-matched systems with crucible temperatures controlled by thermocouple feedback to a rate-proportional controller, a stability of 1/10°C should be achieved.

On the other hand,when using separate effusion cells for growing multinary alloys epitaxial layers, an important lateral composition inhomogeneity is generally observed, specially when large wafers are used, owing to the intensity profile of the beams at the substrate surface. The best way to grow uniform multinary alloys over a large area is to rotate the substrate during the growth [51,52]. However, a sufficient rotation speed (i.e. a complete revolution for one monolayer growth) is needed to avoid periodic modulation of the composition along the growth direction [53].

3.4. Growth parameters

In the growth of epitaxial layers of III-V compounds, the main growth parameters are the growth rate, the growth temperature and the element V/element III flux intensity ratio. Growth rate and growth temperature are not independent parameters: since the growth temperature mainly determines the mobility of atoms on the surface, it should be optimized for each different atom or molecule arrival rate at the surface. The basic rule is that the growth temperature should be increased with the growth rate. It should be also pointed out that the effect of background-gas related impurities in the growth chamber, in particular residual reactive species such as H_2O or CO, increases drastically when the growth rate or the growth temperature are lowered. Therefore, high growth temperatures are generally needed for growing alloys which involve the use of highly reactive materials such as aluminium.

The growth temperature and the element V/element III flux ratio fix together the growth interface stoichiometry. This parameter plays a crucial role in the determination of morphological, electrical and optical properties of the growing layer. It can be controlled in real time by using RHEED which indicates the type of surface reconstruction for a given set of the growth parameters. In the case of GaAs, surface reconstruction change with surface stoichiometry has been extensively studied by high and low energy electron diffraction [54,55], Auger electron spectroscopy [56] and soft X-ray electron spectroscopy [57[]. Results dealing with the relationship between surface structure and surface stoichiometry have been also reported for $Ga_xIn_{1-x}As$ [58,59], $Ga_xIn_{1-x}P$ [58] and $GaAs_xSb_{1-x}$ [59].

The first clear evidence of the crucial role of the growth interface stoichiometry in the incorporation of intentional impurities has been reported

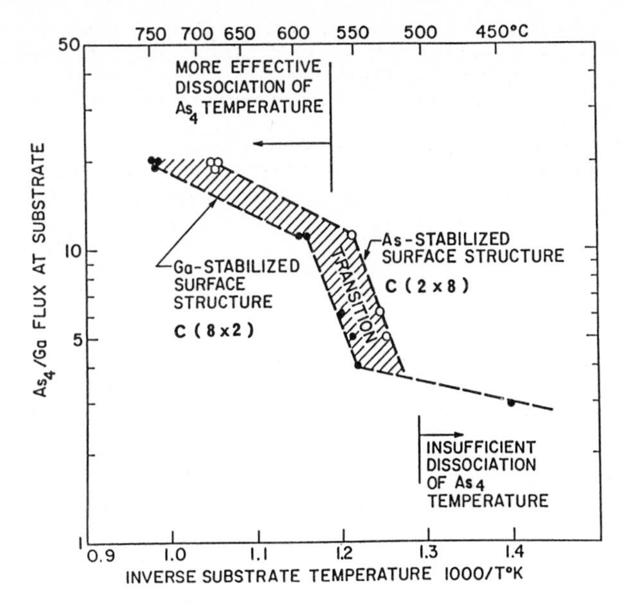

Fig.13 Surface structure of GaAs as a function of the growth temperature and the As_4/Ga flux ratio (growth rate ∿ 1μm h^{-1}) (after ref. 33)

several years ago by CHO and HAYASHI [60]. They have demonstrated that a Ge flux impinning on GaAs surface during growth gives n type doping for As -stabilized growth conditions corresponding to the C(2x8) surface reconstruction, and p type doping for Ga - stabilized conditions with the C(8x2) surface reconstruction. This effect is not restrictive to the Ge incorporation and affects also, to a less extent, other amphoteric dopant incorporation (column IV) such as Si, which is generally used to obtain n type MBE III-V layers [61]. The key role of the growth interface stoichiometry in the incorporation mechanisms of impurities impinning the surface of the growing layer has been confirmed recently on nonintentionally doped GaAs layers [62] : p-type and n-type conduction mechanisms have been respectively observed for low and high values of the As_4/Ga flux ratio. At least in the case of GaAs and $Al_xGa_{1-x}As$ growth, recent results of several laboratories lead to the conclusion that the highest quality materials are grown for As_4/Ga flux ratio giving a growth interface stoichiometry very close to the transition region between the C(2x8) As -stabilized surface structure and the C(8x2) Ga -stabilized surface structure (Fig. 13). For GaAs growth, this transition region corresponds to the onset of the 3x1 surface reconstruction [55].

References

Vacuum and thin films technology

See for example

A. Roth: Vacuum Technology, second edition, North Holland, Amsterdam 1982.

Handbook of Thin Film Technology, Edited by L.I. Maissel and R. Glang, Mc Graw Hill New York 1970.

Reviews articles and books on MBE

- A.Y. Cho and J.R. Arthur: Progr. Solid State Chem., 10, 157 (1975)
- L.L. Chang and R. Ludeke: Epitaxial Growth, p. 37, J.W. Matthews editor, Academic New York (1975)

. J. Massies: P. Etienne and N.T. Linh: Rev. Tech. THOMSON CSF, 8, 5, (1976)
. R.F.C. Farrow: Crystal Growth and Materials, 1, 237, E. Kaldis et H.J. Scheel editors, North Holland Amsterdam (1977)
. B.A. Joyce and C.T.Foxon: Inst. Phys. Conf. Ser. N°32, p. 17, (1977)
. R.Z. Bachrach: Progress in crystal growth and characterization, 2, 115, Pergamon Press (1979)
. C.E.C. Wood: Phys. Thin Films, 11, 35, (1979)
. Molecular Beam Epitaxy, Editor B.R. PAMPLIN, Pergamon, Oxford 1980
. K. Ploog: Crystal : Growth, Properties and Applications, 3, 73, Springer Verlag Berlin (1980)
. R.Z. Bachrach: Crystal Growth, Second Edition, Brian Pamplin Editor, Pergamon (1980)
. K. Ploog: Ann. Rev. Mater. Sci., 11, 171, (1981)
. P.E. Luscher: Thin Solid Films, 83, 125, (1981)
. A.Y. Cho: Thin Solid Films, 100, 291, (1983)
. Molecular Beam Epitaxy and Heterostructures Edited by L.L. CHANG and K. PLOOG, NATO ASI Serie E N° 87, Martinus Nijhoff, Dordrecht 1985

Comprehensive bibliography on MBE

K. Ploog and K. Graf: Molecular Beam Epitaxy of III-V Compounds, Springer-Verlag Berlin 1984

References explicitly cited in the text

1. E. Bloch: Theorie Cinetique des Gaz, Armand Colin, Paris 1921
2. E.H. Kennard: Kinetic Theory of Gases, Mc Graw Hill, New York 1938
3. R.D. Present: Kinetic Theory of Gases, Mc Graw Hill, New York 1958
4. M. Knudsen: The Kinetic Theory of Gases, Methuen, London 1934
5. R.W. Berry, P.M. Hall and M.T. Harris: Thin Film Technology, D. Van Nostrand, Princeton 1968
6. P.E. Luscher and D.M. Collins: in Molecular Beam Epitaxy, B.R. Pamplin Editor, Pergamon, Oxford 1980, p. 15
7. B.B. Dayton: 1956 Vacuum Symposium Transactions, Committee on Vacuum Techniques, Boston 1956, p. 5
8. R.E. Stickney, R.F. Keating, S. Yamamoto, W.J. Hastings: J. Vac. Sci. Technol. 4, 10, (1967)
9. K.G. Gunther: Z. Naturforsch, 13a, 1081, (1958)
10. J.H. Neave, B.A. Joyce, P.J. Dobson, N. Norton: Appl. Phys. A,31, 1, (1983)
11. J.M. Van Hove, C.S. Lent, P.R. Pukite, P.I. Cohen: J. Vac. Sci. Technol. B, 1, 741 (1983)
12. P.R. Pukite, J.M. Van Hove, P.I. Cohen: J. Vac. Sci. Technol. B, 2, 243, (1984)
13. J.H. Neave, B.A. Joyce, P.J. Dobson: Appl. Phys. A, 34, 179, (1984)
14. B.F. Lewis, T.C. Lee, F.J. Grunthaner, A. Madhukar, R. Fernandez, J. Maserjian: J. Vac. Sci. Technol. B, 2, 419, (1984)
15. M.N. Charase, L. Goldstein, Proc. of the "3ème Séminaire Epitaxie par Jets Moléculaires", Carry Le-Rouet, France 1984, Edited by CNRS - Sophia Antipolis, p. 135
16. B.A. Joyce: in "Molecular Beam Epitaxy and Heterostructures" p. 37, ed. by L.L. Chang and K. Ploog, NATO ASI Serie E N° 87, Martinus Nijhoff, Dordrecht 1985
17. J.R. Arthur: Surf. Sci., 43, 449, (1974)
18. C.T. Foxon, B.A. Joyce: Surf. Sci. 64, 293, (1977)
19. C.T. Foxon, B.A. Joyce: Surf. Sci. 50, 434, (1975)
20. B.A. Joyce, C.T. Foxon: Inst. Phys. Conf. Ser. N° 32, p. 17, (1977)
21. C.E.C. Wood, D.V. Morgan, L. Rathbun: J. Appl. Phys. 53, 4524, (1982)
22. B. Goldstein, D. Szostak, Appl. Phys. Lett. 26, 685, (1975)
23. C.T. Foxon, B.A. Joyce: J. Cryst. Growth 44, 75, (1978)
24. D. Bonnevie, D.Huet: J. Phys. (Paris) 43, C5, 445, (1982)
25. J. Massies, J.F. Rochette, P. Delescluse: Proc. Third International Conf. on MBE, San Francisco 1984, to be published in J. Vac. Sci. Technol. B, 3, March-April 1985

26. K. Ploog: Ann. Rev. Mater. Sci., 11, 171, (1981)
27. C.E.C. Wood: in "Molecular Beam Epitaxy and Heterostructures", p. 149, ed. by L.L. Chang and K. Ploog: NATO ASI Serie E N° 87, Martinus Nijhoff, Dordrecht 1985
28. A.Y. Cho, J.R. Arthur: Prog. Solid State Chem. 10, 157, (1975) and references there in
29. D. Delagebeaudeuf, P. Delescluse, P. Etienne, M. Laviron, J. Chaplart, N.T. Linh: Electron. Lett. 16, 667, (1980)
30. P.N. Tung, P. Delescluse, D. Delagebeaudeuf, M. Laviron, J. Chaplart, N.T. Linh: Electron. Lett. 18, 109 and 517, (1982)
31. M. Laviron, D. Delagebeaudeuf, P. Delescluse, P. Etienne, J. Chaplart, N.T. Linh: Appl. Phys. Lett. 40, 530, (1982)
32. J. Massies, J.F. Rochette, P. Delescluse, P. Etienne, J. Chevrier, N.T. Linh, Electron. Lett. 18, 758, (1982)
33. A.Y. Cho: Thin Solid Films, 100, 291, (1983)
34. M. Heiblum, E.E. Mendez, L. Osterling: J. Appl. Phys. 54, 6982, (1983)
35. P. Etienne, J. Massies, N.T. Linh: J. Phys. E, 10, 1153, (1977)
36. A.Y. Cho, J.C. Tracy, Jr: United States Patent 3, 969, 164, (1976)
37. K. Ploog: in Crystals : Growth, Properties, and Applications Vol. 3, p. 73, Springer-Verlag 1980 ; J.C.M. Hwang, H. Temkin, T.M. Brennan, R.E. Frahm: Appl. Phys. Lett. 42, 66, (1983)
38. D. Bonnevie and D. Huet: J. Phys. (Paris) 43, C5-445 (1982)
39. J.P. Contour, J. Massies, A. Saletes: Jpn. J. Appl. Phys. 24, L563, (1985)
40. J. Massies, J.P. Contour: J. Appl. Phys. 58, 806 (1985)
41. J. Massies, J.P. Contour: Appl. Phys. Lett. 46, 1150 (1985)
42. J.P. Contour, J. Massies, A. Saletes, P. Staib: Appl. A 38, 45 (1985)
43. J. Massies, F. Lemaire-Dezaly: J. Appl. Phys. 55, 3136 (1984) and 57, 237, (1985)
44. G.J. Davies, R. Heckingbottom, H. Ohno, C.E.C. Wood, A.R. Calawa: Appl. Phys. Lett. 37, 290, (1980)
45. J. Massies, P. Devoldere, N.T. Linh: J. Vac. Sci. Technol. 15, 1353, (1978)
46. C. Raisin: Private Communication (1984)
47. C.E.C. Wood, D. Desimone, K. Singer, G.W. Wicks: J. Appl. Phys. 53, 4230 (1982)
48. T.A. Flaim, P.D. Ownby: J. Vac. Sci. Technol. 8, 661, (1971)
49. K.Y. Cheng, A.Y. Cho, W.R. Wagner, W.A. Bonner: J. Appl. Phys. 52, 1015 (1981)
50. J. Massies, M. Sauvage-Simkin: Appl. Phys. A, 32, 27, (1983)
51. A.Y. Cho, K.Y. Cheng: Appl. Phys. Lett. 38, 360 (1981)
52. K.Y. Cheng, A.Y. Cho, W.R. Wagner: Appl. Phys. Lett. 39, 607, (1981)
53. K. Alavi, P.M. Petroff, W.R. Wagner, A.Y. Cho: J. Vac. Sci. Technol. B1, 146, (1983)
54. A.Y. Cho: J. Appl. Phys. 47, 2841, (1976)
55. J. Massies, P. Devoldere, P. Etienne, N.T. Linh: in : Proc. 7th Intern. Vacuum Congr. and 3rd Inter. Conf. on Solid Surfaces (Vienna, 1977), R. Dobrozemsky et al. Editors, Vol. 1, p. 639
J. Massies, P. Etienne, F. Dezaly, N.T. Linh: Surface Sci. 99, 121, (1980)
56. P. Drathen, W. Ranke, K. Jacobi: Surface Sci. 77, L 162 (1978)
57. R.Z. Bachrach, R.S. Bauer, P. Chiaradia, G.V. Hansson: J. Vac. Sci. Technol. 19, 335, (1981)
58. C.T. Foxon, B.A. Joyce: J. Crystal Growth 44, 75, (1978)
59. R. Ludeke: J. Vac. Sci. Technol. 17, 1241, (1980)
60. A.Y. Cho, I. Hayashi: J. Appl. Phys. 42, 4422, (1971)
61. R. Nottenburg, H.J. Buhlmann, M. Frei, M. Illegems: Appl. Phys. Lett. 44, 71, (1984)
62. A. Saletes, J. Massies, G. Neu, J.P. Contour: Electron. Lett. 20, 872, (1984)

Molecular Beam Epitaxy of II-VI Compounds

A. Million

CENG, IRDI, LETI/LIR, 85 X, F-38041 Grenoble Cedex, France

The growth of II-VI compounds by vacuum evaporation has been extensively studied [1]. Molecular beam epitaxy (MBE) of these compounds has been developed more recently. The first publication on this subject was reported in 1975 [2]. Since then, most of the studied compounds have been : ZnTe [2-4], ZnSe [2-7], $ZnSe_x\ Te_{x-1}$ [8], ZnS [9-10], CdSe [2], CdS [29] which are interesting alloys for their optical and electroluminescent properties. In the eighties, the development of the MBE of CdTe [11-18] and $Cd_x\ Hg_{1-x}$ Te (CMT) [19-23], mainly known for their applications in infrared detection, started.

In the field of superlattices using II-VI compounds, to our knowledge, only the CdTe-HgTe superlattice has been successfully grown and studied [24-27]. Very recently, the ZnS - ZnSe superlattice has been grown by hot wall epitaxy [28].

In this paper, we will describe the evaporation of these compounds and a very simple model which allows us to evaluate the epitaxial conditions and their evolutions with the different parameters. We will take the case of CMT as an example of the experimental procedure, because growing this semiconductor involves most of the problems encountered in MBE of II-VI compounds.

1. Evaporation of Binary II-VI Compounds

1.1. P-T Diagram

To grow a crystal under vacuum conditions, we have to know the vapor species in equilibrium with the solid and its mode of evaporation.

The different studies of the evaporation of II-VI compounds have shown that they all sublime dissociatively [30-31]

$$MX_{(S)} \rightarrow M_{(g)} + \frac{1}{2} X_{2(g)} \qquad (1)$$

The vapor species over group II are the monomers and over group VI the dimers, except for selenium and sulfur, where a series of polyatomic molecules are present with proportions a function of temperature. The gaseous molecule $MX_{(g)}$ is practically nonexistent.

Following the mass action law, we can write :

$$K_p(T) = P_M . P_{X_2}^{1/2} \qquad (2)$$

The dependence of the value of K_p on the deviation from the stoichiometry of the compounds can be neglected [32]. In a P-T diagram, the partial pressures P_M and P_X describe loops (Fig. 1). The curved lines represent the three phase equilibrium curves and the area within the boundary line represents the pressure values of the element which is in equilibrium with the solid compound. A detailed description of diagrams is given in [31,32]. It can be noticed that, if the compound presents a deviation from stoichiometry toward an enriched side of one element (e.g. M) or the other (e.g. X), the partial pressures of elements P_M and P_X can change by several orders of magnitude at the same temperature. These pressures cannot take any values because they are linked together by the relation (2) and none can be superior to the pressure Po of the pure element :

$$P_M < P_M^o \quad \text{and} \quad P_X < P_{X_2}^o$$

These two conditions imply that they should also be superior to a minimal pressure P^*_M and P^*_X given by the following expressions :

$$P_M^* = K_p.(P_{X_2}^o)^{-1/2} \quad , \qquad P_{X_2}^* = K_p^2 \cdot (P_M^o)^{-2} \tag{3}$$

In general, the pressures above the compound are determined at high temperature (T > 600 °C) and we have no experimental data in the range of growth temperatures used (T < 400°C). Nevertheless, the limiting pressure P_i^o is generally well known and P^*_i is calculated using extrapolated values of K_p. These approximations are satisfactory for MBE application.

1.2. Congruent Sublimation

Except for the mercury compounds, the pure elements of II-VI compounds have vapor pressures relatively close to each other. This particularity leads to most of the P-T loops overlapping in a wider range of temperature than in the case of III-V compounds (e.g. CdTe in Fig.1a). This overlapping explains the following phenomena: let us consider a compound with any stoichiometry to be heated under vacuum and, for example, under pressures such that $P_M > 2P_{X_2}$, in this case the preferential vaporisation of M compared with that of X_2 changes the stoichiometry of the compound which takes the new equilibrium partial pressures P', such that $P_M > P_M > 2P_{X2} > 2P_{X2}$. This system will evolve toward a steady state where the vapor phase will have the same composition as the solid phase, within the stoichiometry range. This condition defines what is called the congruent sublimation. In this condition, inside a quasi-closed cell, at equilibrium, the pressure P_M and P_{X2} are related by the relation $P_M = 2P_{X2}$ [4]. Combining with relation (2), it is easy to show that in this case the total pressure $P = P_M + P_{X2}$ is minimum. For an evaporation from an open cell (e.g.,MBE cell), the steady state is reached when the vapor fluxes J are equal : $J_M = 2J_{X2}$ [5]. The fluxes are a function of the atomic mass of elements m :

$$J_M = 2\pi(m_M\ kT)^{-1/2}.P_M \tag{6}$$

(k = Boltzmann's constant), so from (5), (6) we have

$$m_M^{-1/2}.P_M^C = 2m_{X_2}^{-1/2}.P_{X_2}^C \tag{7}$$

Combining with (2), we can determine the pressures for the congruent sublimation :

$$P_M^C = 2^{1/3}\ (m_M/m_{X_2})^{1/6}\ K_p^{2/3} \tag{8}$$

$$P_{X_2}^C = 2^{-2/3}(m_M/m_{X_2})^{-1/3}\ K_p^{2/3} \tag{9}$$

These values are represented by straight lines in a P-T diagram (e.g. CdTe Fig. 1a)

With II-VI compounds (except mercury compounds), the congruent sublimation extends up to temperatures close to the melting temperature (see Fig. 1). So, for growing these compounds, it is easier to use only one cell charged with the compound itself, rather than two cells, charged with pure elements M and X, which requires the adjustment of two fluxes.

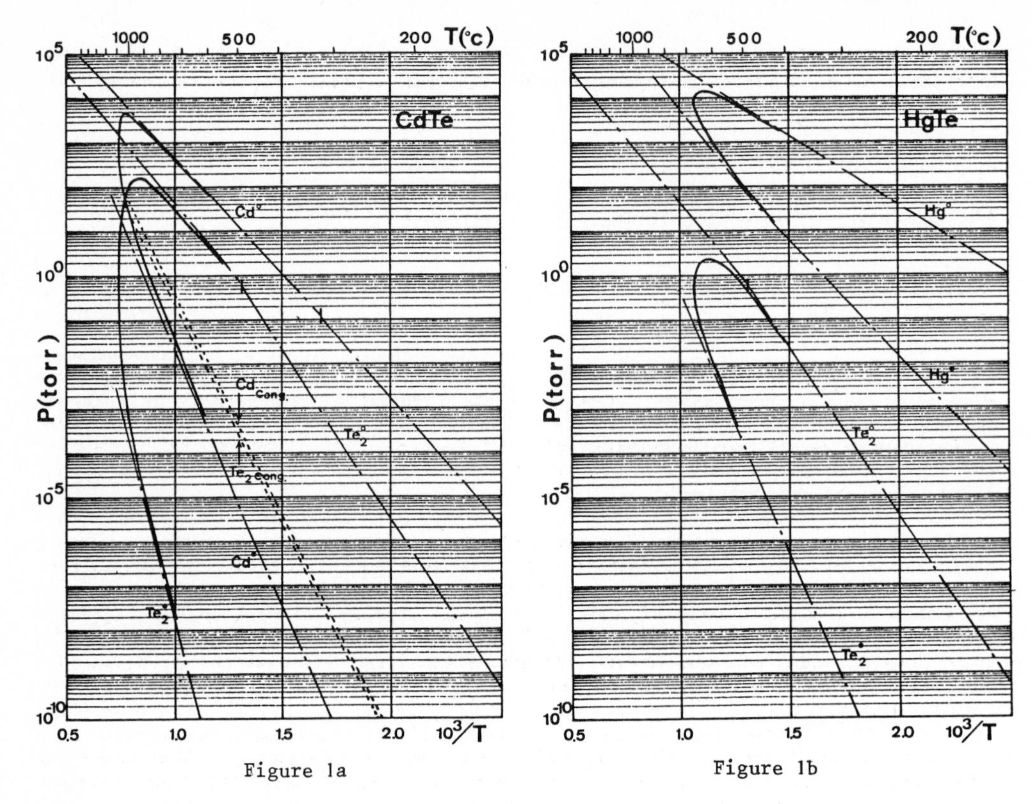

Figure 1a

Figure 1b

Fig.1 P-T diagram of CdTe (a) and HgTe (b)

1.3. Noncongruent Evaporation

We have a noncongruent evaporation when the partial pressure of one element is always superior to that of the other element, i.e. when the pressure loops don't overlap. This is the case for HgTe (Fig. 1b). The heating of HgTe leads rapidly to a depleted Hg surface consisting of a porous Te layer [33]. So the evaporation rate of Hg is controlled by the diffusion of Hg through this porous layer.

For HgSe, evaporation is congruent between 450°C and 600°C [38].

2. Growth Model

This growth model, developed by J.P. GAILLIARD [34], based only on the thermodynamical properties of materials, allows us to calculate the epitaxial conditions for a new material without any empirical kinetic consideration. This model has shown its ability to predict epitaxial parameters for noncongruently evaporating ternary compounds such as CdHgTe.

For a binary, using the relation (2) and the expression (6) of the flux J of elements which evaporate from the surface of the compound MX, we can transform the mass action law for the pressures into an equivalent law for the fluxes

$$K_J \ (T_S) = J_M \cdot J_{X2}{}^{1/2} \qquad (10)$$

where T_S is the substrate temperature.

If we express the growth rate R of a compound by the quantity of M atom (or X) incorporated in the layer for unit time and surface, we have :

$$dM/dT = F_M - J_M, \tag{11}$$

$$dX/dT = 2F_{X_2} - 2J_{X_2} \tag{12}$$

where $F_M(F_X)$ is the flux coming from the cells and $J_M(J_{X2})$ the evaporating flux from the substrate surface.

The conditions of existence of the compound MX impose the relations :

$$R = dM/dt = dX/dt \tag{13}$$

$$J_M < J_M < J_M^o \quad \text{(similarly } X_2\text{)} \tag{14}$$

J^* and J^o are calculated from (6) starting with the values P^* and P^o respectively. In the case of materials comprising selenium or sulfur, which have polyatomic vapor species, all the relations of dissociation in the vapor phase should be taken into account.

Acting on the experimentally controllable parameters F_M, F_X, and T_S, we can solve (10-13) and verify whether the conditions (14) are satisfied or not. For ternary compounds, the calculation is more complex and requires a computer; it is developed in [35].

The results obtained for $Cd_xHg_{1-x}Te$ are shown in Fig. 2, which represents the three cell temperatures (Cd, Te, Hg) as a function of the substrate temperature at a given composition and growth rate.

A direct application of this model is the calculation of the sticking coefficient of one element, which is defined as the ratio of the incorporated quantity of the element to the impinging quantity :

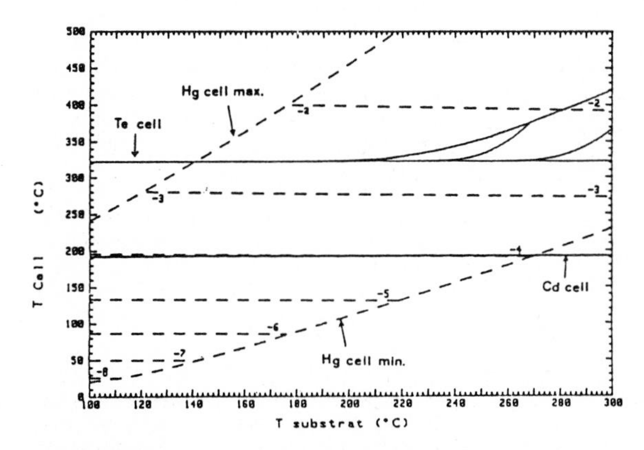

Fig.2 Temperature of the different cells as a function of the substrate calculated with the model, for the growth rate of 1 monolayer per second of $Cd_{0.2}Hg_{0.8}Te$.
The field between the two max. and min. dashed lines represents the possible Hg temperatures. The horizontal lines are the isopressure lines of mercury, the numbers are the values of $\log_{10} P_{Hg}$ (atm) over the CMT layer.

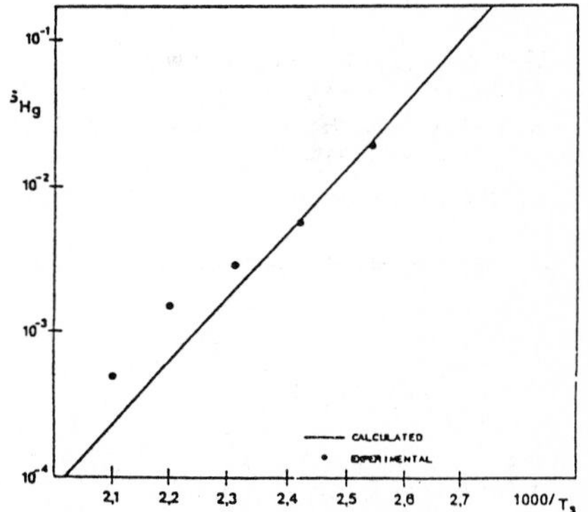

Fig.3 Sticking coefficient of mercury as a function of the substrate temperature for the epi taxy of CMTx = 0.20. Solid line is calculated, the points are experimental points from [20].

$$s_M = \frac{F_M - J_M}{F_M} = \frac{R}{F_M} \; . \qquad (15)$$

Figure 3 represents the sticking coefficient of mercury as a function of the substrate temperature in the case of $Cd_x\,Hg_{1-x}\,Te$ ($x = 0.20$). It shows the good agreement between the model and the published experimental results [20].

3. Apparatus

Usually, we adapt the apparatus used for III-V compounds to the specific problems of II-VI compounds.

The cell temperatures are much lower in the case of II-VI compounds than in the case of III-V compounds, so, the temperature regulation is more difficult. Moreover, the materials to evaporate are generally solid, so it is an inconvenience to have a stable and reproducible evaporation.

For mercury compounds, the presence of mercury, which has a very high vapor pressure (10^{-3} torr at 25°C) requires the adaptation of the mercury cell to the ultra high vacuum environment of MBE. Because of the high Hg flux required, a large quantity of Hg is evaporated during each run (several ten grams per hour). As a consequence, the Hg cell needs to be refilled very often. Another drawback is that no mercury should be present in the chamber in order to have a low residual pressure in the chamber or to be able to bake it.

Different solutions are used, one consists of a movable cell which can be introduced in the chamber during the deposition process and then drawn out of the chamber. A gate allows the cell to be refilled without breaking the vacuum in the growth chamber [20,44]. Other solutions are proposed in which the mercury is introduced as a gas phase [45], or is tranferred from a bellows into the cell [22].

Concerning the mercury pumping, a cryogenic pumping system, which can be isolated from the chamber when the mercury is removed, seems to be the best way.

4. Growth of CdHdTe Systems

4.1. Compounds $Cd_xHg_{1-x}Te$ ($x = 0$ to 1)

Taking into account the congruent sublimation of CdTe and the noncongruent sublimation of HgTe, we use a CdTe source to grow CdTe , two cells mercury and tellurium for the growth of HgTe ; and three pure element cells for the growth of $Cd_x\,Hg_{1-x}\,Te$. Usually, the substrate is a (111) oriented single crystal of CdTe, but because of its relatively poor crystal quality, several others substrates have been tried (sapphire [13], GaAs [15], InSb [12, 17, 36, 37]).

The CdTe substrates are etched in a bromine methanol solution followed by a thermal annealing at 300°C under vacuum. The growth temperature is around 300°C for CdTe and around 200°C for HgTe or CdHgTe. Several studies have shown that the cristallinity of these materials becomes very poor when the substrate temperature is lower [20].

The growth rates range between 1 and 8 $\text{Å}\,s^{-1}$ up to a temperature of 210°C. The growth model (Fig. 2) shows that the Te flux governs the growth rate and that the flux ratio Cd/Te controls the composition x. Beyond 210°C, we should take into account some reevaporation of Te, which fact has been experimentally verified.

As for the mercury flux, it must be superior to a minimal flux for a given substrate temperature (P_{Hg} min. Fig. 2) and theoretically inferior to a maximum (P_{Hg} max. Fig. 2), the intermediate fluxes allow for a change of the stoichiometry of the compounds. Practically, at Ts = 200°C only the Hg flux near the minimum flux is materially possible. This limits the possibility to change the stoichiometry as grown. In the case of CMT, the compound obtained are p type due to doping by the Hg vacancies. Nevertheless, under 200°C we can achieve as grown p type or n type materials, just by changing the relative values of the mercury flux and the substrate temperature [39, 40]. Electrical characteristics of CMT MBE layers can be as good as those of bulk material [39, 40], and photovoltaic devices have been made [20] on these layers.

4.2. Superlattice CdTe-HgTe

The growth of superlattice CdTe-HgTe takes place within a substrate temperature range of 180-200°C. The structure is obtained by moving simultaneously the shutters in front of the CdTe cell, and the mercury and tellurium cells [24]. Table I gives some examples of superlattices which have been achieved :

Table 1

TS [°C]	Number of Periods	Thickness [Å] CdTe	HgTe	Ref.
160	13 1/2	150	400	[24]
180	112	40	50	[27]
200	100	44	180	[24]

In these materials, one of the most crucial problems is in controlling the interdiffusion depth between CdTe and HgTe layers. With the classical growth techniques such as LPE or MOCVD the thickness of the interdiffusion zone is about 0.1μm in MOCVD and 1 to 2 μm in LPE. These relatively high values make impossible the growth temperature (400-500°C). By MBE, the interdiffusion depth, measured from SIMS and AES profile, is lower than 40 Å (Fig. 4) [24]. Observations by TEM on a

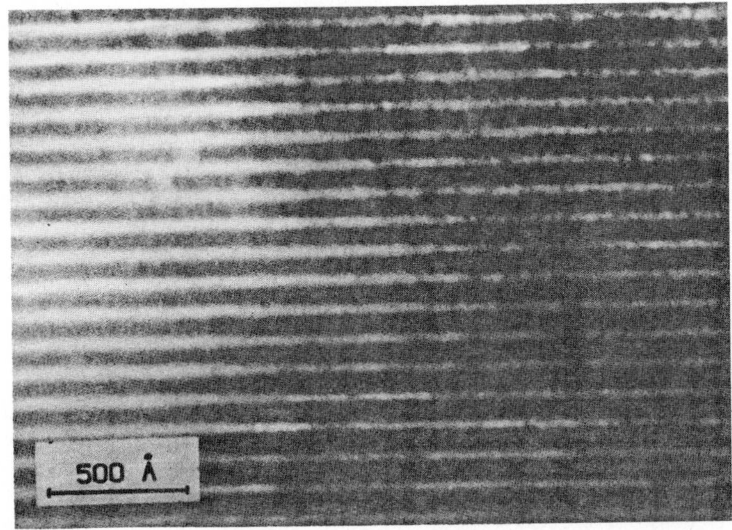

Fig.4 TEM observation of a CdTe-Hg superlattice,in white, CdTe (40 Å thick), in black, HgTe (55 Å thick)

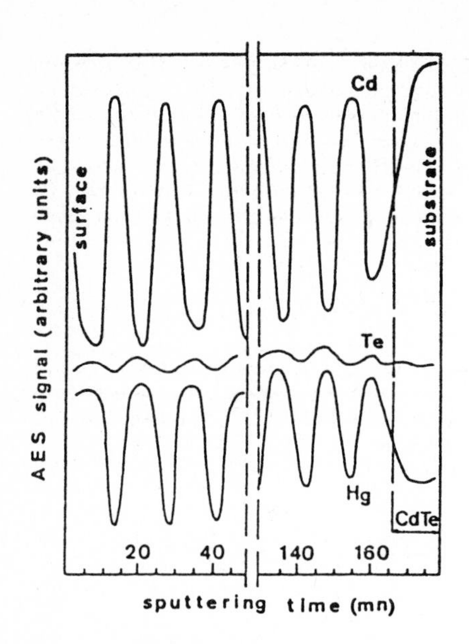

Fig.5 Compositional profiles of a CdTe-HgTe superlattice (150 Å CdTe-400 Å HgTe) analysis by AES (sputter rate is 25 Å/mn for CdTe and 50 Å/mn for HgTe)]24]

superlattice (40 Å CdTe - 55 Å HgTe) have shown a very abrupt interface between the layers (Fig. 5) [41].

Characterizations and properties showing the superlattice effect are described elsewhere in this conference [43].

5. Conclusion

Molecular beam epitaxy of II-VI compounds is presently in full expansion, specially for the compounds $Cd_x Hg_{1-x} Te$ and ZnSe. The first results obtained on the superlattices CdTe-HgTe are encouraging, and open up the possibility of realizing such structures in systems which are similar, such as semimagnetic compounds $Cd_x Mn_{1-x} Te$ [46] or $Hg_x Mn_{1-x} Te$.

Acknowledgments

The presented work is the result of a group effort. The author would like to thank the members of the MBE group of Laboratoire Infrarouge -LETI, J.P. GAILLIARD, J. PIAGUET, C. FONTAINE, Y. DEMAY, P. MEDINA for their fruitful collaboration, as well as M. DUPUY from LETI for TEM observations.

References

1. D.B. Holt: Thin Solid Films 24, 1 (1974)
2. D.L. Smith: V.Y. Pickhardt, J. Appl. Phys. 46, 2366 (1975)
3. T. Yao, S. Amano, Y. Makita, S. Maekawa: Japan Appl. Phys. 15, 1001 (1976)
4. F. Ketagawa, T. Mishima, K. Takahashi: J. Electrochem. Soc. 127 (4), 937 (1980)
5. T. Yao, Y. Makita, S. Maekawa: Jpn J. Appl. Phys. 16 Suppl., 451 (1977)
6. T. Niina, T. Minato, K. Yoneda: Jpn J. Appl. Phys. 21, L 387 (1982)
7. R.M. Park, N.M. Salansky: Appl. Phys. Lett. 44, 249 (1984)
8. T. Yao, Y. Makita, S. Maekawa: J. Crystal Growth 45, 309 (1978)
9. T. Yao, S. Maekawa: J. Crystal Growth 53, 423 (1981)
10. K. Yoneda, T. Toda, Y. Hishida, T. Niina: J. Crystal Growth 67, 125 (1984)
11. J.P. Faurie, A. Million: J. Crystal Growth 54, 577 (1981)
12. R.F.C. Farrow, G.R. Jones, G.M. Williams, I.M. Young: Appl. Phys. Lett. 39 954 (1981)

13. T.H. Myers, Y. Lo, R.N. Bricknell, J.F. Schetzina: Appl. Phys. Lett. 42, 247 (1983)
14. Y. Lo, R.N. Bricknell, T.H. Myers, J.F. Schetzina: J. Appl. Phys. 54, 4238 1983)
15. H.A. Mar, K.T. Chee, N. Salansky: Appl. Phys. Lett. 44, 237 (1983)
16. R.N. Bricknell, R.W. Yanka, N.C. Giles, J.F. Schetzina: Appl. Phys. Lett. 44, 313 (1984)
17. K. Sugiyama: Jpn J. Appl. Phys. 21, 665 (1982)
18. T.H. Myers, Y. Lo, J.F. Schetzina: J. Appl. Phys. 53, 9232 (1982)
19. J.P. Faurie, A. Million: J. Crystal Growth 54, 582 (1981)
20. J.P. Faurie, A. Million, R. Boch, J.L. Tissot: J. Vac. Sci. Technol. A1, 1593 (1983)
21. P.P. Chow, O.K. Greenlaw, D. Johnson: J. Vac. Sci. Technol. A1, 562 (1983)
22. C.J. Summers, E.L. Meeks, N.W. Nox: J. Vac. Sci. Technol. B2, 224 (1984)
23. K. Nishitani, R. Ohkata, T. Murotani: J. Elec. Mat. 12, 619 (1983)
24. J.P. Faurie, A. Million, J. Piaguet: Appl. Phys. Lett. 41, 713 (1982)
25. Y. Guldner, G. Bastard, J.P. Vieren, M. Voos, J.P. Faurie, A. Million: Phys. Rev. Lett. 51, 907 (1983)
26. N.P. Ong, G. Kote, J.T. Cheung: Phys. Rev. B28, 2289 (1983)
27. J.P. Faurie, M. Boukerche, S. Sivananthan, J. Reno, C. Hsu: Proc. Int. Conf. Superlattices, Microstructures and Microdevices - Urbana (1984)
28. H. Takahashi, K. Mochizuki, Y. Nakanishi, G. Shimaoka, H. Kuwabara: 3rd Int. Conf. on MBE - San Francisco (1984)
29. A. Bosacchi, S. Franchi, P. Allegri, V. Avanzini: Collected Papers of 2nd Int. Symp. on MBE and Clean Surface Techniques - Tokyo 223 (1982)
30. P. Goldfinger, M. Jeunehomme: Trans. Faraday Soc. 59, 2851 (1963)
31. M.R. Lorentz: "Physics and Chemistry of II-VI Compounds", ed. by M. Aven, J.S. Prener (North-Holland, Amsterdam (1967)
32. R.F. Brebick: in Progress in Solid State Chemistry, Vol. 3, ed. by H. Reiss (Pergamon, Oxford 1967) Chap.5
33. R.F.C. Farrow, G.R. Jones, G.M. Williams, P.W. Sullivan, W.J.O. Boyle, J.T.M. Watherspoon: J. Phys. D, 12, L117 (1979)
34. J.P. Gailliard: 3rd French Meeting on MBE - Carry le Rouet (1984)
35. J.P. Gailliard: to be published
36. C. Fontaine, Y. Demay, J.P. Gailliard, A. Million, J. Piaguet: 3rd Int. Conf. on MBE - San Francisco (1984)
37. S. Wood, J. Greggi, R.F.C. Farrow, W.J. Takei, F.A. Shirland, A.J. Noreika: J. Appl. Phys. 55, 4225 (1984)
38. R.F. Brebick: J. Chem. Phys. 43, 3846 (1965)
39. J.P. Faurie, A. Million: Appl. Phys. Lett. 41, 264 (1982)
40. J.P. Faurie, A. Million, J. Piaguet: J. Crystal Growth 59, 10 (1982)
41. M. Dupuy: J. Microsc. Spectrosc. Electron., 9, 163 (1984)
42. A. Million, G. Fontaine, J.P. Gailliard, J. Piaguet: unpublished
43. Y. Guldner: "Superlattices of II-VI compounds", this conference
44. MBE News, Vol. 2, n° 1, ISA Riber documentation (1983)
45. V.G. Semicon. documentation (1983)
46. L.A. Kolodziejski, T. Sakamoto, R.L. Gunshor, S. Datta: Appl. Phys. Lett. 44 799 (1984)

Dynamic Aspects of Growth by MBE

C.T. Foxon

Philips Research Laboratories, Redhill, Surrey, RH1 5HA, England

This article discusses the practical application of time-dependent RHEED measurements for the control of growth by MBE. In practice it is convenient to study the time-dependent intensity of electrons specularly reflected from the surface. Following an interruption to growth, the intensity of various features in the RHEED pattern, including the specular beam, oscillate with a period equal to the time taken to deposit one monolayer of material. This provides a direct in-situ measure of the growth rate and can be used to control film thickness and as a means of calibrating the beam-monitoring ion gauge used in most MBE systems. By predepositing Ga atoms, it is also possible to measure the As flux using the same technique. The temperature dependence of the intensity of the specular beam reflected from the surface provides a means of estimating substrate temperature for systems where the indicated and actual temperature differ. The method can also be used to grow structures of known thickness by counting the number of layers deposited; examples of superlattice structures grown by this method will be presented.

1. Introduction

There have been many excellent reviews dealing with the growth and properties of heterojunctions and superlattices by MBE [1-3] and many of the lectures in this Workshop will discuss such matters in great detail. The RHEED technique has been used extensively to study the MBE growth process, and this has resulted in a better knowledge of the chemistry involved. A fairly detailed understanding of the As stable 2x4 reconstructed surface has emerged, and the c 4x4 has been shown to be due to trigonally bonded As chemisorbed at sub-monolayer coverage onto the As stable surface. This article will concentrate on the application of time-dependent RHEED measurements for controlling the growth of high quality heterojunction, quantum well and superlattice structures in the AlGaAs system. No attempt is made to review completely the literature on this subject. The examples chosen are taken from work carried out at Philips Research Laboratories, Redhill, England.

For a simple heterojunction device such as the TEGFET we would like to know as accurately as possible the thicknesses of the GaAs buffer layer, the Al concentration in the AlGaAs film, the thickness of the undoped spacer layer, the sharpness of the GaAs-AlGaAs interface and the doping concentration in the AlGaAs. All of these, except for the last item, can be measured precisely using time-dependent RHEED measurements, as will be shown below. For more complex Quantum Well Lasers, the precise thickness of the well will influence the wavelength of operation and should therefore be determined as accurately as possible, for Multiple Quantum Well Lasers it is clearly desirable to know whether the wells are all of identical width. For superlattice structures it is obvious that we need to know how reproducible are the thicknesses of the various layers. For such structures time-dependent RHEED measurements can be invaluable in assessing the structure of the films grown by MBE.

It was recognised quite early in the development of MBE by WOOD [4] that the features of a RHEED pattern changed with time, especially at the beginning of growth or following a growth interruption. The significance of this observation was recognised later by HARRIS et al. [5] who showed that the periodic variations in the various features of the RHEED pattern corresponded to the growth of monolayers of GaAs. Since then this technique has developed rapidly, and much additional information can be obtained about the growth process from such measurements, as will be shown below.

2. Equipment Considerations

Fig. 1 shows a general view of an MBE system constructed at PRL which incorporates many of the features found in most commercial systems. In order to grow high quality material, particularly when Al is involved it is essential to have a leak-free system equipped with a vacuum interlock. What is less obvious is that in order to grow superlattice structures with layers only a few monolayers thick, it is essential to have shutters which close or open as rapidly as possible. Since typical growth rates are 1-2 monolayers s^{-1} it follows that the shutters should open or close in less than 0.1 s.

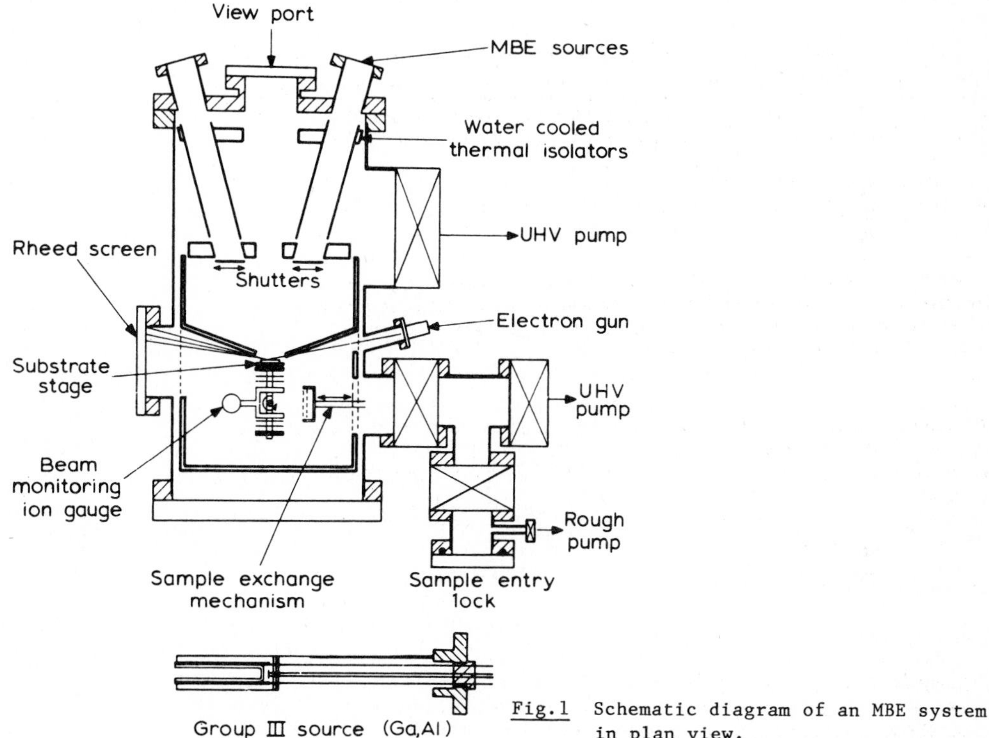

Fig.1 Schematic diagram of an MBE system in plan view.

Since the shutter is in most commercial systems a flat sheet mounted some distance in front of the furnace, when closed it will reflect heat into the cell which will be lost when the shutter is opened. This results in a difference in heat input required to maintain a particular temperature when the shutter is open or closed. The usual 3-term (PID) temperature controller used for MBE will compensate for the difference, but there will be a transient change in flux until the original temperature is restored. This is a serious problem when growing superlattice structures, the magnitude of which we can also assess using the RHEED techniques, to be discussed below. Problems associated with transients of this type can be avoided bv introducing a fourth term into the control algorithm; this simply applies the additional heat required for the furnace whenever the shutter is open. In a software controlled system such as the one developed by the author at PRL this is readily implemented, since the additional heat required is simply the difference in power levels at steady state with shutter open and closed.

At present most MBE systems are equipped with an ion gauge which can be placed in the sample position in order to measure the flux emerging from the various sources. Such an arrangement is quite adequate for establishing the relative fluxes of group III elements,but day to day variations in the sensitivity of the gauge make it less ideal as an absolute measure of intensity. It is also difficult to use this method to estimate the ratio of group V to group III fluxes,since the relative ionisation efficiencies are unknown. There is an additional problem in many commercial systems where this gauge is shielded in such a way as to prevent its use as a single pass density detector. For the group V fluxes the signal is time-dependent,and can be influenced by previous deposits of group III material. The RHEED measurements to be described below can be used to calibrate the beam-monitoring ion gauge and to determine the relative sensitivities to the various fluxes.

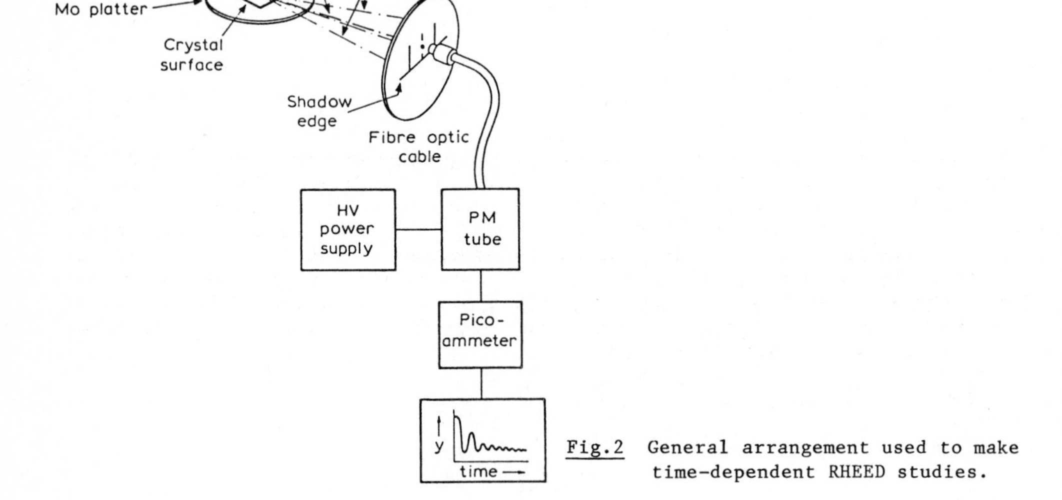

Fig.2 General arrangement used to make time-dependent RHEED studies.

The general arrangement usual to make time-dependent RHEED measurements is illustrated in Fig. 2.A simple collimator consisting of a pair of apertures 1 to 2mm diameter about 10 mm apart is used to collect light from a single feature of the RHEED pattern. The time-dependent intensity is measured using an optical fibre, photomultiplier and Y-t recorder. A simple X-Y manipulator can be mounted externally on the RHEED screen flange to position the collimator over the feature of interest.

Finally,whilst discussing equipment it is essential to use a substrate rotation stage in order to obtain uniformity for large area samples. Several problems then arise, first the speed must be high enough to ensure one rotation per monolayer deposited. This is particularly important for alloy films to avoid fluctuations in composition through the layer. Since the properties of films are influenced strongly by the substrate temperature,it is essential to determine and control this as accurately as possible. With many commercial systems the indicated temperature is far from the actual temperature,and the relation between the two can change as a function of time,due to variations in the emissivity of the platten during growth. The RHEED measurements to be described below can also be used to obtain an independent estimate of the substrate temperature as a function of time.

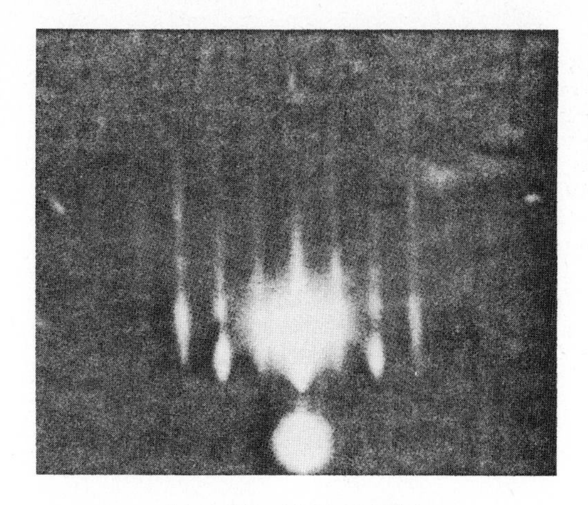

Fig.3 [010] azimuth RHEED pattern from a GaAs (001) 2x4 reconstructed surface showing a strong specular beam suitable for measurements of growth rate.

[110] AZIMUTH
GaAs (001) – 2X4

3. Time-Dependent RHEED Studies.

For most practical purposes it is convenient to measure the intensity of the electrons specularly reflected from the surface under conditions where they are not coincident with any other diffraction feature along the 00 rod. Such a situation is illustrated in the RHEED pattern shown in Fig. 3. With the collimator positioned over the specular spot, if the growth is interrupted and then restarted the intensity oscillates as shown in Fig. 4 [6]. The oscillations gradually decay but can reappear if additional beams are incident on the surface or if the growth is interrupted once more. The best conditions under which to observe such oscillations will be discussed in more detail below.

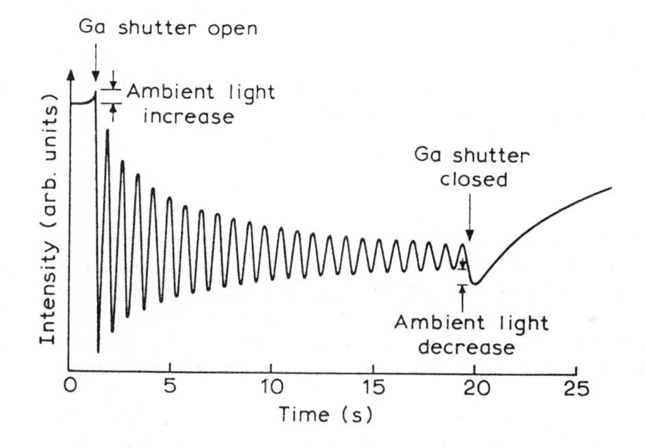

Fig.4 Intensity oscillations of the specular beam in the RHEED pattern from a GaAs (001) 2x4 reconstructed surface using a [110] azimuth.

The observation of periodic variations in thin film growth is not new and has always been explained in terms of a layer by layer growth process. The mechanism giving rise to the oscillations observed in the RHEED features has been discussed by a number of groups. Whilst there are differences in detail, the general features of the currently accepted model are illustrated in Fig. 5 [6]. The intensity of the specular 00 beam relates to changes in surface roughness. It is generally observed that the static surface has a high reflectivity. Upon commencing growth, a number of islands nucleate and grow horizontally. Since the wavelength of the electrons incident on the surface is small (∿ 0.1 A) compared to the height of a GaAs

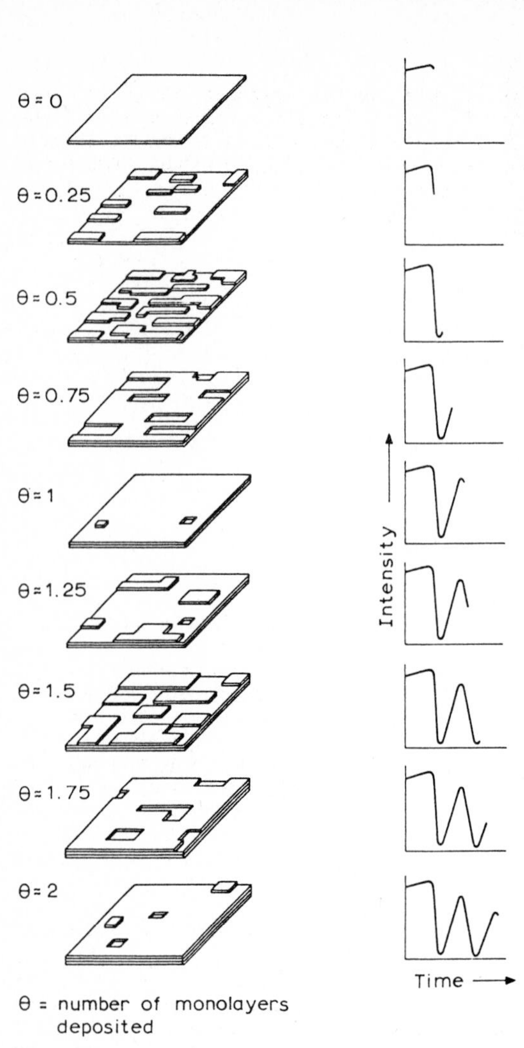

Fig.5 Idealised model of the growth process on an initially smooth surface together with the expected intensity of the specularly reflected beam.

monolayer (2.8 A) some of the electrons will be scattered diffusely, resulting in a loss in intensity in the specular beam. The maximum scattering will occur when the surface coverage is about 0.5 and thereafter the intensity in the specular beam will increase again and would reach its original value if the layer were completed before nucleation occurred for the second layer. Statistically this is unlikely, and therefore in general we see a damped series of oscillations as the surface becomes distributed over several incomplete monolayers. The important practical point is that the period of the oscillation corresponds precisely to the time taken to deposit one monolayer of material. For compound semiconductors such as GaAs this corresponds to a layer of Ga plus a layer of As atoms (2.8 A) in the [100], for elemental semiconductors such as Si and Ge however, the period of the oscillation will correspond to a single atomic layer of the material (1.4 A) for ge in the [100].

A number of attempts have been made to analyse the time-dependent intensity data using kinematical calculations of diffraction intensities, the results obtained may be qualitatively correct but are suspect because, even for the specular beam, dynamical many-beam calculations are required to correctly predict the observed behaviour.

Since under As stable conditions at relatively low temperatures the growth rate of GaAs is determined by the arrival rate of Ga atoms to the surface, it is obvious that we have an absolute in situ means of determining the flux of Ga atoms arriving at the surface and hence of calibrating the beam-monitoring ion gauge. The same technique can also be used to determine the relative fluxes of Ga and Al arriving during the growth of AlGaAs and hence to determine the Al content of the mixed binary compound. In such measurements it is possible to determine the Ga flux during the growth of the GaAs buffer layer. The Al flux can be obtained by measuring the growth rate of the AlGaAs layers, since on opening the Al shutter the oscillations will in general reappear. Measurements of the Al fraction made in this way agree within experimental error with those made after growth by other techniques, and with the relative measurements of flux made in situ using a calibrated ion gauge as discussed above.

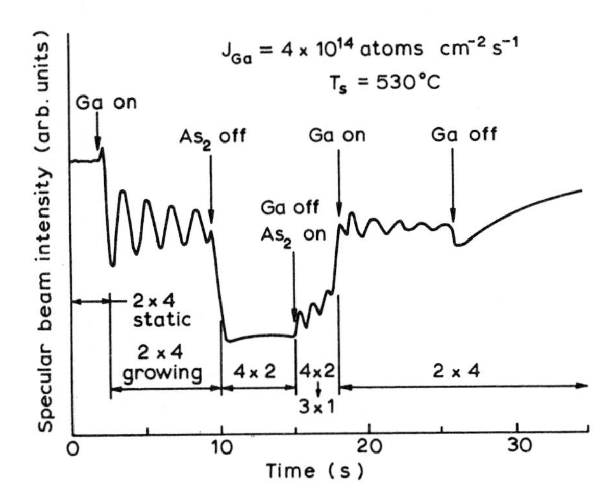

Fig.6 Time-dependence of the intensity of the specular beam following Ga predeposition on a GaAs (001) surface. Note that the oscillations occur at higher frequency and decay once all of the surface Ga atoms are consumed.

It is also possible to determine the magnitude of the group v fluxes using the RHEED oscillation method [7]. To do this the As flux reaching the surface is interrupted, and a known quantity of Ga deposited. The Ga flux is then turned off and the As flux started. Oscillations are then observed whose frequency relates to the As beam intensity, and they disappear when all of the Ga is consumed. Fig. 6 illustrates this method for GaAs grown from Ga and As_2. It is probably better to use this procedure as a means of calibrating the beam-monitoring ion gauge, since layers grown with the As flux interrupted may not have optimum properties. In using this technique with As4 it should be recalled that for this species only 50% of the incident As atoms are consumed, the rest being lost by desorption [2].

At high substrate temperature Ga is lost by re-evaporation from the growing surface, leading to an effective decrease in the growth rate. This can also be seen using the RHEED oscillation method, since the net growth rate is measured by this method. The results of such a study are shown in Fig. 7, from which it can be seen that above 650 C it is important to take this factor into account in predicting both the thickness and composition of AlGaAs layers grown by MBE. It has also been possible to determine the relative rates of loss of Ga from GaAs and AlGaAs using the method illustrated in Fig. 8. The growth rates of GaAs, AlAs and AlGaAs are all measured. The difference between the growth rate of the AlGaAs and the sum of the growth rates of GaAs and AlAs gives the difference in re-evaporation rates directly. The results of this study are also shown in Fig. 7 and it can be seen that there is little difference in the re-evaporation rates from GaAs and AlGaAs contrary to some reports in the literature. The solid line shown in Fig. 7 is not a best fit to the data but a theoretical calculation of the expected re-evaporation rate based upon a very simple model. The thermodynamic vapour pressure of Ga over Ga is used to estimate the Ga loss rate, taking into account the fact that the

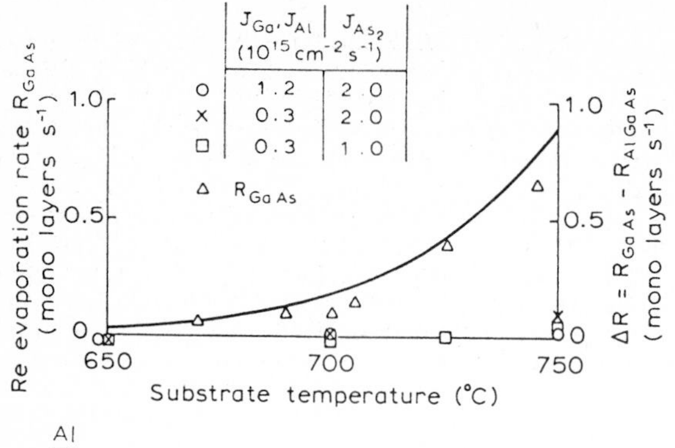

Fig.7 Measurements of the rate of re-evaporation of Ga from GaAs and of the difference in re-evaporation rates of Ga from GaAs and AlGaAs as a function of substrate temperature.

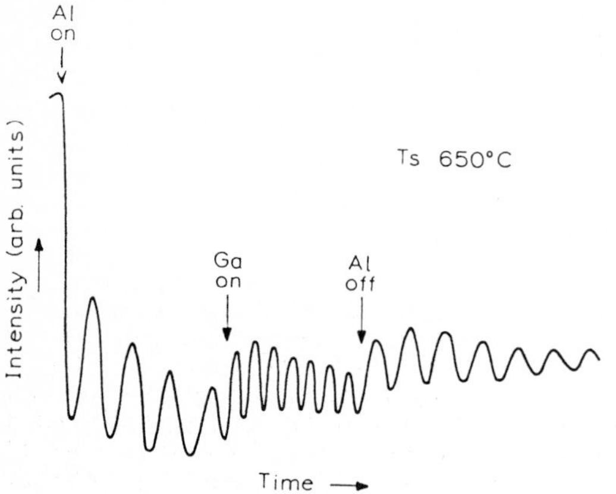

Fig.8 Method used to measure the difference in re-evaporation rates of Ga from GaAs and AlGaAs.

density of Ga atoms in a Gaas surface is lower than that of Ga atoms in a liquid Ga surface. It can be seen that within experimental error the agreement between the data and the calculation is quite good.

The intensity of the specular beam changes both with orientation and with substrate temperature for both static and growing surfaces as shown in Fig. 9. In order to observe strong oscillations in intensity at the commencemnt of growth,it is necessary that there is a large difference between the static and growing surface. At low temperatures where a 2x4 reconstruction is observed,the [110] or [010] orientations will show strong oscillations,but in the [-110] orientation they will be difficult to detect. At higher temperatures, where a 3x1 reconstruction occurs, the [010] or [-110] may be better than the [110]. At very high temperatures under Ga stable conditions where a 4x1 or 4x2 reconstruction is present it is difficult to observe oscillations under any conditions for GaAs alone. In the measurements shown in Fig. 9 the substrate temperature was measured quite accurately using a thermocouple passing through the molybdenum block, to which the substrate on a separate platten was attached using liquid indium to conduct the heat. The temperature was also measured using a radiation pyrometer, and good agreement was obtained between the two methods. It follows that we can use the known temperature dependence as a measure of substrate temperature for situations where with commercial rotation stages the indicated temperature is far from the true value. The data shown in Fig. 9 depends critically upon the As flux used for the experiment, and it is essential therefore to use the same value. As the As flux is increased the transitions move to a higher temperature,as would be expected since for any given temperature the surface is more As stable. As an additional cross check we can measure the growth rates at low and high temperatures and estimate the temperature from the Ga re-evaporation rate.

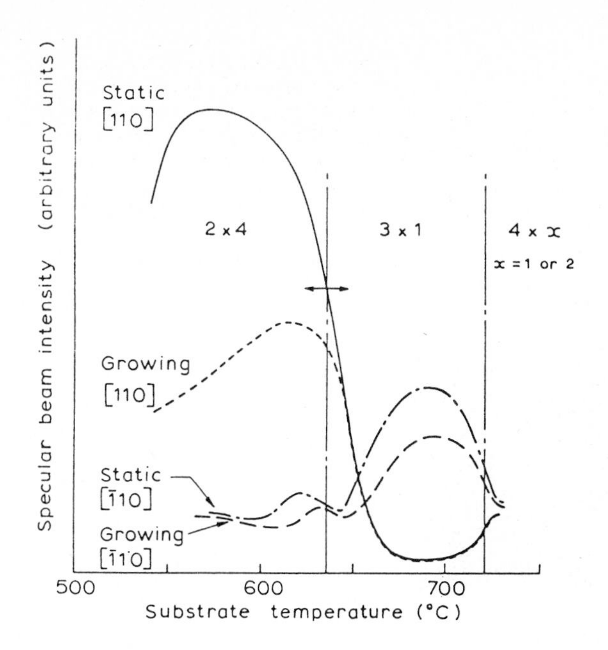

Fig.9 Dependence of the specular beam intensity on the substrate temperature from static and growing surfaces for (001) GaAs in the [110] and [-110] azimuths. The surface reconstruction observed is also indicated. J As_2 ∿ $8x10^{14}$ $mol.cm^2$ s^{-1}, J Ca (during growth) = $6x10^{14}$ cm^{-2} s^{-1}.

In practice it is convenient to measure the growth rate during the deposition of the buffer layer. If the surface after thermally desorbing the oxide is not clean, however, the specular beam will be very weak and oscillations may be difficult to detect at this stage. Under such conditions, it is better to grow a small quantity of GaAs and then interrupt the growth briefly to establish the growth rate. The strength of oscillations observed after thermal desorption of the oxide is a good qualitative measure of the chemical process used to prepare the surface.

There is one other important consideration when using a rotation stage. The growth rate must be measured at a position where the Ga flux is equal to the mean value obtained when rotating the platten. Under such conditions good agreement will be obtained between the measured growth rate with the sample static and the layer thickness using rotation. The only part of the structure in which there will be significant variations in thickness will be the GaAs buffer layer, where for part of the time no rotation was used. Since it is difficult to know precisely where the RHEED beam strikes the surface, it may be convenient to keep a sample within the preparation chamber for RHEED oscillation studies prior to growing the required layers. If such a procedure is adopted, a relatively small sample may be placed on the platten in a position corresponding to the average growth rate.

The RHEED technique may also be used directly to grow samples of known thickness by counting the number of monolayers of GaAs or AlGaAs deposited. The difficulties with this procedure are knowing precisely when the shutters open and close relative to the RHEED data, especially when the shutter speed is slow compared with the monolayer growth rate. There is additionally a difficulty in defining precisely where the frequency changes from that associated with GaAs to AlGaAs, as can be seen from Fig. 8. This uncertainty can be removed by stopping the growth between layers. There may be an additional benefit in using growth interrupts since the recovery in reflectivity is thought to be associated with a smoothing of the surface, and hence may lead to sharper interfaces under suitable conditions. Fig. 10 shows a portion of a time-dependent RHEED trace taken for samples where alternate monolayers of GaAs and AlAs were deposited using this method. Fig. 11 shows part of a cross-sectional TEM image of a 3x3 monolayer GaAs AlAs superlattice structure grown in this manner. The accompanying diffraction pattern clearly shows the additional periodicity associated with the superlattice. The main practical limitation of this method arises from the variations in flux across the substrate surface, which

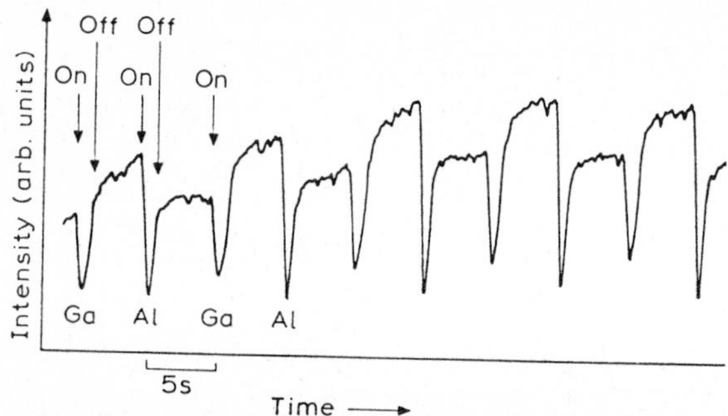

Fig.10 RHEED intensity measurements for a 1x1 GaAs Alas superlattice grown using interrupts.

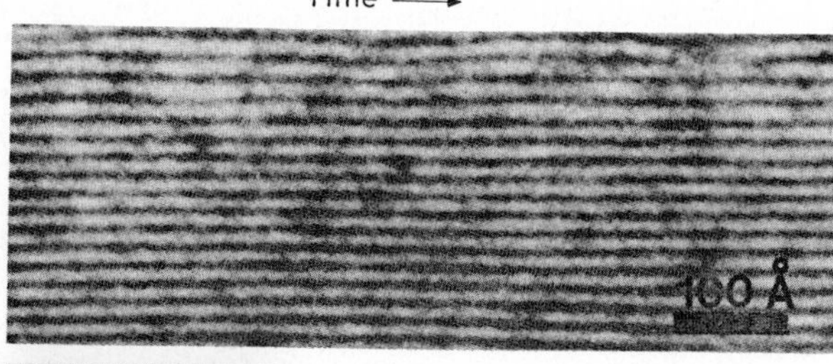

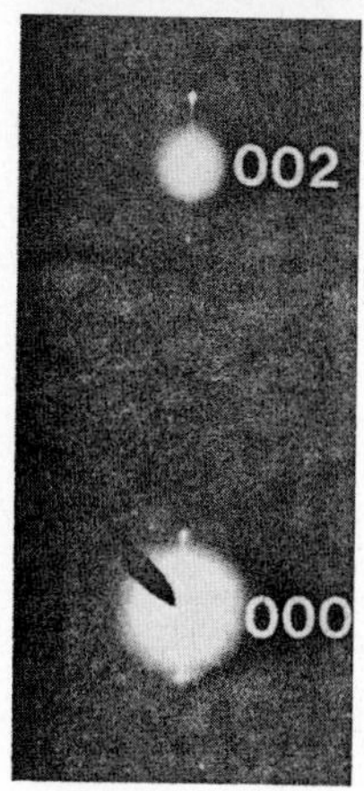

Fig.11 TEM cross-section for a 3x3 monolayer GaAs AlAs structure grown using RHEED oscillations to control the thickness. The additional reflections in the diffraction pattern arise from the superlattice periodicity.

limits the sample size. Nevertheless it has been possible to grow Quantum Well samples and other special structures in this way with precisely known thicknesses for the various layers.

Conclusions

Time-dependent RHEED studies of the GaAs surface have led to a better understanding of the nature of the MBE process and have provided a practical method of determining in situ the growth rate of the films. Such measurements can be used in turn to calibrate the beam-monitoring ion gauge fitted to most MBE systems. It is also possible to use this technique to estimate both the substrate temperature during growth and the reevaporation rate of Ga at high temperatures.

Acknowledgements

The author wishes to thank his colleagues at PRL who have supplied many of the examples used in this article, and in particular J.P. GOWERS for the TEM pictures which have not been published elsewhere.

References

1 A.C. Gossard : Treatise on Materials Science and Technology, Vol. 24 (Academic (1982) p. 13

2 K. Ploog : M. Crystals. Growth, Properties and Applications, Vol. 3 (Springer, Berlin, Heidelberg) (1980) p. 73

3 C.T. Foxon, B.A. Joyce : Current Topics in Materials Science, Vol. 7 (North-Holland (1981) p. 1

4 C.E.C. Wood, Surf. Sci. 108, L441 (1981)

5 J.J. Harris, B.A. Joyce, P.J. Dobson : Surf. Sci. 104, L90 (1981)

6 J.H. Neave, B.A. Joyce, P.J. Dobson, N. Norton : Appl. Phys. A31 (1983)

7 J.H. Neave, B.A. Joyce, P.J. Dobson : Appl. Phys. A34 179 (1984)

An Introduction to OMVPE of III-V Compounds

P.M. Frijlink

L.E.P. Limeil Brevannes, France

Of the four existing methods mainly used for epitaxial growth of III-V compounds - Liquid Phase Epitaxy, Chloride transport epitaxy, organometallic vapour phase epitaxy and Molecular Beam Epitaxy - OMVPE has been used to grow the largest variety of possible III-V compounds. The non-equilibrium character with respect to the deposition of the group III element adds much to the easy application of the method to grow binary, ternary and quaternary alloys, with precise control of composition and thickness. In some materials systems, virtually monolayer control can be achieved when switching from one compound to another. Doping profiles can be reealized with transitions of two decades over distances of the order of 50 (switch on) to 100 (switch off) angstroms.

The achievable purity, which is reflected by the electrical and optical quality of the materials, is practically state-of-the-art for most alloys. For the best-known compounds involving Al, Ga, In, As, P, it is mainly limited by the purity of the (commercially available) starting products, which has considerably improved over the last five years. These possibilities make the method attractive for application to the epitaxial growth of advanced structures.

1. Historic

MOVPE is a relatively young method, which shows a boom of interest in recent years. A short historic outline :

1968-1971	Early work by H. Manasevit III-V and II-VI compound semiconductors on insulators
1972	ICGC Marseille The Manasevit growth technique is acknowledged by the crystal growth community [1]
1975	High purity GaAs demonstrated by Y. Seki [2] First report by S.J. Bass of device quality GaAs for reflection mode photocathodes and MESFET's [3]
1976	(Al,Ga)As - GaAs heterostructures for transmission mode photocathodes by J.P. Andre et al [4]
1977	- D.H. lasers by R.D. Dupuis and P.D. Dapkus [5] - Intense activity in several laboratories - First report on GaAs growth by low-pressure MOVPE by J.P. Duchemin [6]
1978	- Quantum well (Al,Ga)As-GaAs heterostructure lasers by R.D. Dupuis, P.D. Dapkus and N. Holonyak [7] - InP growth by LP MOVPE by J.P. Duchemin [8]
1980	- Quaternary laser by J.P. Hirtz [9]
1981	- First international conference on Metal Organic Vapour Phase Epitaxy
1981-NOW	Stormy development of MOVPE worldwide.

2. Basic growth method

Smooth epitaxial growth on a crystal requires a minimum temperature for the crystal to order its growing surface at a given growth rate. MOVPE mostly involves a reaction between the crystal surface, group III organometallics, and group V organometallics or hybrides, all of them partially decomposed in the vapour phase.

In a variant of the method, the group III metals are provided by adducts of group III and group V organometallics. The best known example is the case of GaAs epitaxy from trimethylgallium (TMG) and arsine, for which the overall reaction in the presence of a standard hydrogen atmosphere might be described as :

Ga(CH3)3+AsH3 ---> GaAs + 3 CH4

Although indeed GaAs and CH4 are formed, this simple formula does not at all describe what really happens. In fact, if we would provide arsine in the same molar fraction as TMG, the result would at normal growth temperatures be of very poor crystalline quality. At normal growth temperatures, the vapour pressure over the alloy of the group III elements is negligibly small, but the group V elements are very volatile, except for antimony, which makes its case a special one. To ensure sufficient surface coverage of the group V elements or complexes, the group V element must be provided in excess from the vapour phase. In normal growth conditions, the input flow of group III organometallics will totally govern the growth rate. If more than one group V element is present, a competition between these elements for surface coverage takes place, which is describable by a quasi thermodynamic equilibrium. Fig. 1 shows the typical behaviour of the residual doping as a function of III-V ratio for GaAs and (Ga,Al)As. The transition to n-type material coincides usually with the obtention of good crystalline quality, reflected by surface morphology and carrier mobility. A similar transition is observed as a function of growth temperature (Fig. 2).

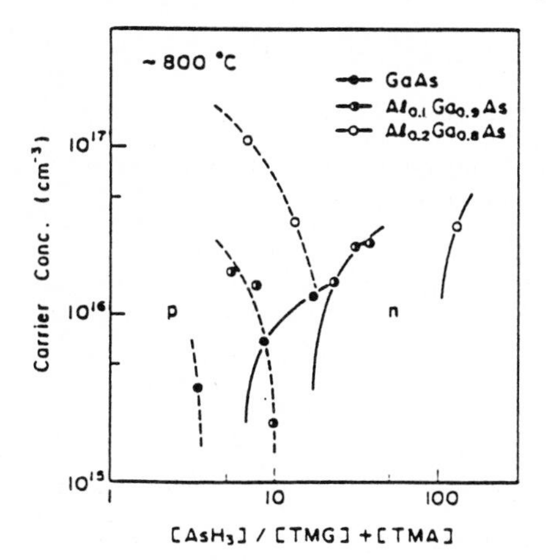

Fig.1 Carrier concentration versus III/V ratio for GaAs and (Ga,Al)As after [10]

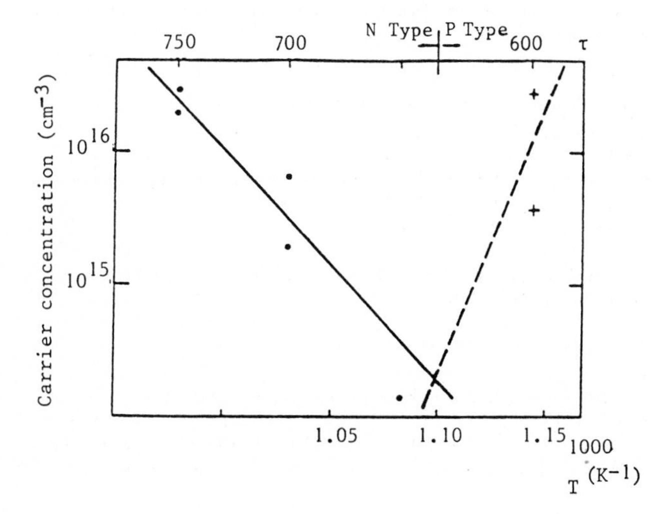

Fig.2 Variation of the background doping (type and carrier concentration) as a function of the deposition temperature after [11].

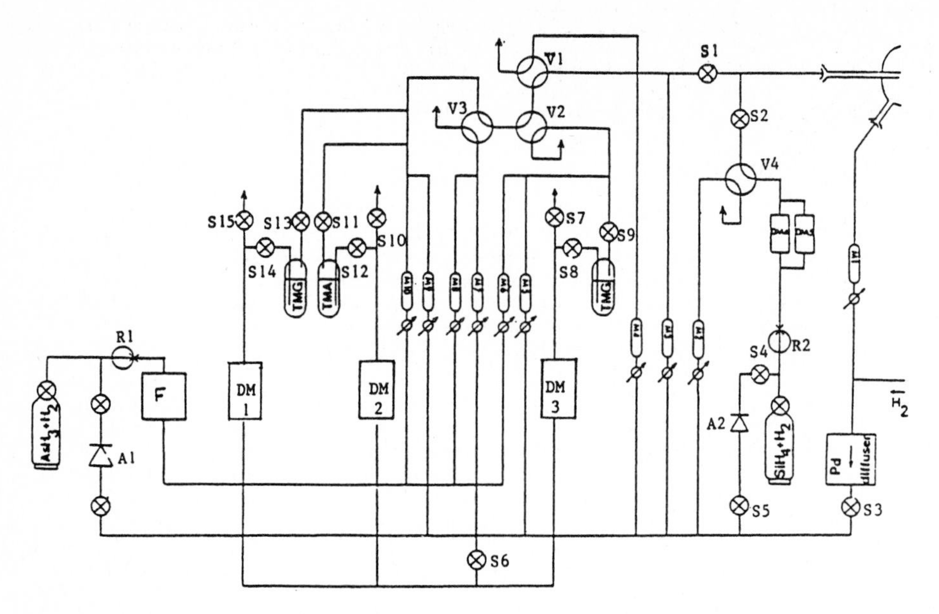

Fig.3 MOVPE reactor system for abrupt GaAs-(Ga,Al)As heterostructures. F is a drying unit.

Fig. 3 is a basic drawing of an MOVPE reactor system for GaAs/(Ga,Al)As heterostructures. The group III organometallics are contained in thermostated stainless steel bubblers. The flowrates through these (D_{TMG}, D_{TMA}) bubblers determine the growth rate, together with the equilibrium vapor pressure of the organometallics. For the case of undoped $Ga_{1-x}Al_xAs$ the growth rate and the obtained aluminium percentage in group V rich conditions can be expressed as :

$$\left.\begin{aligned} R &= A.D_{TMG}P_{TMG} + 2B\ D_{TMA}P_{TMA} \\ x &= \frac{2B\ D_{TMA}P_{TMA}}{A.D_{TMG}P_{TMG} + 2B\ D_{TMA}P_{TMA}} \end{aligned}\right\} \qquad \text{experimental : } A = B$$

The constants A and B depend mostly on the mass transport in the reactor chamber. These expressions take into account that TMA behaves like a dimer in the bubbler, i.e. the molecule is Al2(CH3)6. Fig. 4 shows the vapour pressures of relevant organometallic compounds.

The gas mixing system is made in such a way that switching from GaAs growth to (Ga,Al)As growth and vice-versa comes down to switching between established flows of organometallic vapour mixtures. Secondary entrainment vectors provide a limitation of transient problems upon switching. The growth rate will normally be of the order of 5 Angstrom/sec. The necessary TMG mole fraction for this average growth rate is for this reactor about 2.4*10-5 at a total flow of 5L/min. The growth of indium-containing compounds from trimethyl-indium or triethyl-indium is complicated by the possible formation of adducts between group V hybrides and indium-alkyls. These adducts will split off alkyl groups and polymerise to form a deposit on relatively cold parts of the reactor tube. To avoid or diminish these elimination reactions, several solutions are possible :

1) Pre-cracking of the group V hybride. Adduct formation will be avoided.

2) The use of reduced pressure (10-100 mBar) increases the practical volumetric flowrate and increases the stability of the flow with respect to turbulence.

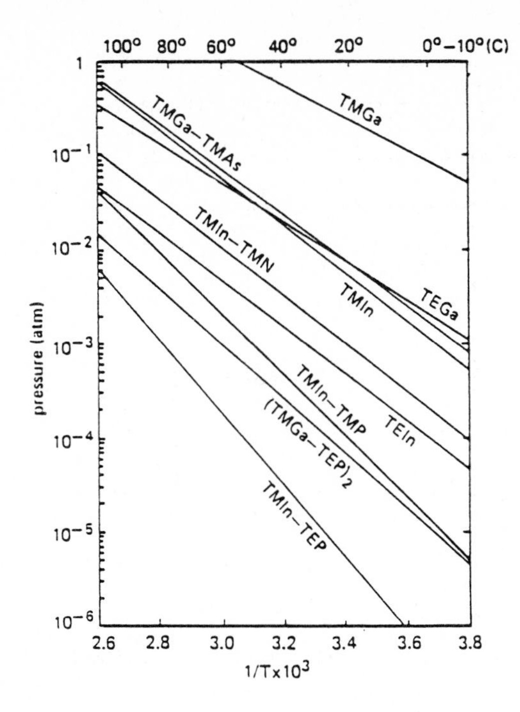

Fig.4 Vapour pressures of group III source materials. (after R. Moss)

3) The use of group V alkyls, instead of group V hybrides. The possible adducts are volatile and apparently less susceptible to elimination.

4) The use of volatile preformed adducts between group III alkyls and group V alkyls.

Fig. 5 is an example of a reactor for the growth of InP and (Ga,In)As lattice-matched to InP at reduced pressure.

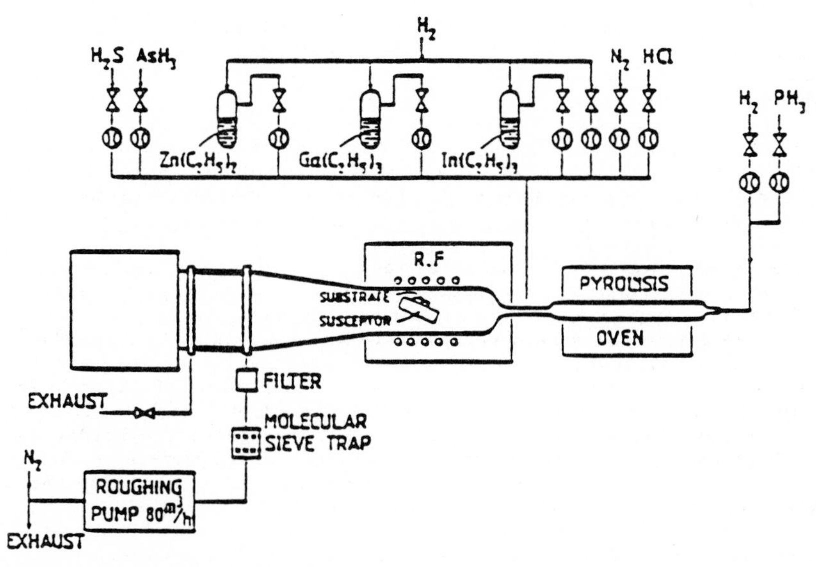

Fig.5 Low-pressure MOVPE reactor for InP-(Ga,In)As after M. Razeghi, M.A. Poisson, J.P. Larivain and J.P. Duchemin.

In general, the actual growth procedure in the case of a GaAs or InP substrate is :

1) A chemical etch of the substrate in aqueous or alcohol-based solution to withdraw surface contaminants and form a standard native (hydrated) oxide layer, followed by a rinse in ultrapure water.

2) Loading into the reactor and heating to growth temperature under group V pressure to avoid decomposition of the surface. The oxides of Ga, In, As, P are all volatile at normal growth temperature. This step will provide an essentially clean and reconstructed crystal surface. Some authors propose additional cleaning by in situ etching with HCl or AsCl3. For most applications this is not necessary.

3) Epitaxial growth.

3. Growth kinetics

The reaction kinetics of MOVPE growth is subject to controversy. Due to the fact that the overall reaction in terms of the causal dependence between input flows in the reactor and the resulting epitaxial layers is in first order simple in the case of GAAs - (Ga,Al)As - the first materials system used extensively -, in which device results could quickly be obtained, the growth kinetics has at first received relatively little attention. That even GaAs growth cannot be described only by diffusion towards the crystal surface with zero concentration in the gas on the surface is immediately evident from Fig. 6 after Reep and Gandhi [12]. We can distinguish three different regions :

- Mostly kinetic limitation of the growth rate below 600°C
- Mostly mass transport limitation of the growth rate between 600°C and 850°C
- A region of decreasing growth rate above 850°C which the authors attribute to GaCH3 desorption from the surface. Elsewhere, this region has been attributed to homogeneous gas phase nucleation,and also to depletion of the gas before it reaches the wafer.

The turnover points on the temperature scale depend of course on the reactor geometry and flowrates.

A number of models have been presented by different authors. One model described the process as diffusion towards and adsorption on the crystal surface of the entire group III metal organic molecules and group V hybrides, complex formation on the crystal surface and subsequent successive demethylation [13]. Leys and Veenvliet [14] have shown by an infrared absorption technique in which gas is sampled from the reaction chamber by a capillary tube and led into a cold space for infrared absorption, that at normal growth temperatures TMG cannot reach the crystal surface without being at least partially cracked. They also found that arsine decomposition is catalized by the presence of GaAs, and to an even stronger degree by decomposing TMG. TMG all by itself in hydrogen atmosphere, does not fully decompose down to free gallium until 800°C [15]. The bond energies of TMG are [15]:

Ga(CH3)3 ⟶ Ga(CH3)2	59.5 Kcal/mol
GA(CH3)2 ⟶ Ga(CH3)	35.4 Kcal/mol
Ga(CH3) ⟶ Ga	77.5 Kcal/mol

The first decomposition step might very well be the bi-molecular reaction :

Ga(CH3)3 + H2 ⟶ Ga(CH3) + 2 CH4 + 9 Kcal/mol

This step is exothermal, and its activation energy might be reasonably low due to

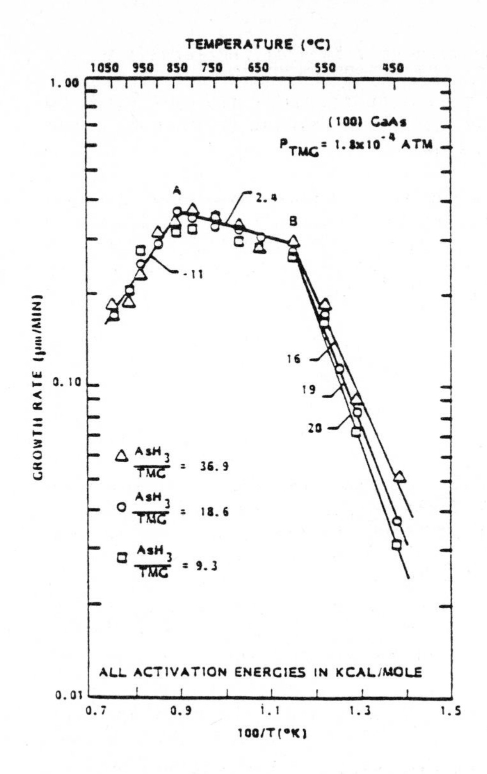

Fig.6 Growth rate vs. reciprocal temperature for (100) GaAs epitaxial layers, as a function of P(AsH3) and for P(TMG) = 1.5×10^{-4} atm.(after [12])

the favorable steric configuration : H2 can simply place itself between two methyl groups. Succesful collisions are possible on all sides of the molecule. H2 is present in abundance, so the rate of this bi-molecular reaction will be appreciable. The high dissociation energy of the last bond and the fact that only one methyl group is available blocks further decomposition if we consider the usual short residence time in a well-designed reactor system, unless the temperature is very high. A tri-molecular reaction would again have a lower enthalpy, but its frequency of occurrence based on collision rates in the gas phase will be much lower. The authors who did IR measurements [14] [16] did not see monomethyl-gallium, but this compound might very well not exist at room temperature, and it is not known if the resulting products upon cooling down are transported all the way to the analysis chamber. The last methyl group will only be lost after adsorption of the mono-methyl gallium on the crystal surface.

For the decomposition of arsine, a similar mechanism may be proposed, the first step being :

$$AsH3 \longrightarrow AsH + H2$$

instead of :

$$AsH3 \longrightarrow AsH2 + H$$

as was proposed so far by several authors. The last step :

$$AsH \longrightarrow As + H$$

is so unfavorable because of the energy-consuming formation of free H, that it will only occur at higher temperatures. The bi-molecular reaction :

$$2\ AsH \longrightarrow As2 + H2$$

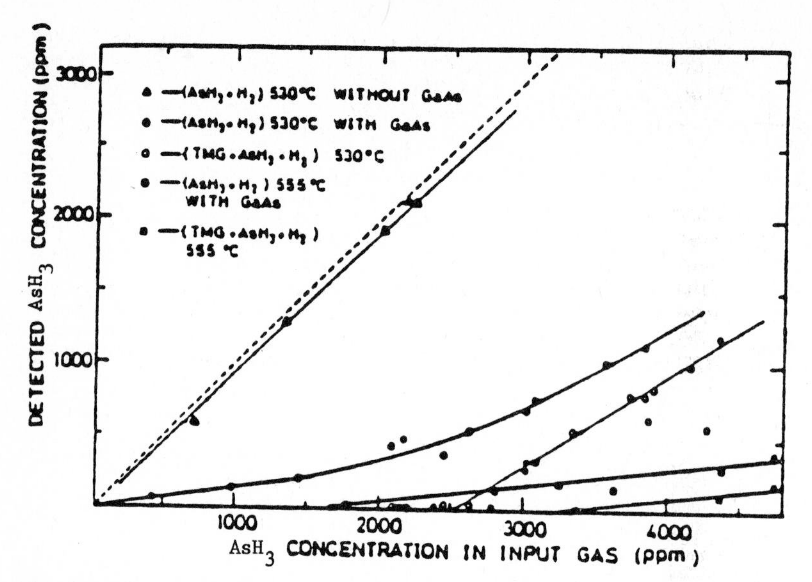

Fig.7 AsH3 concentration depends on TMG addition and GaAs existence. Broken line indicates AsH3 concentration in the input gas (after [16])

is much more favorable but needs a collision of two AsH molecules, and will therefore have a much lower rate except for the case of undiluted arsine. However, if a GaAs surface is present, the AsH molecules will be collected on it and the reaction rate will be much higher. This explains the observation that a GaAs surface catalizes arsine decomposition (fig. 7).

Following these considerations, in the case of GaAs epitaxy the reaction between adsorbed species on the crystal surface is between AsH*, its reaction product As*, Ga(CH3)* and its reaction product Ga*. A number of reactions occur between these adsorbed molecules on adjacent sites :

$$2AsH^* \longrightarrow 2As^* + H2$$

$$AsH^* + Ga(CH3)^* \longrightarrow As^* + Ga^* + CH4$$

$$2Ga(CH3)^* + H2 \longrightarrow 2Ga^* + 2CH4$$

$$2Ga(CH3)^* \longrightarrow 2Ga^* + C2H6$$

The possibility of obtaining atomically smooth interfaces with MOVPE, which will be demonstrated below, indicates strongly that there is an ordering process of adsorbed species which can be described as lateral growth or step growth, and that for strongly disoriented substrates (2 to 6 degrees off the <100> direction) there is even a mechanism which keeps the distance between adjacent steps about equal. This mechanism may simply arise from the fact that the position of a gallium atom INSIDE a step is energetically more favorable than a position on the rest of the surface, but that the position ON TOP of the step is energetically less favorable than any other position on the surface. The position on top will act as a potential barrier for adsorbed species diffusing on the surface.

This means that the steps are only fed from ONE side, and that if during growth two steps approach one another, i.e. one step starts to catch up with the other, it will statistically get less feeding because the area of the surface between it and the step it is catching up, from which it collects incoming atoms, is decreasing (Fig. 8). Fig. 9 summarizes the reaction kinetics for MOVPE of GaAs.

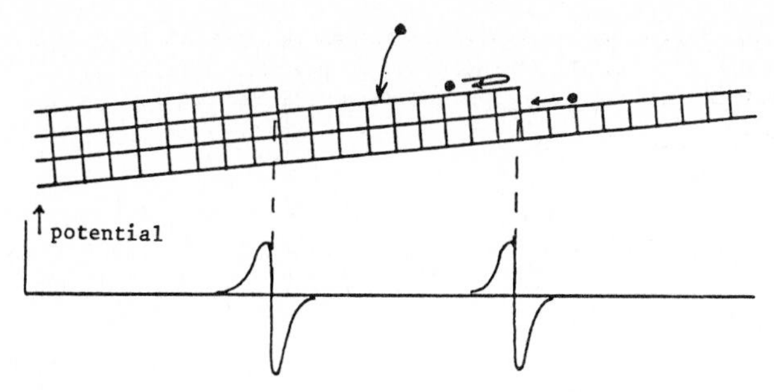

Fig. 8 Why a disoriented surface remains flat and ordered. A possible reason for the stability of interstep distances.

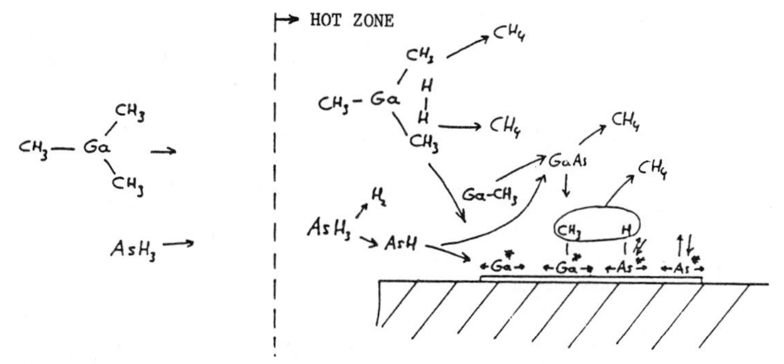

Fig.9 GaAs growth by MOVPE

The reaction mechanism for MOVPE of (Ga,Al)As is similar, but the Al Al bearing species may not be of the same form. Fig. 10 shows the composition profile obtained in the reactor cell of fig. 10c with heating of the entrance region to 400°C. There is a thickness variation over the wafer due to depletion in the gas phase of almost one in three. The composition of the (Ga,Al)As remained however 35+1% throughout the wafer. These profiles depend on the diffusion coefficient of the species diffusing towards the surface. The diffusion coefficient of a molecule depends in first approximation only on its mass and of the molar mass of the carrier gas, especially if we compare molecules with the size/mass ratio of the same order of magnitude.

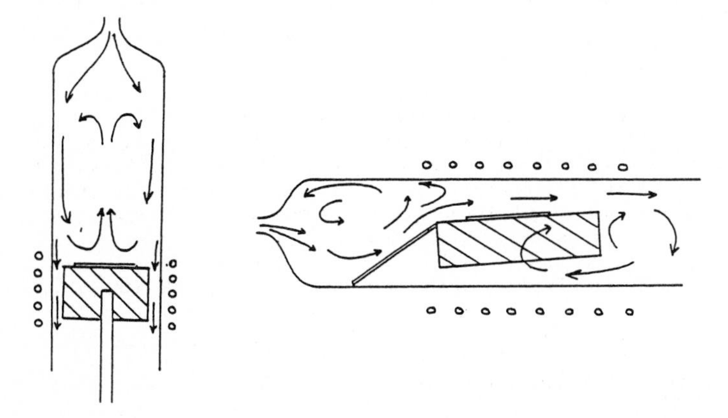

Fig. 10 Two different reactor chamber arrangements among the many currently used.

The diffusing gallium species is in this case with preheating only monomethylgallium (MMG). The molecular mass of MMG is 85. The molecular mass of MMA is 42. An alternative for MMA could be Al2(CH3)2, obtained from TMA by dissociation of the weakest bonds via :

$$CH_3\text{—}Al\langle {}^{CH_3}_{CH_3} \rangle Al\text{—}CH_3 \;+\; H_2 \;\longrightarrow\; Al\langle {}^{CH_3}_{CH_3} \rangle Al \;+\; 2CH_4$$

The molecular masses are :

Al2(CH3)2	84 a.m.u.
Ga(CH3)	85 a.m.u.

The calculation of Fig. 11 shows however, that the growthrate profile is too insensitive to the diffusion coefficient to conclude which species diffuses to the surface. This insensitivity is of great practical importance for MOVPE growth of ternary and quaternary materials like (Ga,In)As, of which the lattice match to the substrate depends critically on the composition of the ternary.

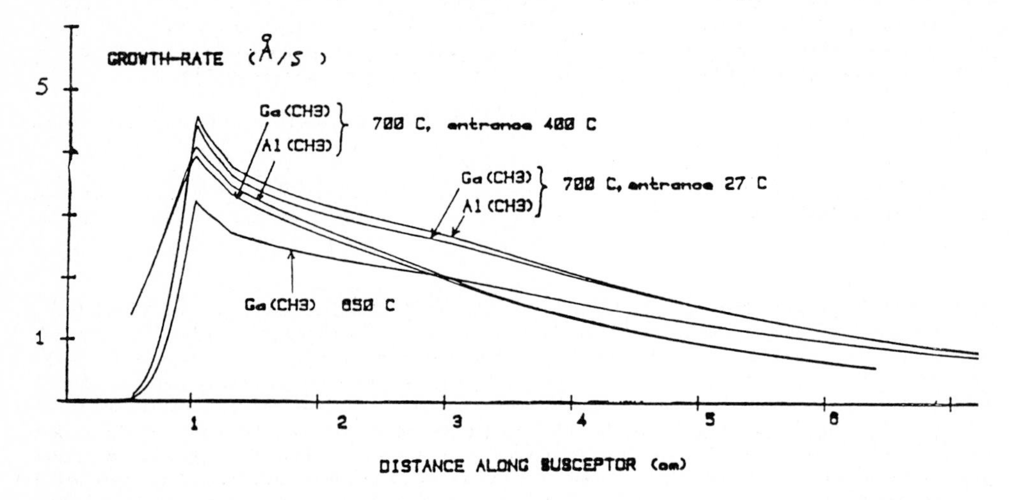

Fig. 11 Calculated growthrate versus distance along susceptor.

4. Mass transport

Fig. 10 shows several geometries currently used in MOVPE. Several of these show important stationary vortices. If we like to grow very (atomically) sharp interfaces in continuous growth, or if we want to realize a sharp decrease in a doping profile, then we must be able to change the concentrations of reactant gases over the wafer in a time of the order of 0.1 sec, if the growthrate is 5 A/sec. The mass transport in and out the vortices is by diffusion.

The tracer diffusion coefficient of a gas with mass Ma in a gas with mass Mb is approximately [17] :

$$D = 6.10^{-4}\, T^{1.67} \frac{(M_a + M_b)^{1/2}}{(M_a M_b)^{0.54}(M_a^{0.12} + M_b^{0.12})^2}$$

A few values : [F = 1 atm]

molecule	D in H2 [cm2/s]
AsH3	0.622
SiH4	0.745
Ga(CH3)3	0.579
Ga(CH3)	0.615
Ga(C2H5)	0.597
Al2(CH3)6	0.580
Al(CH3)	0.708
In(CH3)	0.565
In(C2H5)	0.554

The characteristic group for diffusion is : $x/\sqrt{Dt}$ in which x is the characteristic distance and t the characteristic time. If the pressure is 1 atm and the vortex is in the cold part of the reactor (D=0.6),it will empty by diffusion with a characteristic time of about 15 seconds. At 400°C, it is about 4 seconds. At 5 Å/sec these times correspond with thicknesses of 75 Å resp. 20 Å. In general, a few times the characteristic time are needed to approach zero conveniently.

Let us now study the limit of MOCVD in the case of an ideal growth chamber which would not show stationary vortices.

The type of the flow - laminar or turbulent - depends on the ratio of the inertial forces and the viscous forces in the gas, indicated by the Reynolds number :

$$Re = \frac{U_m \chi \rho}{\eta}$$

U_m = mean spured $\quad$ η = viscosity

χ = char. distance $\quad$ ρ = density

For Re > 5000, the flow is turbulent. For Re < 2000 the flow is laminar. We think that Re = 20 is achievable under practical condition. If, Re = 20 at room temperature, it can be expected to be even lower at high temperature. The flow will be a perfect streamline flow in which the streamlines will only be bent by the temperature profile. This situation can be treated numerically [18]. Fig. 11 shows the calculated profiles, based on the above considerations.

REFERENCES

1. H.M. Manasevit: J. Cryst. Growth, 13/14, (1972), 306
2. Y. Seki, K. Tanna, K. Iida and E. Ichiki: J. Electrochem. Soc., 122, (1975), 1108
3. S.J. Bass: J. Cryst. Growth, 31 (1975), 172
4. J.P. Andre, A. Gallais and J. Hallais: Inst. Phys. Conf. Ser. 33a
5. R.D. Dupuis and P.D. Dapkus: Appl. Phys. Lett., 31, (1977), 466
6. J.P. Duchemin, M. Bonnet and Huyghe: Revue Technique Thomson-CSF, 9 (1977), 685
7. N. Holonyak Jr., R.M. Kolbas, W. Laidig, B.A. Vojak, R.D. Dupuis and P.D. Dapkus: Inst. Phys. Conf. Ser. 45, (Institute of Physics, London 1979), 387
8. J.P. Duchemin, M. Bonnet, C. Benchet and F. Koelsch: Inst. Phys. Conf. Ser. 45, (Institute of Physics, London 1979), 10

9. J.P. Hirtz, J.P. Duchemin, P. Hirtz and B. De Cremous: Electr. Lett., 16 (1980), 275
10. Y. Mori, M. Ikeda, H. Sato, K. Kaneko, and N. Watanabe: Inst. Phys. Conf. Ser., 63 (1981), 95
11. J. Hallais, J.P. Andre, P. Baudet, D. Boccon-Gibod: Inst. Phys. Conf. Ser. 45, (Inst. of Phys., Bristol 1979), 361
12. D.H. Reep, S.K. Ghandhi: J. Electrochem. Soc., 131, (1984), 2697
13. D.J. Schlyer, M.A. Ring: J. Electrochem. Soc., 124, (1977), 569
14. M.R. Leys, H. Veenvliet: J. Cryst. Growth 55, (1981), 145
15. M.G. Jacko, S.J.W. Price: Can. J. Chem., 41 (1963), 1560
16. J. Nishizawa, T. Kurabayashi: J. Electrochem. Soc., 130, 2 (1983), 416
17. M.L. Hitchmann: Progress in Crystal Growth and Characterization, 4, 3 (1981), 249
18. P.M. Frijlink: to be published

Part V

Applications

The Two-Dimensional Electron Gas Field Effect Transistor

B. Vinter

Thomson-CSF, Laboratoire Central de Recherches, Domaine de Corbeville, BP 10, F-91401 Orsay, France

A field effect transistor (FET) is a semiconductor device in which the current between two contacts - source and drain - is controlled by the voltage on a third contact - the gate. The role of such devices in electronic circuits is simple in principle. In logic circuits it functions as a switch - depending on the gate voltage the connection between source and drain is broken or closed - and in analogue circuits a small time-varying signal on the gate yields a time-varying current between source and drain, and since the gate current ideally is purely a displacement current, a very small input power can be amplified.

There are several kinds of FETs. By far the most commonly applied is the Metal-Oxide-Semiconductor FET (MOSFET) where the metal gate is insulated from the Si semiconductor by a SiO_2 layer. The current between source and drain is carried by a thin channel of electrons near the insulator-semiconductor interface, formed by applying a positive voltage to the gate. Another type is the MESFET (Metal-Epitaxial-Semiconductor FET) in which the source-drain current is carried in a thin highly doped epitaxial semiconductor layer. The current is controlled by a gate which forms a Schottky barrier on the semiconductor and therefore - depending on the applied gate voltage - depletes more or less the semiconductor layer of electrons under the gate. For transistors based on GaAs the MESFET technology is the most developed at present.

Here we shall discuss another type of FET. In principle, one would like to make a MOSFET structure with GaAs replacing Si as the semiconductor. The reason for this is that intrinsically GaAs conduction electrons have a considerably lower effective mass than those of Si and therefore a higher mobility, and, for most electric fields, a higher drift velocity. However, it has not been possible to find an insulator which forms a sufficiently good interface on GaAs to form a conducting channel. Instead, a solution has been found in the form of the TEGFET (or HEMT or MODFET or SDHT depending on manufacturer) which as we shall see has features in common with both MOSFETs and MESFETs.

The basic structure as shown in Fig. 1 consists of a semiinsulating substrate on which is first grown a buffer layer of undoped (to the extent possible) GaAs; on top of this is grown a thin ($\sim$ 500 Å) layer of $Al_xGa_{1-x}As$ part or all of which is rather heavily ($\sim 10^{18} cm^{-3}$) n-type doped. The gate metal forms a Schottky barrier to

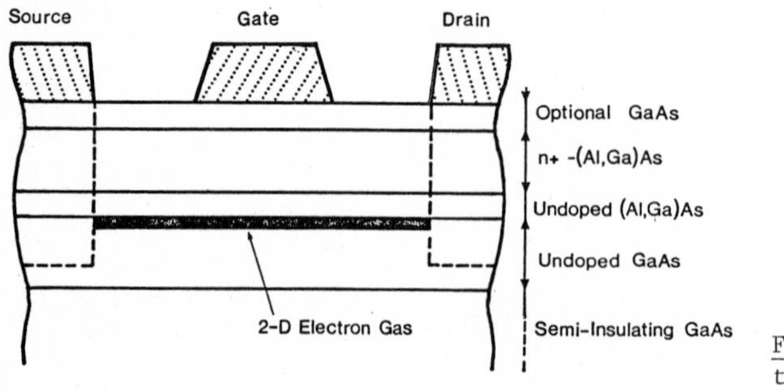

Fig.1 The basic TEGFET structure

the AlGaAs and by making the ternary layer thin enough, the gate can completely deplete the AlGaAs of electrons. Then the density of electrons on the GaAs side of the AlGaAs/GaAs heterojunction is controlled by the voltage applied to the gate. Consequently the current between the source and drain contacts can be controlled by the gate voltage.

The simplest description of normal transistor operation can now be given. With gate length L_g and gate width Z the gate/AlGaAs/channel structure acts as a capacitor with capacitance $C \cong \varepsilon_r \varepsilon_o Z L_g / d_{AlGaAs}$ so that the number of electrons in the channel depends linearly on gate voltage

$$e\, n_s\, L_g\, Z \cong C(V_G - V_{th}) \ , \tag{1}$$

where n_s is the electron density per surface area. Normally the drain voltage is so large that roughly all electrons move at saturation velocity v_s ($\sim 10^7$ cm/s) independent of drain voltage so that the current between source and drain is given by :

$$I_{DS} \cong n_s\, Z\, e\, v_s \tag{2}$$

so that I_{DS} is also roughly linear in gate voltage. A very important parameter is the (intrinsic) transconductance

$$g_{mo} \equiv \left.\frac{\partial I_{DS}}{\partial V_G}\right)_{V_{DS}}$$

which in this simple model is given by (1) and (2) as

$$g_{mo} \cong Z\, eV_s \frac{dn_s}{dV_G} = C \frac{V_s}{L_g} \ , \tag{3}$$

from which we find

$$\frac{g_{mo}}{C} = \frac{V_s}{L_g} = \frac{1}{\tau} \ , \tag{4}$$

where τ is the transit time for an electron under the gate. Clearly, for fast operation it is advantageous to reduce the gate length as much as technologically possible; the limit at the moment is at $L_g \cong 0.25$ μm.

To make a slightly better model let us assume that the velocity of electrons as a function of electric field has the dependence shown in Fig. 2. Below a critical field ε_c the velocity is proportional to the field $v = \mu E$ and above it is constant. Furthermore let us assume that the voltage in the channel varies along the channel, $V_{ch} = V(x)$, but that the density of electrons follows the law described in (1). The current is then given by

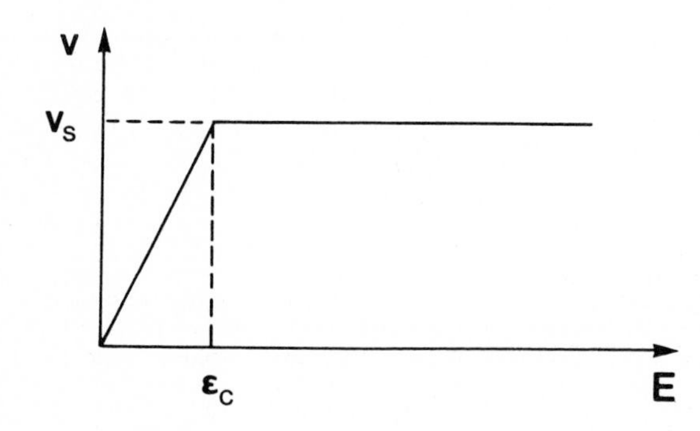

Fig.2 Velocity-field characteristic employed in the simplified model. $\mu = v_s/\varepsilon_c$.

$$I = Zen_s(x)v(x) = \frac{C}{L_g}[V_G - V_{th} - V(x)]\ \mu \frac{dV}{dx} \tag{5}$$

as long as $dV/dx \leqslant \varepsilon_c$. Since the current is conserved we can integrate the equation by separation :

$$I(x-x_s) = \mu \frac{C}{L_g} \int_{V(x_s)}^{V} (V_G - V_{th} - V')dV' \tag{6}$$

where x_s is the position of the entrance to the channel on the source side. If we define the saturation current to be the current for which the field on the drain side of the channel $x_s + L_g$ just reaches ε_c we find the implicit equation for I_{sat} :

$$I_{sat} = g_{mo}\{\sqrt{[(V_G - V_{th} - R_s I_{sat})^2 + \varepsilon_c^2 L_g^2]} - \varepsilon_c L_g\} \quad \text{where} \tag{7}$$

$$g_{mo} = \frac{C\,\mu\,\varepsilon_c}{L_g} = \frac{C\,V_s}{L_g} \tag{8}$$

Here R_s is the resistance between the source contact and x_s, and the source voltage is at 0.

It is instructive to consider the two limits of (7)

(I): $\varepsilon_c L_g << V_G - V_{th} - R_s I_{sat}$, short, highly conductive channel.

$$I_{sat} \simeq \frac{g_{mo}}{1 + R_s g_{mo}} (V_G - V_{th} - \varepsilon_c L_g) \tag{9}$$

i.e. linear behaviour as in (2) but with an extrinsic transconductance $g_m = g_{mo}(1 + R_s g_{mo})^{-1}$.

(II): $\varepsilon_c L_g >> V_G - V_{th} - R_s I_{sat}$, long, channel near pinch-off

$$I_{sat} \simeq \frac{g_{mo}}{2\varepsilon_c L_g} (V_G - V_{th})^2 \ , \tag{10}$$

i.e. a quadratic dependence on V_G.

In Fig. 3 we show experimental measurements of $I_{sat}(V_G)$ for four TEGFETs on the same wafer but with different gate lengths. The limiting behaviour corresponding to (9) and (10) is evident, and indeed one can fit all four curves quite well with (7) using basically the same values for g_{mo}, V_{th}, and R_s. So we see that for normal operation even this crude model, which is actually a standard MOSFET model, describes quite well the basic functioning of the TEGFET.

In equivalent circuit representation the simple small-signal model is as shown in Fig. 4 where the current generator is controlled by the voltage across the gate capacitor. Simple circuit theory then leads to the expression for the current gain

$$h_{21} = \frac{I_D}{I_G} = \frac{g_{mo}}{j\omega C} \equiv \frac{f_T}{jf} \ , \tag{11}$$

where the cut-off frequency $f_T = g_{mo}/2\pi C$. This again shows the importance of keeping g_{mo}/C as large as possible; the model is too simple to calculate the available power gain, which falls off with frequency as $(f_{max}/f)^2$, where f_{max} is slightly higher than f_T.

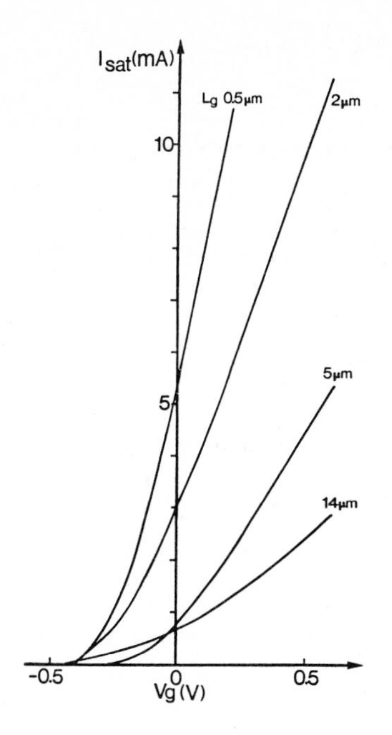

Fig.3 Experimental drain saturation current vs gate voltage for transistors of different gate length L_g. The gate width Z = 150 µm.

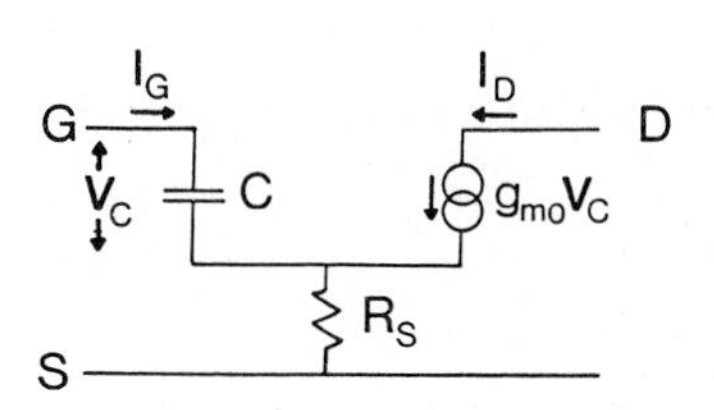

Fig.4 The simplest small signal equivalent circuit for a (TEG)FET.

In the following we shall look in more detail at some of the physics and limitations of the simple model. The discussion is in a natural way divided between the charge control, i.e. the behaviour transverse to the heterojunction, and the parallel transport in the channel.

Charge control

A charge control model must describe the relation between the charge (and its distribution) in the channel and the applied gate voltage. To understand the problem, assume that the $Al_xGa_{1-x}As$ layer is doped with N_D donors/cm^3 in a layer of thickness d_2, followed by an undoped spacer layer of AlGaAs of thickness d_3. If we furthermore assume that we work at a gate voltage for which the AlGaAs is completely depleted of electrons, the conduction band diagram will look schematically as in Fig. 5. The field at the heterojunction interface on the AlGaAs

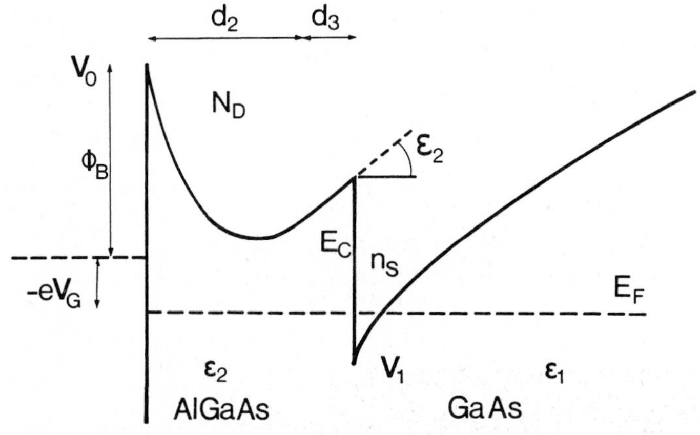

Fig.5 Conduction band diagram for the gated heterojunction. Broken lines: Fermi level

side ε_2 will be given by

$$\mathcal{E}_2 = \frac{\varepsilon_1}{\varepsilon_2}\,\mathcal{E}_1 = \frac{e}{\varepsilon_2}\,(n_s + N_{depl}) \cong \frac{e}{\varepsilon_2}\,n_s\,, \tag{12}$$

where ε_2 (ε_1) is the permittivity of AlGaAs (GaAs), n_s is the charge in the channel, and N_{depl} is the effective depletion charge in the GaAs. If we introduce the band bending necessary to deplete a layer of doping N_D and thickness d_2, $V_{p2} = e^2N_Dd_2^2/2\varepsilon_2$, expressing the value of V_0 in two ways gives the equation

$$V_{p2} - e\mathcal{E}_2\,(d_2+d_3) + \Delta E_c + V_1 = E_F + \phi_B - eV_G\,, \tag{13}$$

where ϕ_B is the built-in Schottky barrier ($\sim$ 1eV), V_G is the applied gate voltage, and E_F is the Fermi energy. Combining (12) and (13) we obtain

$$en_s = \frac{\varepsilon_2}{e(d_2 + d_3)}\,[eV_G - (E_F - V_1) + \Delta E_c + V_{p2} - \phi_B] \tag{14}$$

and further progress now depends on describing the dependence of $E_F - V_1 = f(n_s)$ as a function of n_s. Differentiating (14) we find the capacitance

$$\frac{C}{ZL_g} = \frac{d(en_s)}{dV_G} = \frac{\varepsilon_2}{d_2 + d_3 + \dfrac{\varepsilon_2}{e^2}\dfrac{df}{dn_s}}\,, \tag{15}$$

which shows that C is constant, as assumed in the model of the first section, if df/dn_s is constant, but that the value is slightly different from the capacitance of the AlGaAs layer alone; in practice a value for $(\varepsilon_2/e^2)df/dn_s$ of about 80 Å is often used.

For the function $f(n_s)$ several simpler models exist in the literature; we shall base our discussion on a fairly complete treatment of the problem, which also brings in evidence other features of charge control in the TEGFET structure.

In that model, which is purely numerical, we solve self-consistently the Schrödinger equation and Poisson's equation. Thus for the electron potential we have

$$\frac{d^2V}{dz^2} = \frac{e^2}{\varepsilon}\,[N_D^+(z) - n_{el}(z)]\,, \tag{16}$$

$$V(d_2 + d_3\,-) = V(d_2 + d_3\,+) + \Delta E_c\,, \tag{17}$$

$$V(z = 0) = -\,eV_G + \phi_B\,, \tag{18}$$

where the charges in the system consist of the ionized donors $N_D^+(z)$ and the electrons $N_{el}(z)$. The distribution of ionized donors is given by $N_D^+(z) = N_D(z) - N_D^0(z)$, where N_D is the donor and N_D^0 is the neutralized donor density determined by Fermi statistics

$$N_D^0\,(z) = \frac{N_D}{1 + \exp\,(E_F - [V(z) - E_B]/k_BT)} \tag{19}$$

where E_B is the donor binding energy.

The electron density $n_{el}(z)$ is found by solving the Schrödinger equation

$$- \frac{\hbar^2}{2m} \frac{d^2 \zeta_n}{dz^2} + V(z) \zeta_n (z) = E_n \zeta_n (z) , \qquad (20)$$

where $\zeta_n(z)$ are the subband wavefunctions, E_n the subband bottom energies, and m the effective mass of electrons in gaAs (in this calculation we take permittivity and electron mass to be the same in AlGaAs). The electron density is then given in the Hartree approximation as

$$n_{el} (z) = \sum_n | \zeta_n (z)|^2 N_n \qquad (21)$$

where

$$N_n = \frac{k_B T m}{\pi \hbar^2} \ln [1 + \exp (\frac{E_F - E_n}{k_B T})] \qquad (22)$$

is the subband occupation number.

Because the charge distribution depends on the potential (19 , 20) and the potential depends on the charge distribution (16), the solution must be found in a self-consistent way. Calculations of this kind have been presented earlier at this Winter School,so we want to point out the following differences from those other calculations. The Fermi energy is kept constant (=0) throughout the semiconductor structure, and the independent parameter is therefore the gate voltage V_G rather than n_s; the electrons of the system are not only on the GaAs side of the heterojunction, they can penetrate into the ternary, they can even tunnel through the barrier, and they can become bound on donor atoms in the AlGaAs; and the calculations are made at finite temperature, so that a sufficient number of subbands must be included.

Results of such self-consistent calculations are shown in Figs. 6-8 based on the parameters given in Table I. Device 3468 is "normally on" or works in the "depletion mode", i.e. for $V_G = 0$ the channel is conducting and a negative V_G must be applied to deplete the channel of electrons, whereas 3469 is "normally off" or works in "enhancement mode", i.e. a positive V_G must be applied to create a conducting channel. If we consider device 3468 in Fig. 6 we see the charge control. At lower gate voltage $V_G = -1.5$V there are no electrons in the AlGaAs and four subbands in the channel are shown; their separation is such that about 50-60% of the electrons are in the lowest and about 20% in the first excited subband (this can be taken as the justification for the name TEGFET). As the gate voltage is

Table I. Parameters of the two devices used in the calculations.

Device nr.		3468	3469
x	Al concentration in AlGaAs	0.26	0.28
N	Doping in AlGaAs (10^{18} cm^{-3})	1.3	0.6
d_2	Doped AlGaAs thickness (Å)	550	400
d_3	Spacer thickness (Å)	75	65
d_1	GaAs thickness (μm)	1.0	0.8
ΔE_c	Conduction-band discontinuity (meV)	260	280
E_B	Donor binding energy (meV)	50	50
V_0	Potential at $d_2 + d_3 + d_1$ (eV)	1	1

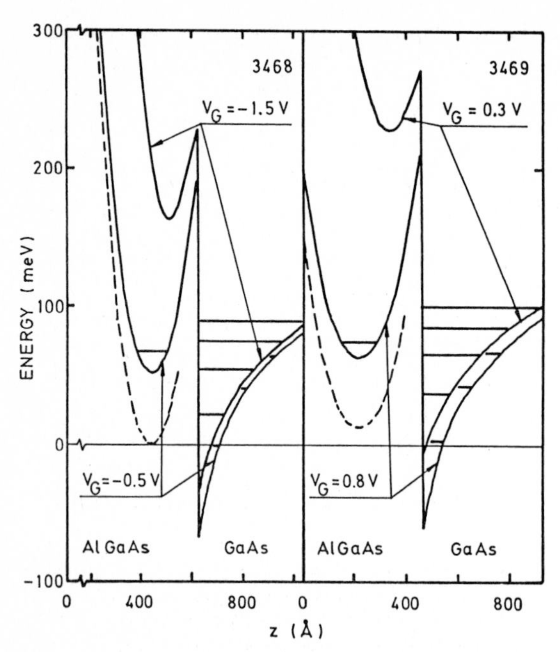

Fig.6 Self-consistent potentials for conduction electrons in the two TEGFET's for two gate voltages V_g. left part : normally-on device; right part : quasi-normally-off device. The Fermi level $E_1 = 0$. Horizontal lines : bottom energy of the lowest four subbands. The energies of the donor levels are shown broken for the higher gate voltage.

increased, the potential and the subbands are lowered and gradually some of the donors in AlGaAs become neutralized and free electrons begin to appear in AlGaAs. In Fig. 7 the electron density in the system is shown as a function of the gate voltage : below $V_G = -2V$ the device is pinched off, there are no electrons in the system; with increasing gate voltage the electrons begin to go into the channel in gaAs (curve a) and the density increases almost linearly with V_G as in the simple model (1) around $V_G = -1.5V$, however, the channel density tends to saturate because the additional electrons now either neutralize donors in AlGaAs (curve d) or enter the AlGaAs as free electrons (curve b). This means that it becomes less and less possible to control the density of useful electrons, viz. those that are in the high-mobility channel, and it has an important influence on capacitance and transconductance,as can be inferred from Fig. 8 in which we show the derivatives of the various densities with respect to gate voltage. By definition, the derivative for all electrons in the system (curve a) gives the gate capacitance. Note that it is actually not very constant : From pinch-off it increases rapidly, tends to flatten out a little around -1.5V and then increases again, above all because of the electrons that neutralize the donors in AlGaAs. The curves corresponding to the channel electrons (c) and all the free electrons (b) show a different behaviour, namely first an increase, a maximum around -1.5V followed by a decrease. Since current from source to drain can only be carried by free electrons, we can understand from these curves that qualitatively the transconductance must show a behaviour similar to curve (b); the decrease above the maximum is called transconductance degradation by some authors, but we see that it is a natural intrinsic property of the heterostructure. These features are seen experimentally as shown in Fig. 9 for the capacitance and Fig. 10 for the transconductance.

It should be mentioned that some of the parameters used in Table 1 are somewhat controversial; the accepted value of ΔE_c has varied over the years, a consensus seems to build up around $\Delta E_C \cong 0.7x$eV (where x is the Al concentration) rather than

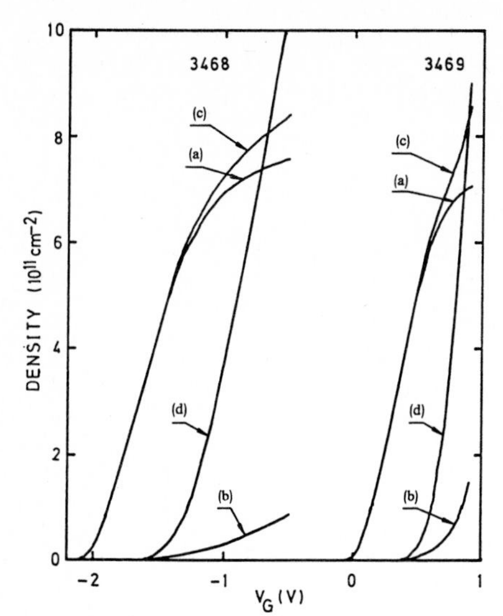

Fig.7 Calculated electron density vs gate voltage; (a) Free electron density in GaAs channel; (b) free -electron density in AlGaAs; (c) total density of free electrons; (d) density of neutralized donors in AlGaAs.

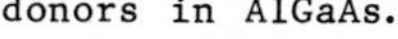

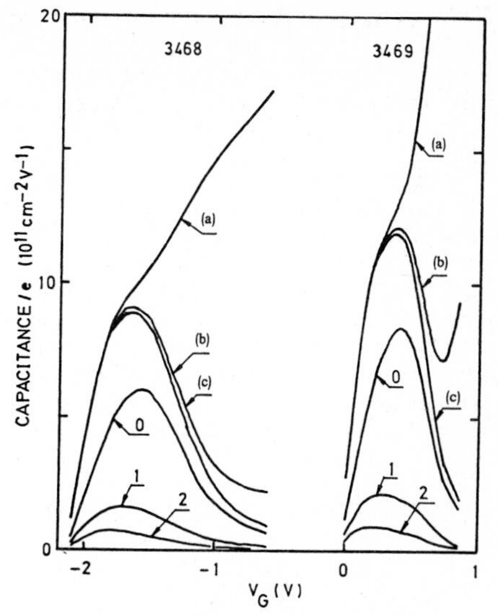

Fig.8 Calculated capacitance edN_1/dV_G of (a) all electrons in system, (b) all free electrons, (c) all electrons in the channel in GaAs; and (0,1,2) electrons in lowest three channel subbands.

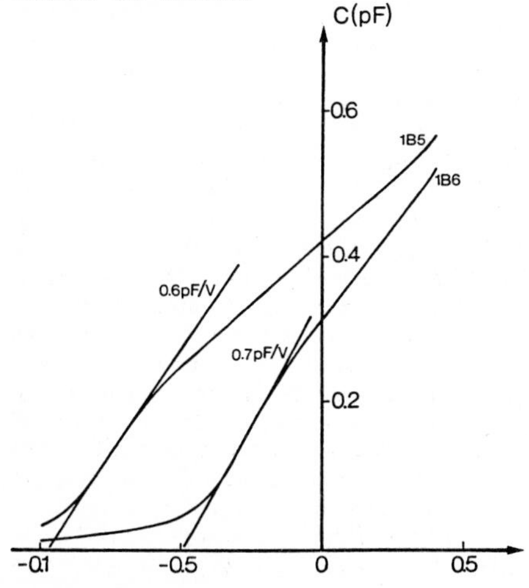

Fig.9 Capacitance vs gate voltage for two transistors

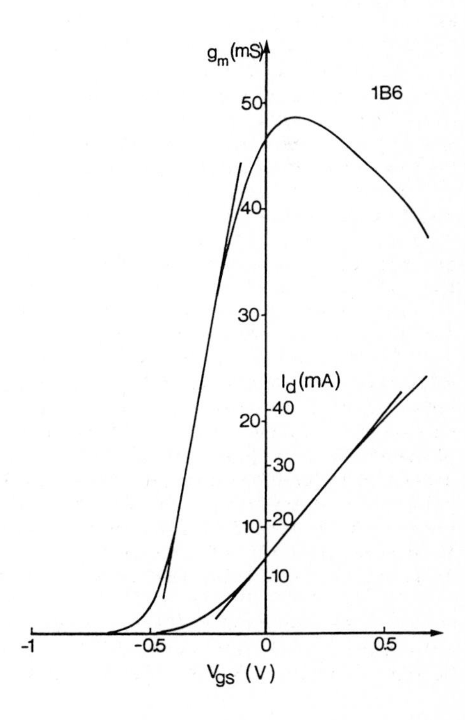

Fig.10 Transconductance g_m and drain current I_D vs gate voltage V_G for transistor 1B6. Gate length L_g =0.5 μm, gate width Z =300 μm.

the value used here; moreover, donor atoms in AlgaAs give rise to a much more complex level structure than the single fairly deep donor assumed in the model. We shall not discuss this point in these notes; a few references on that subject are given at the end. These are fine points that do not change the results of the calculation in any serious way.

Parallel Transport

In our simple modelling of the saturated current in the TEGFET we already used the most simple representation of velocity of electrons versus field described in Fig. 2, and we saw that a quite reasonable description of the behaviour of the transistor in normal saturated operation could be obtained. In fact, theories for transport in the channel are still very much in a beginning stage, so we can only describe those elements of a transport theory that are known with reasonable certainty, and indicate speculations on where new effects of the special quasi-two-dimensional structure might possibly play a role.

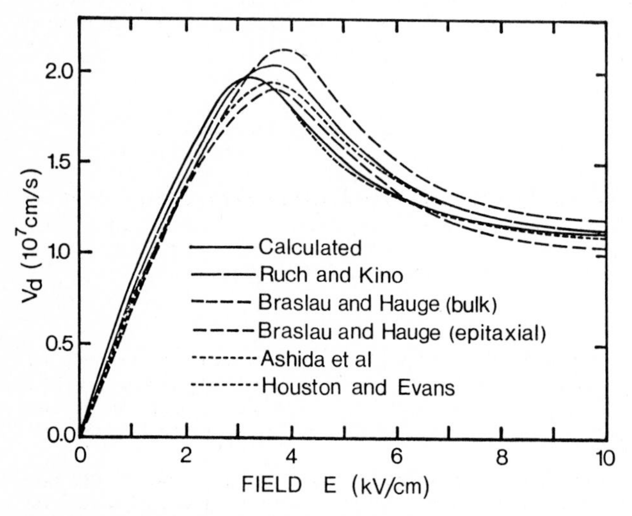

Fig.11 Drift velocity of conduction electrons in GaAs at 300K, for the 0-10kV/cm range, after Pozela and Reklaitis [3]. The solid curve was generated by Pozela and Reklaitis in a Monte Carlo calculation, and compared with experimental data of Ruch and Kino [4], Braslau and Hauge [5], Ashida et al [6] and Houston and Evans [7].

As a starting basis, consider the dependence of electron velocity on electric field in bulk GaAs, shown in Fig. 11. At low field the velocity is proportional to the field, where by definition the proportionality constant is the low-field mobility. This parameter has been heavily studied both experimentally and theoretically in heterojunction structures. Part of this subject has been treated in earlier lectures at this School, and we shall return to the phonon-scattering contribution to the mobility below. It must be borne in mind, however, that in transistors in operation the parallel field is considerable, several kV/cm, so that hot-electron transport is of decisive importance in the channel. What happens in bulk GaAs is that the electrons gain more kinetic energy from the electric field than they can dissipate rapidly, so that the mean electron energy increases from the thermal equilibrium value $(3/2)k_BT$. With increasing energy the scattering rates, especially those due to high-energy phonons, increase, so that an energy balance is established but with a drift velocity which is no longer proportional to the field. The particularity of the maximum followed by a region of decreasing velocity originates from the scattering of hot electrons from the low-mass Γ -valley in the center of the Brillouin zone to heavy-mass, low-mobility valleys lying about 300 meV above the Γ valley; this process is only possible for electrons that have obtained a kinetic energy close to the valley separation. Since scattering on ionized impurities becomes less efficient with increasing electron energy, it is understandable from these qualitative considerations that the low-field mobility depends much on impurity concentration, whereas the high-field behaviour of the electrons is much less determined by the impurities.

We now show how the quasi-two-dimensional structure of the electron system in a TEGFET modifies the electron-phonon interaction in comparison with the bulk. We use as illustration the interaction with polar optical phonons. The principles for other interactions are the same and pertinent expressions can be found e.g., in [1].

A longitudinal optical phonon of wave vector $\vec{q}$ and frequency ω_o creates (in a heteropolar crystal like GaAs) a polarization wave. Associated with the polarization wave is an electric potential on which the conduction electrons can be scattered. In three dimensions the matrix element for scattering of an electron in state $|\vec{k}\rangle \Omega^{-1/2}\exp(i\vec{k}.\vec{r})$ to another state $|\vec{k}'\rangle$ with annihilation of one phonon is given by (e.g., [2]) :

$$<\vec{k}'\, n_g -1\, |H_{int}|\, n_g\, \vec{k}> = i\, g\, \frac{n_g^{1/2}}{|\vec{g}|}\, \delta_{\vec{k}',\vec{k}+\vec{g}} \tag{23}$$

where $n_g = [\exp(\hbar\omega_o/k_BT) - 1]^{-1}$ is the phonon occupation number and g is a coupling constant which can be calculated as (S.I. units) :

$$g^2 = \frac{e^2\hbar\omega_o}{2\Omega}\left(\frac{1}{\varepsilon_\infty} - \frac{1}{\varepsilon_s}\right) \tag{24}$$

In these equations ω_o is the LO phonon frequency, Ω is the volume of the sample, ε_∞ and ε_s are the permittivities far above and far below ω_o respectively. The momentum conservation implied in Eq. (23) arises from the matrix element of the $e^{i\vec{g}.\vec{r}}$ dependence of the phonon potential between the two electron states.

In the quasi-two-dimensional system the potential created by the phonon is not changed if we neglect the difference of interatomic forces in AlGaAs and GaAs. On the other hand, the electrons are no longer in plane-wave states but in subband states characterized by subband index n and wave vector parallel to the interface $\vec{K}$: $|n\vec{K}\rangle = A^{-1/2}\zeta_n(z)\exp(i\,\vec{K}.\vec{R})$. Then the matrix element of the electron-phonon interaction corresponding to (23) becomes (we use capital letters for vector components parallel to the interface) :

$$\begin{aligned}<n'\vec{k}'|H_{int}|n\vec{K}> &= ig\,\frac{n_g^{1/2}}{\sqrt{Q^2+g_2^2}}\,\frac{1}{A}\int dz\,\zeta^*_{n'}(z)\zeta_n(z)e^{ig_2z}\int d\vec{R}\,\exp(i(\vec{K}+\vec{Q}-\vec{K}').\vec{R})\\ &= ig\,\frac{n_g^{1/2}}{\sqrt{Q^2+g_2^2}}\,\delta_{\vec{K}',\vec{K}+\vec{Q}}\int dz\,\zeta^*_{n'}(z)e^{ig_2z}\zeta_n(z)\end{aligned} \tag{25}$$

from which we see that momentum conservation is only fulfilled parallel to the interface, since the electrons no longer have a well-defined momentum perpendicular to the interface. The scattering rate from state $|n\vec{K}\rangle$ to $|n'\vec{K}'\rangle$ by absorption of a phonon $\vec{g}$ is given by Fermi's golden rule

$$S_{\vec{g}}\,(n\vec{K}\rightarrow n'\vec{K}') = \frac{2\pi}{\hbar}\,|<|H_{int}|>|^2\,\delta(\varepsilon_{n'\vec{k}'} - \varepsilon_{n\vec{k}} - \hbar\omega_o)\ , \tag{26}$$

where the δ-function conserves energy in the scattering process. Since the phonon energy is constant it is possible to sum over processes involving phonons with the same $\vec{Q}$,

$$S(n\vec{K}\to n'\vec{K}') = \sum_{g_z} S_{\vec{g}} = \frac{\Omega}{A}\int \frac{dg_z}{2\pi} S_{\vec{g}} = \tag{27}$$

$$\frac{2\pi}{\hbar} g^2 \frac{\Omega}{A} \frac{n_g f_{n'n}(Q)}{2Q} \delta_{\vec{K}',\vec{K}+\vec{Q}} \delta(\varepsilon_{n'\vec{K}'} - \varepsilon_{n\vec{K}} - \hbar_{\omega o}) ,$$

where the form factor is

$$f_{n'n}(Q) = \iint dz\, dz' \, \zeta^*_{n'}(z)\zeta_n(z)e^{-Q|z-z'|}\zeta^*_n(z')\zeta_{n'}(z') . \tag{28}$$

Comparing this expression with the corresponding three-dimensional result obtainable by Fermi's rule on (23), we see that the scattering rate has effectively been reduced to two dimensions; the difference is that the Q-dependence has been changed, and that we now have scattering processes in which the electron stays in the same subband ($n' = n$) and intersubband scattering processes in which the electron is scattered to a different subband ($n' \neq n$), subject to conservation of parallel momentum and energy.

For processes involving emission of an optical phonon, the scattering rate is as in (27) with a replacement of n_q by $n_q + 1$ to account for spontaneous emission and a change of sign for $\hbar\omega_o$ in the energy conservation.

The scattering rate out of a state $|n\vec{K}>$ follows from (27) by summing over final states for each process :

$$\tau^{-1}_{n\vec{K}} = \sum_{n'\vec{K}'} S\,(n\vec{K}\to n'\vec{K}') = \sum_{n'} A \int \frac{dK'}{(2\pi)^2} S\,(n\vec{K}\to n'\vec{K}') . \tag{29}$$

This rate is shown in Fig. 12 for electrons in the lowest subband n = 0. If we concentrate on part (d) of the figure we see that the most pronounced feature is

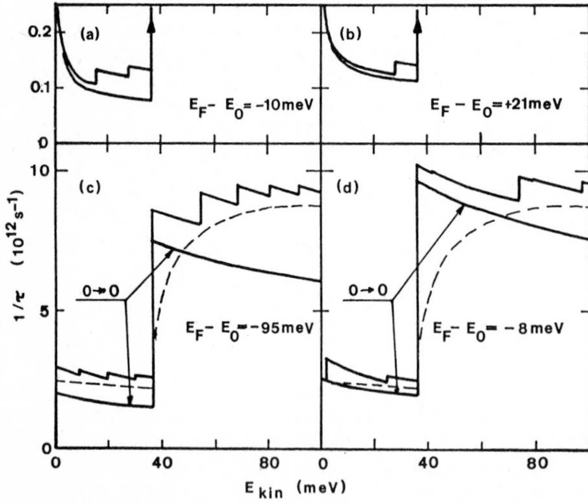

Fig.12 Scattering rates vs kinetic energy out of states in the lowest subband for (a) T = 77K,n, = $4\times10^{(1)}$ cm^{-2}, (b) T = 77K,n, = $6.5\times10^{(1)}$ cm^{-2}, (c) T = 300K,n, = $4\times10^{(1)}$ cm^{-2}, (d) T = 300K,n, = $6.2\times10^{(1)}$ cm^{-2}. Intrasubband scattering ($0 \to 0$) alone and the three-dimensional polar optical phonon scattering rate (dashed) are also shown. $E_V - E_0$ is the Fermi energy relative to the bottom of the lowest subband.

the discontinuity at a kinetic energy of 36 meV = $\hbar\omega_o$, which is due to the possibility of scattering by emission of an optical phonon above this energy; comparing with the corresponding polar optical phonon scattering rate in bulk (shown broken), we see that the discontinuity is a result of the two-dimensionality of our system, in which the density of final states for the scattering by emission of a phonon remains constant; in 3D the density of states goes to zero as $E_{kin}^{1/2}$. The other discontinuities in the scattering rate are due to onset of scattering into other subbands; e.g. the discontinuity very close to E_{kin} = 0 corresponds to scattering into the first excited subband n' = 1 by absorption of an LO phonon.

For lower density the scattering rate shown in part (c) of the figure is seen to approach the scattering rate for bulk. This is understandable,since in that case the subband separations are small and the channel becomes fairly wide. Note that intersubband scattering is responsible for this trend towards bulk.

Since the polar optical phonon scattering is neither isotropic [it depends on Q in (27)] nor elastic, it is not possible to define a momentum relaxation time and one cannot use the relaxation time approximation to solve the Boltzmann equation to obtain the mobility. Instead, one must resort to numerical methods; this is so in three dimensions as well, and the methods developed for that case [8] can be adapted to our quasi-two-dimensional case. To describe these methods is beyond the scope of this course, so we shall only show the results for the low field mobility in the TEGFET channel calculated with full account of subband structure and the relevant phonon scattering mechanisms but without any impurity scattering. The results therefore represent the upper limit for the mobility obtainable in a TEGFET at finite temperature.

Figure 13 shows that at room temperature the mobility almost does not depend on the density of electrons in the channel,and that its value is very close to the mobility in very pure bulk GaAs (~8500 cm^2/Vs). Thus the only reason for an

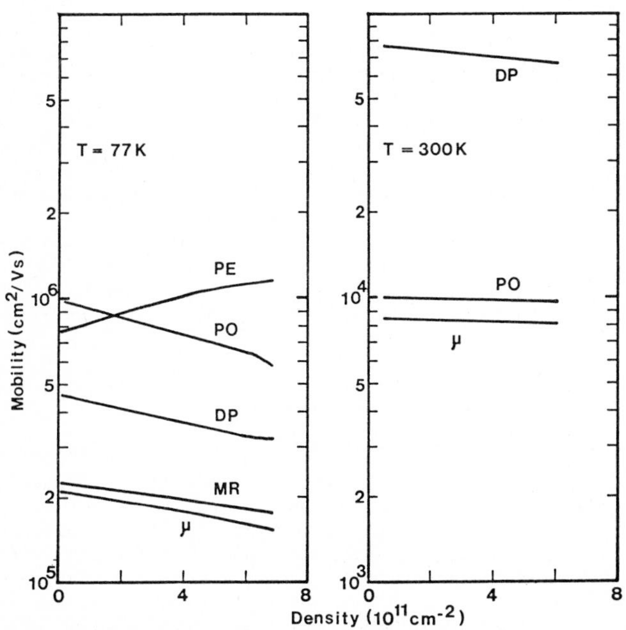

Fig.13 Calculated mobility μ vs density for T = 77K and 300K. Also shown are the mobilities due to each of the processes alone,(PO) polar optical phonon, (PE) piezoelectric, and (DP) deformation potential acoustic phonon scattering. MR shows the mobility obtained from Matthiessen's rule.

enhanced mobility in a TEGFET at room temperature is the reduction in impurity scattering obtained by having the electron channel in undoped GaAs, separated from the ionized donors in the AlGaAs. This may seem a trivial result, but by looking at Fig. 12 again one can convince oneself that it is not at all obvious from the scattering rates; the result is a balance of subband occupancies and intrasubband and intersubband scattering processes. If intersubband scattering did not exist, electrons in the lowest subband would have a mobility of 11 000cm^2/Vs at the highest density but 16500 cm^2/Vs at the lowest density; with intersubband scattering the mobility of electrons in the lowest subband is 8400 cm^2/Vs and 9400 cm^2/Vs at the highest and lowest densities respectively.

At 77K we see that the confinement plays a more significant role; the mobility decreases with increasing density by about 25%. At this temperature, however, the impurity scattering is relatively more important than at room temperature, so in most samples one observes a mobility increasing with density.

As demonstrated earlier, transport in a TEGFET under normal operation is not describable in terms of a low-field mobility; the field parallel to the interface easily reaches several kV/cm so it is important to understand high-field transport. Unfortunately, this is a complicated subject which has not been investigated with much rigour in TEGFETs, so we shall only indicate some of the effects that have been suggested to play a special role in TEGFETs. The problem with these ideas is that they are very difficult to demonstrate experimentally in a TEGFET where many different effects, whose importance is not well known, are working at the same time and cannot be easily eliminated.

1. By looking at Fig. 12 one may imagine that the strongly enhanced scattering probability just above the threshold for optical phonon emission would change the velocity-field relationship compared with that of bulk GaAs, when the field increases and the electrons aquire a higher kinetic energy. It has even been suggested that negative differential mobility might result from this process.

2. In bulk GaAs the negative differential mobility seen in Fig. 11 is essentially due to scattering of electrons into a higher-lying valley in the conduction band,where they have a much lower mobility. One might envision a competing effect in the TEGFET structure : Since the heterojunction barrier is not very high, electrons heated to a sufficiently high kinetic energy in the channel can traverse the barrier and enter the AlGaAS where they would have a much lower mobility. While this "real space transfer" effect has been clearly demonstrated and utilized in other devices, its role is probably much less important in the TEGFET; the electrons are hottest near the drain end of the channel where the gate-to-channel voltage is also relatively high,so that the electrons tend to be pushed away from the interface [see (5) and Fig. 8].

3. Overshoot or quasi-ballistic transport. The velocity-field curve shown in Fig. 11 is the stationary curve in which the power acquired by the electrons from the electric field equals the power dissipated by scattering. Roughly speaking there is a time constant - the energy relaxation time - associated with the establishment of the stationary state. If the energy relaxation time is considerably greater than the scattering time, the drift velocity of the electrons accelerated by the field can be larger than the stationary value for times shorter than the energy relaxation time. If the channel length is made so short that the time for passing under the gate is comparable or smaller than the energy relaxation time, one would obtain a higher effective drift velocity and therefore a higher transconductance according to (8). A similar utilization of transient transport can be envisioned in a channel which is so short that even though the field parallel to the interface is high, the voltage drop is so small that the electrons cannot acquire energy

enough to reach the stationary state corresponding to the field; e.g. if the voltage drop is smaller than 300 mV the electrons will not be able to scatter into the higher-lying valleys. These effects are not limited to TEGFETs, but the quantitative question of how short the channel must be for observing them is still debated.

4. In the theory described in the first part it was assumed that the field parallel to the channel is much smaller than the field which confines the electrons. This is a good approximation for the case where it was used, viz. up to the critical field ε_c. For higher source-drain voltages this approximation must break down, and in principle we can no longer use the essentially one-dimensional charge control model. Instead, two-dimensional simulations would have to be developed, as for MESFETs and MOSFETs.

As mentioned earlier these four points only indicate qualitative speculations and projects for further research.

For deeper study the following list quotes recent, generally more comprehensive articles in which further references to the very many shorter communications on heterojunction FET structures can be found

General reviews :

N.T. Linh: In Festkörperprobleme (Advances in Solid State Physics) XXIII, p. 227, P. Grosse (ed.) (Vieweg, Braunschweig 1983)

H.L. Störmer: Surf. Sci. 132, 519 (1983)

Subband calculations :

T. Ando : J. Phys. Soc. Jpn 51, 3893 (1982)
B. Vinter : Appl. Phys. Lett. 44, 307 (1984)
F. Stern , S. Das Sarma : Phys. Rev. B 30, 840 (1984)

Impurities in AlGaAs :

T. Ishikawa, J. Saito, S. Sasa, S. Hiyamizu : Jpn. J. Appl. Phys. 21, L675 (1982)
H. Künzel, K. Ploog, K. Wünstel, B.L. Zhou : J. Electr. Mater. 13, 281 (1984)
E.F. Schubert, K. Ploog : Phys. Rev. B 30, 7021 (1984)

Transport theory :

F. Stern : Appl. Phys. Lett. 43, 974 (1983)
P.J. Price : Surf. Sci. 113, 199 (1982)

Transport experiment :

E.F. Schubert, K. Ploog, H. Dämbkes, K. Heime : Appl. Phys. A 33, 66 (1984)

Device modelling :

T. Mimura, K. Joshin, S. Kuroda : FUJITSU Sci. Tech. J. 19, 243 (1983)

References

1. P.J. Price : Surf. Sci. 113, 199 (1982)
2. J.M. Ziman : Electrons and Phonons (Oxford University Press, Oxford 1960) p. 211
3. A. Pozela, A. Reklaitis : Solid-State Electron. 23, 927 (1980)
4. J.G. Ruch, Kino : Phys. Rev. 174, 921 (1968)
5. N. Braslau, P.S. Hauge : IEEE Trans. ED-17, 616 (1970)
6. K. Ashida, M. Inoue, J. Shirafuji, Y. Inuishi : J. Phys. Soc. Jpn. 37, 408 (1974)
7. P.A. Houston, A.G.R. Evans : Solid-State Electron. 20, 197 (1977)
8. D.L. Rode : Semicond. Semimetals 10, 1 (1975)

Index of Contributors

Springer Series in Solid-State Sciences

Editors: **M. Cardona, P. Fulde, K. von Klitzing, H.-J. Queisser**

Volume 63

Electronic Properties of Polymers and Related Compounds

Proceedings of an International Winter School, Kirchberg, Tirol, February 23–March 1, 1985
Editors: **H. Kuzmany, M. Mehring, S. Roth**
1985. 267 figures. XI, 354 pages. ISBN 3-540-15722-0

Volume 62

Theory of Heavy Fermions and Valence Fluctuations

Proceedings of the Eight Taniguchi Symposium, Shima Konko, Japan, April 10–13, 1985
Editors: **T. Kasuya, T. Saso**
1985. 106 figures. XII, 287 pages. ISBN 3-540-15922-3

Volume 61

Localization, Interaction, and Transport Phenomena

Proceedings of the International Conference, August 23–28, 1984, Braunschweig, Federal Republic of Germany
Editors: **B. Kramer, G. Bergmann, Y. Bruynseraede**
1985. 125 figures. IX, 264 pages. ISBN 3-540-15451-5

Volume 60

Excitonic Processes in Solids

By **M. Ueta, H. Kanzaki, K. Kobayashi, Y. Toyozawa, E. Hanamura**
1986. 307 figures. XII, 530 pages. ISBN 3-540-15889-8

Volume 59

Dynamical Processes and Ordering on Solid Surfaces

Proceedings of the Seventh Taniguchi Symposium, Kashikojima, Japan, September 10–14, 1984
Editors: **A. Yoshimori, M. Tsukada**
1985. 89 figures. XII, 195 pages. ISBN 3-540-15108-7

Volume 58

The Recursion Method and Its Applications

Proceedings of the Conference, Imperial College, London, England, September 13–14, 1984
Editors: **D. G. Pettifor, D. L. Weaire**
1985. 42 figures. VIII, 179 pages. ISBN 3-540-15173-7

Volume 57

Polycrystalline Semiconductors

Physical Properties and Applications
Proceedings of the International School of Materials Science and Technology at the Ettore Majorana Centre, Erice, Italy, July 1–15, 1984
Editor: **G. Harbeke**
1985. 159 figures. VIII, 245 pages. ISBN 3-540-15143-5

Volume 56
T. Moriya

Spin Fluctuations in Itinerant Electron Magnetism

1985. 98 figures. X, 239 pages. ISBN 3-540-15422-1

Volume 55
D. C. Mattis

The Theory of Magnetism II

Thermodynamics and Statistical Mechanics
1985. 40 figures. XII, 177 pages. ISBN 3-540-15025-0

Volume 54

Magnetic Excitations and Fluctuations

Proceedings of an International Workshop, San Miniato, Italy, May 28–Juni 1, 1984
Editors: **S. W. Lovesey, U. Balucani, F. Borsa, V. Tognetti**
1984. 114 figures. IX, 227 pages. ISBN 3-540-13789-0

Springer-Verlag
Berlin Heidelberg New York
London Paris Tokyo

Volume 53

Two-Dimensional Systems, Heterostructures, and Superlattices

Proceedings of the International Winter School Mauterndorf, Austria, February 26–March 2, 1984
Editors: **G. Bauer, F. Kuchar, H. Heinrich**
1984. 231 figures. IX, 293 pages. ISBN 3-540-13584-7

Volume 52

Superconductivity in Magnetic and Exotic Materials

Proceedings of the Sixth Taniguchi International Symposium, Kashikojima, Japan, November 14–18, 1983
Editors: **T. Matsubara, A. Kotani**
1984. 106 figures. XII, 211 pages. ISBN 3-540-13324-0

Volume 51

Phonon Scattering in Condensed Matter

Proceedings of the Fourth International Conference, University of Stuttgart, Federal Republic of Germany, August 22–26, 1983
Editors: **W. Eisenmenger, K. Laßmann, S. Döttinger**
1984. 321 figures. XIV, 472 pages. ISBN 3-540-12954-5

Volume 50
Sh.-L. Chang

Multiple Diffraction of X-Rays in Crystals

1984. 152 figures. XI, 300 pages. ISBN 3-540-12955-3

Volume 49

Organic Molecular Aggregates

Electronic Excitation and Interaction Processes
Proceedings of the International Symposium on Organic Materials at Schloss Elmau, Bavaria, June 5-10, 1983
Editors: **P. Reineker, H. Haken, H. C. Wolf**
1983. 113 figures. IX, 285 pages. ISBN 3-540-12843-3

Volume 48

Magnetic Phase Transitions

Proceedings of a Summer School at the Ettore Majorana Centre, Erice, Italy, 1–15 July, 1983
Editors: **M. Ausloos, R. J. Elliott**
1983. 103 figures. VII, 269 pages. ISBN 3-540-12842-5

Volume 47

Statics and Dynamics of Nonlinear Systems

Proceedings of a Workshop at the Ettore Majorana Centre, Erice, Italy, 1–11, 1983
Editors: **G. Benedek, H. Bilz, R. Zeyher**
1983. 117 figures. VIII, 311 pages. ISBN 3-540-12841-7

Volume 46

Topological Disorder in Condensed Matter

Proceedings of the Fifth Taniguchi International Symposium, Shimoda, Japan, November 2–5, 1982
Editors: **F. Yonezawa, T. Ninomiya**
1983. 158 figures. XII, 253 pages. ISBN 3-540-12663-5

Volume 45
B. I. Shklovskii, A. L. Efros

Electronic Properties of Doped Semiconductors

Translated from the Russian by S. Luryi
1984. 106 figures. XII, 388 pages. ISBN 3-540-12995-2
Originally published in Russian by Nauka Publishing House, Moscow 1979

Volume 44
I. A Kunin

Elastic Media with Microstructure II

Three-Dimensional Models
1983. 20 figures. VIII, 272 pages. ISBN 3-540-12078-5

Springer-Verlag
Berlin Heidelberg New York
London Paris Tokyo